Structural Restoration of Masonry Monuments

Structural Restoration of Masonry Monuments

Arches, Domes and Walls

George G. Penelis and Gregory G. Penelis

CRC Press
Taylor & Francis Group
Boca Raton London New York

CRC Press is an imprint of the
Taylor & Francis Group, an **informa** business

CRC Press
Taylor & Francis Group
6000 Broken Sound Parkway NW, Suite 300
Boca Raton, FL 33487-2742

Printed on acid-free paper

International Standard Book Number-13: 978-0-367-10947-9 (Hardback)

Visit the Taylor & Francis Web site at
http://www.taylorandfrancis.com

and the CRC Press Web site at
http://www.crcpress.com

Contents

List of abbreviations

ADRS	Acceleration Demand Response Spectra
API	Application Programming Interface
ASCE/SEI	American Society of Civil Engineers/Structural Engineering Institute, USA
ATC	Applied Technology Council
AUTH	Aristotle University of Thessaloniki
BAR	British Archaeological Reports, UK
BRAU	Biennial of Architectural and Urban Restoration
BSI	British Structural Institute, UK
CAMBA 98	International Seminar on the Conservation of Byzantine Masonry
CEN	European Committee for Standardization
CIAS	Centro Internazionale di Aggiornamento Sperimentale-Scientifico
CICOP	Network of International Centres for the Conservation of Architectural Heritage
CLT	Cross Laminated Timber
CSIR	Council of Scientific & Industrial Research (India)
DEM	Descrete Element Method
DICAT	Dipartimento di Ingeneria de Construzioni, dell' ambiente e dell Territorio
DIN	Deutsches Institut für Normung, Germany
DOF	Degrees of Freedom
DT	Destructive Test
DTA	Differential Thermal Analysis
EAK	Greek Aseismic Code (in Greek)
E.C.8-1/EN 1998-1	Eurocode 8-1
ECEE	European Association for Earthquake Engineering
EERC	Earthquake Engineering Research Centre (UC Berkeley, California)
EN	European standard
ENR	Engineering News-Record
FEM	Finite Element Method
FRP	Fiber-Reinforced Plastic
GEM	Global Earthquake Model
IABSE	International Association for Bridge and Structural Engineering

ICCROM	International Centre for the study of the Preservation and Restoration of Cultural Property
ICE	Institution of Civil Engineers, UK
ICOMOS	International Council of Monuments and Sites
IESEE	Institute of Engineering Seismology and Earthquake Engineering
ISPRA	Instituto Superiore per la Protezione e la Ricerca Ambientale (Italian National Institute for Environmental Protection and Research)
LOI	Loss of Ignition
LVDT	Linear Variable Differential Transformer (Sensors)
MAFEA	Masonry Finite Element Analysis
MARE 14	Materials and Techniques for Restoration of Monuments
MASW	Multichannel Analysis of Surface Waves
MEP	Mechanical, Electrical, Plumbing
NDT	Nondestructive Test
NTUA	National Technical University of Athens
OPCM	Ordinanza del Presidente del Consiglio dei Ministri (Prime Ministerial Decree)
PGA	Peak Ground Acceleration
R/C	Reinforced Concrete
RILEM	Reunion Internationale des Laboratoires d'Essais et de Recherches sur les Materiaux et le Constructions
SAHC	Structural Analysis of Monuments and Historical Construction
STREMAH	Structural Studies, Repairs and Maintenance of Heritage Architecture
UNDP	United Nations Development Programme
UNESCO	United Nations Educational, Scientific and Cultural Organisation
UNIDO	United Nations Industrial Development Organisation
URM	Unreinforced Masonry
UV	Ultra Violet
WCEE	World Conference of Earthquake Engineering
WIT	Wessex Institute of Technology, UK

Foreword

The scientific–formative interest of the textbook *Structural Restoration of Masonry Monuments: Arches, Domes and Walls* lies both in the *completeness of the arguments presented*, regarding restoration interventions on deteriorated masonry structures or damaged by calamitous events, and in the *distinction of the treatment in common masonry buildings and monumental buildings.*

From the point of view of the completeness of the topics covered in the text, it ranges between norms and methods to evaluate the mechanical performance of the materials constituting the structures and the evaluation of the carrying capacities of the individual structural elements and of the structures as a whole, both those simplified and those more sophisticated methods, before and after restoration/reinforcement.

The completeness of the treatment of the text becomes even more incisive and valuable, since the authors constantly refer to the *in-situ* and laboratory experiments that they themselves have carried out over the past decades and are known to be avant-garde experimenters in the field of structural restoration and in particular in postearthquake interventions. In fact, their experiments have enriched the Greek technical literature and that of other countries, such as neighbouring Italy.

The authors further point out what is set out in this textbook through their restoration projects applied to important monumental buildings in Greece and in various other countries in the Mediterranean area in which they operated. A comparison therefore also emerges with different socio-economic and procedural realities, through which the text is enriched, confirming its great applicability and operational flexibility, which is invaluable for young architects and engineers wishing to pursue professional careers outside their country of origin.

Therefore, this book by engineers Penelis and Penelis is not only a qualified academic book – which in any case would already be valuable by itself – but also a unique educational and operational offer.

The *distinction between common and monumental building structures* clarifies to professionals the different level of operational complexity of the two categories of structural intervention and therefore the need to acquire specific specialist knowledge that is not always obtained during the course of university studies, but obtained only by attending postgraduate specialization courses.

The authors firmly point out this distinction, highlighting the specific points of differentiation of the intervention process. In particular, they deal with the very difficult topic of *interdisciplinarity,* which since the eighties has been a topic of great debate in the specific sector of structural restoration interventions, still unresolved today. Translating into simple application formulations, even of a semiempirical nature, the complex link between the fundamental requirements of 'structural safety' and those of 'integrity of values', 'durability', 'reversibility' and 'compatibility' is by itself already complex, if for no other reason, because there are infinite solutions, for each specific case of intervention, which can all be satisfactory, as well as 'safe'. It is therefore necessary to search among these infinite solutions, all of which are structurally valid, that solution which at the same time minimises the 'global costs' of the intervention.

In this research, obviously, the structural engineer in charge of the intervention, engineer or architect needs to be interfaced and necessarily confronted with experts from other disciplines intervening in the same intervention process (architecture historians, technologists, geologists, economists etc.). And if, on the one hand, this relationship of comparison is a cultural stimulus for all the 'actors' of the intervention, it presents, on the other hand, considerable operational difficulties – which often slow down the approval process of the restoration projects in which they are involved. A part of these difficulties resides in the lack of a common technical terminology that each discipline uses in its own semantic way.

Other difficulties result from the assignment of the role of the 'coordinator' of the intervention, who interfaces with the institution that engages in the project and with all the other 'actors' of the intervention process. Some regulations, such as the Italian one, entrust the role of coordinator to an architect, only in the case of intervention on monumental buildings and especially in emergency conditions.

The authors highlight and deepen, with particular 'technological sensitivity', the importance of the knowledge of materials and of 'historical' technologies of monumental buildings. Intervening in the structures of buildings of historical-monumental value, the intervention process – be it for the restoration of accidental damage or for the restoration of pathologies in progress, for seismic adaptation or improvement or finally for adaptation to the new functional standards – will be all the more optimal, the more the 'project of knowledge' of the architectural good will be deepened. In fact, it is not a common and simple diagnostic project but an accurate and difficult cognitive project, which in the most complex cases goes hand in hand with the operational phases of the restoration and is modified step by step with it.

This specific professional learning is offered in the text of the authors, and it is the result of their direct experience in the field of structural restoration, deriving from their experimental studies and from interventions directed by them in buildings of monumental prestige.

Enjoy the reading

Prof. Arch. Nina Avramidou
President of the International Centre for the Conservation of Architectural Heritage, Firenze, 2019 (CICOP Italia, CICOP Net)

Preface

In the whole procedure of preservation, conservation and restoration of monuments and historical buildings, *structural restoration* is considered an undesirable parameter, since it implies to interventions on a large scale that might harm the authenticity of a culturally protected building.

However, this type of intervention is an inevitable action, because it pertains to the safety of the building and, most important, to the safety of the users. In this respect, authenticity and safety are two concepts contradicting each other, and they are ensured by professionals of different origin and philosophy, namely archeology and architecture on one side and science and technology on the other. In case of structural restoration, these professionals are required to find a common space of coexistence so that the results of restoration satisfy both concepts, authenticity and safety.

This book is meant primarily for structural engineers involved in structural restoration and postgraduate students involved in the restoration of monuments and historical buildings.

The content of this book has been guided primarily by the above thoughts and from some basic issues that distinguish structural restoration of monuments and historical buildings from repair and strengthening of conventional ones. These issues are as follows:

1. There is a need for thorough knowledge of materials, forms and techniques of construction used in the past. In fact, in a period of more than five millennia, various materials and techniques of construction have been developed as a result of big revolutions that have taken place in the history of human civilisation. A deep knowledge of all the above is necessary for the suitable choice of present-day techniques and materials that might be used in structural restoration where principles of *reversible* or *irreversible* techniques play a vital role together with *compatibility* and *durability* of intervention materials.
2. The assessment of the bearing capacity of monuments and historical buildings, in relation to the expected effects of calamitous events cannot rely only on analytical procedures. Thanks to the vast development of computer science, very reliable models have been developed for elastic or inelastic, static or dynamic analysis of complicated continuous or discontinuous space structures. In fact, the low reliability of material properties of the past, existing cracks, splits and drifts in the body of the structures from the past and, finally, uncertainties in the prediction of effects of calamitous events for a long term period in the future, particularly those of the seismic events, influence seriously the reliability of the results of analysis and verification. Therefore, the basic guide for the structural assessment should be

a historical documentation of damages and the present condition of the structure (cracks, splits, drifts), this information being a basic indication for the verification of the reliability of the results of analysis.
3. According to what has been presented in the beginning, there is a need for close cooperation mainly between archeologists and architects on one side and structural engineers on the other in the framework of restoration activities. The authors of this book having participated in the structural restoration of numerous prominent monuments and historical buildings, in various supreme councils for the administration and approval of restoration activities, and in the formulation of postgraduate studies in restoration, are aware that the collaboration between the above professionals may be more effective and fruitful, if each has some understanding of the other's aims, needs and approaches.
4. Last but not least, it is very important for a structural engineer who is in the beginning of his involvement in structural restoration to have in mind some integrated case studies of structural restoration of not only prominent monuments but also less important historical buildings in order to use them as a guide to his activities.

We hope that the structure of our book covers our intentions to address the above issues properly.

Here we would like to express our gratitude to Nina Avramidou, former professor of architecture in Florence and president of CICOP (Network of International Centres for the Conservation of Architectural Heritage), Italy, for her kindness in writing a preface for our book and approaching our work from the architectural point of view.

Finally, we express our thanks to the following collaborators for their contributions to the preparation of this book:

- *Panagiotis Savvas* for the elaboration of the figures
- *Melissa Ponias* for the typewritten preparation of the text
- *Dr Sotiria Stefanidou,* Civil Engineer, for her critical reading and correction of the text
- *Tony Moore,* CRC Press senior editor for encouraging us in our attempt to write this book.

George G. Penelis
Gregory G. Penelis
Thessaloniki, Greece, July 2019

Authors

George G. Penelis (MSc, PhD, Dr. Habil) is emeritus professor at the Aristotle University of Thessaloniki, Greece. He is an ordinary member of Academia Pontaniana, Italy. He has published more than 250 technical papers and is a co-author of two books on earthquake-resistant concrete structures (*Earthquake Resistant Concrete Structures*, Chapman & Hall, 1997; *Concrete Buildings in Seismic Regions*, CRC Press, 2014). He was the convenor of 'Repair and strengthening of historical monuments and buildings in urban nuclei' (UNDP/UNIDO, 1984). He has been involved in many projects regarding the seismic assessment and restoration of some of the most imposing monumental buildings in eastern Mediterranean basin.

Dr. Gregory G. Penelis (MSc, DIC, PhD) has been involved in many research projects regarding the seismic assessment of listed and monumental buildings as well as urban nucleus. His team has developed several masonry analysis tools which are in use worldwide. As principal of Penelis Consulting Engineers, he has been involved in the design or review of monumental and listed buildings throughout Europe and Middle East. He has been awarded the gold medal by the Patriarch of Alexandria and All Africa for the restoration of the 2000-year-old Roman tower and church (Mar Girgis) in Cairo, which also received the 2017 ENR Global Best Project Award. He is the co-author of *Concrete Buildings in Seismic Regions* (CRC Press, 2014).

Chapter 1

Introduction

1.1 GENERAL

The conservation of monuments and historical buildings is one of the most important expressions of the contemporary cultural activities. These constructions are all those which merit special care on account of their individual historical or architectural importance or their significance as surviving representatives of earlier tradition (UNDP/UNIDO, 1984).

The structural restoration of monuments and historical buildings, being a part of their conservation, has only been recognised, as a special field of the construction technique, in the last few decades.

Main purpose of this book is to outline the techniques, the materials and the design procedures in use for the structural restoration of masonry monuments and historical buildings generally and particularly in seismic regions.

Both historical buildings and contemporary ones are approached from a structural point of view with the same scientific background, namely the 'structural mechanics' (theory of structures, strength of materials and earthquake engineering). In both cases, the structural engineer in charge is concerned primarily with giving the structures the ability to withstand gravity loads and seismic actions without excessive damage, safeguarding primarily the life safety. However, the problem of *structural restoration* of monuments and historical buildings differs radically from *repair* and *strengthening* of conventional contemporary buildings damaged by accidental effects like seismic action. The aim of repair and strengthening of conventional buildings *is to keep them in use at a specified safety level* (Penelis, G. and Penelis, Gr., 2019). On the other hand, for monumental buildings, emphasis is given to the preservation of their *aesthetic* and *historical values*, while the task to remain in use may be considered of secondary importance and, in any case, as consequence of the effort at fulfilment of the main task (Penelis, G., 2002).

More precisely, the features and the values that have to be considered for the characterisation of a building as a monument or historical building may be summarised (UNDP/UNIDO, 1984) as follows:

- A symbolic value that has led to the original construction. A temple, a church or a mosque has such a value from the beginning. Other buildings may have acquired their symbolic value afterwards, as a result of historical events during the life of the structure.
- An aesthetic value, as a work of art deliberately created, or as an important feature in the landscape or the townscape.

- An indirect value based on contents or decorations that are themselves of aesthetic value (mosaics, frescos paintings etc.).
- A 'documentary value' of the architectural or structural creativity of a particular time and region.
- A secondary 'documentary' value in relation to particular historical or religious events.
- A social function as a focus of community life.
- A value, as part of a group of structures which, together, have one or more of the above values.

The decision for characterising a building or an urban nucleum as a monument or as a historical building is the result of extensive investigation that is carried out by archaeologists, historians and architects, and in this respect is beyond the scope of this book. It should be noted, however, that the final decision for the characterisation belongs to the relevant authorities of each country, or of UNESCO.

1.2 RESTRICTIONS UNDER CONSIDERATION FOR THE STRUCTURAL RESTORATION

The constraints for a conventional contemporary building that should be taken into account for repair and strengthening are basically the following (Penelis, G. and Penelis, Gr., 2019):

- The specified use function of the building after retrofitting.
- The aesthetics of the building.
- The cost of retrofitting.
- The lifetime of the building after retrofitting.

On the other hand, for the *structural restoration* of a monument, basic restrictions that should be taken into consideration may be resumed as follows (UNDP/UNIDO, 1984; Penelis, G., 2002):

- Respect for the original material and authentic documents.
- Respect to the valid contributions of all periods to the building.
- Replacement of missing parts must integrate harmoniously with the whole but, at the same time, must be distinguishable from the original.
- Additions cannot be allowed, except in cases where they do not distract from the interesting parts of the building, its traditional setting and its relations with the environment.
- The use of traditional techniques and materials are *clearly preferable* for structural restoration.
- Modern techniques and materials are *admissible* where adequate capacity cannot be ensured by traditional techniques and materials. In this case, *durability* and *compatibility* of the interventions should be adequately proven; otherwise the modern techniques and materials should be used only in a manner that will permit easy corrective action at a later date if necessary (*reversible interventions*).
- Measures are necessary to prevent and safeguard frescos and mosaic decoration. This may exclude the use of some strengthening techniques that may cause damage.

All above restrictions or rather *principles* or *rules* are included in *Venice Charter*, a document which was elaborated at 11th International Conference of Venice, in 1964, an international meeting of experts aiming at the protection of the cultural heritage (Feilden, 2003; UNDP/UNIDO, 1984). Particularly, articles 9, 10, 11, 12, 13 and 16 of this document refer to 'Restoration' and include all above-mentioned constraints.

1.3 AUTHENTICITY AND LIFE SAFETY

From what has been presented so far, the preservation of authenticity, in the framework of a restoration, plays a crucial role. However, it is the authors' opinion that it is also of major importance to ensure many other values, as for example, the use or the visit availability of the monumental building and, most important, *the protection of human life*, which is directly connected with the safety of a structure against collapse (Penelis, G. and Penelis, Gr, 2019). In this respect, a close cooperation must be established, mainly between *archaeologists*, *architects* and *structural engineers* in charge of the restoration of a monument, so that a common place of understanding may be found, satisfying both main tasks, namely, *authenticity* and *life safety*.

The collaboration between the above professionals may be more effective and fruitful, if each has some understanding of the others' aims, needs and approaches, as well as, the choices open to them. The above constitutes a demanding task that calls for considerable flexibility of mind of all components in defining objectives, seeking, interpreting and assessing evidence and choosing schemes to be put into effect.

It is interesting to note here that the recent publication of ASCE 41-17 (ASCE 41-17, 2017), referring to evaluation and retrofitting of buildings against earthquakes, includes in its scope (see commentary of Chapter 1 – general requirements) the structural evaluation and the retrofitting of historical buildings too. However, it poses serious restrictions for the applicability of the actions that could be taken to evaluate and retrofit conventional contemporary buildings. In this respect, there is, even in this standard, a complete compliance with the concerns presented in detail in the above paragraphs.

1.4 THE STRUCTURE OF THIS BOOK

As it is already noted above, the purpose of this book is the presentation of techniques, materials and design procedures in use for the structural restoration of masonry monumental and historical buildings.

However, for the accomplishment of this task, it is necessary to precede with a detailed presentation of materials and techniques used in their original construction, the mechanics of masonry, being the basic structural material, their structural forms classified in groups and their behaviour to conventional loads and seismic actions. Furthermore, an insight approach should be made to the philosophy behind of all these for the formation of safe paths for load transfer from the structure to the foundation.

Additionally, before any intervention, it is necessary to obtain a deep knowledge of the texture of a monument under consideration, the materials used for its construction, its pathology and, finally, its residual strength. So, the structure of this book is formatted as follows.

In the first three Chapters (2, 3 and 4), an overview of the main characteristics of the existing historical masonry building stock and its structural response to various loading actions is presented. Particularly a short presentation of the materials and the techniques of the past is made, the masonry mechanics is reviewed and the behaviour of each characteristic structural form under gravity loads and seismic actions is analysed basically.

In the next Chapter (5), the following issues are examined: the restoration philosophy (reversible–irreversible techniques, local–general types of intervention), the intervention techniques and materials in use (Tassios, 2001).

In Chapters 6, 7 and 8, the following procedures are presented: the collection of all necessary information of a monumental building under consideration for structural restoration, i.e. *in-situ* investigations and laboratory tests, the procedure for static or dynamic analysis and design and the assessment, i.e. diagnosis and safety evaluation together with the choice of the structural restoration scheme.

Finally, in Chapter 9, a series of case studies of some prominent monumental buildings, which the authors have restored during their professional career, is presented.

Chapter 2

Building materials and construction techniques of monuments

2.1 GENERAL

Building materials available for structural construction up to 18th century, when the industrial revolution began, are compared (see Figure 2.1) (Mainstone, 1975) with those available up to recent dates. From this comparison, it is concluded that the materials in use for building construction before 1800 were mainly

- Stones
- Bricks
- Tiles
- Mortars
- Wood.

In very special cases, iron items had been used to ensure special connections under tension, protected from corrosion with a lead coating.

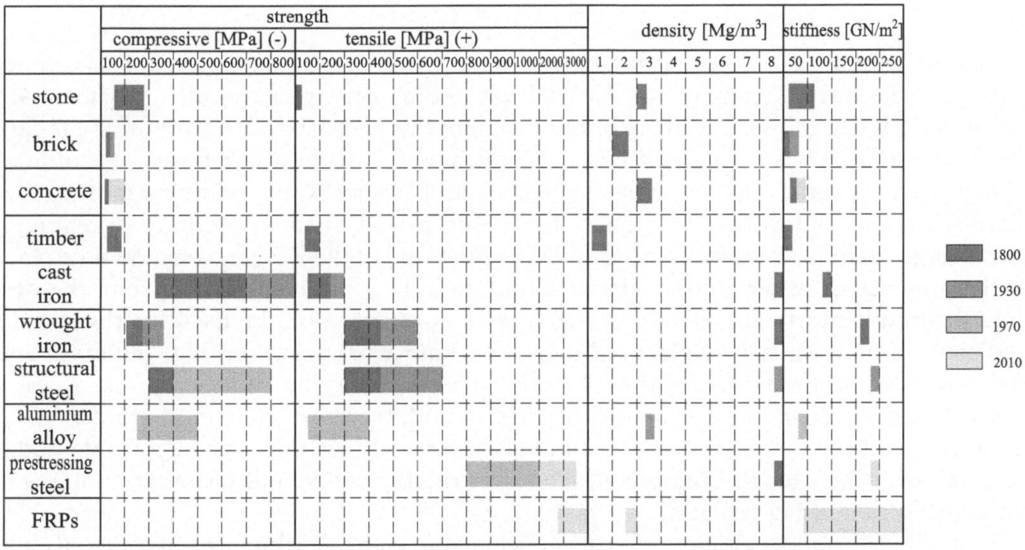

Figure 2.1 Approximate ranges of strength, density and stiffness for structural materials up to 2010. (Adapted from Mainstone, R. (1975). *Developments in structural form*. Penguin Books Ltd, England.)

Masonry construction, as it was composed of stones, bricks and mortars was characterised by medium but reliable compressive strength (1.0–10.0 MPa) but a very low and unreliable tension capacity (0.2–1.0 MPa). Therefore, it was used for walls, arches, vaults and domes, since these structural elements, as it will be discussed later, are subjected mainly to compression under gravity loads.

Wood, having a reliable tensile as well as compressive strength of the order of 5–120 MPa, was used for structural elements under bending or under axial loading, in other words for construction of decks, (joists) roofs (trusses), ties etc. After the first half of 19th century, wood elements were replaced by steel members, in almost all serious works, and only recently wood has begun to be used in modern structures again as a structural material, due to its low carbon imprint.

In a period of almost ten thousand years of human history, the development of building construction has passed through some critical milestones that have influenced radically the structural forms and techniques. These milestones are related to the building materials and are summarised as follows.

Unshaped stone	(8000 B.C.)
Sun-dried mud bricks	(8000 B.C.)
Mud mortars	(8000 B.C.)
Burnt bricks	(4000 B.C.)
Lime mortars	(4000 B.C.)
Roman concrete	(1st century B.C.)
Iron items	(4th–3rd century B.C.)
Steel	(19th century A.D.)
RC concrete	(19th century A.D.)
Fiber-reinforced plastics (FRPs)	(21st century A.D.)

It is the authors' opinion that for *masonry construction*, the most critical milestones were first the introduction of the *burnt bricks* combined with the *lime mortars* (4th millennium B.C.) and second the *Roman concrete* (1st century B.C.). Both of them gave the freedom for constructing easily at a relatively low cost desirable forms of buildings by setting of mortar or concrete on site. Setting and strengthening of a lime mortar take place when exposed to the atmosphere for a long period of time, resulting in relatively low strength at mortar joints. Setting and strengthening of the Roman concrete take place in the presence of water, and its strength and durability are much higher than the first ones (hydraulic mortars). Therefore, the Roman concrete could be used for thick walls and piers, even in water (e.g. piers of bridges, harbour works etc.) for huge buildings and civil works of imposing dimensions. In this respect, it may be considered that *Roman concrete* caused a *real revolution* in construction and, therefore, may be characterised as the ancestor of *modern concrete*. It should also be noted that this type of concrete dominated all over the Roman Empire and later over the Byzantine and Ottoman Empires in the construction of big projects.

From all of the above, it can be easily concluded that the development of materials and construction techniques moved side by side. Needs of the society in construction accelerated the massive use of new or known materials from the past, and these gave the opportunity to develop new forms and techniques in construction.

2.2 MATERIALS

2.2.1 Stones

Cut stone in drums or monolithic single blocks or in fully shaped orthogonal blocks (ashlars) for construction of columns, unmortared walls and facings is the main material for the articulated monuments. The orthogonal blocks were also used for the construction of the two faces of thick walls filled with Roman concrete (*opus caementicium*). In this case, blocks were mortared at their joints, and so there was no need for a high level of smoothness at the contact sides of the stone, as in the case of the unmortared walls or columns (Figure 2.2).

The raw materials are usually marble, tuffstone or more rarely limestone. In Table 2.1 the types of stones used in various monuments and their mechanical properties may be found (Robertson, 1982).

Rubble, used mainly as mortared filling of thick masonry walls or piers, may be of any kind of raw stone, easily found in quarries and in the surrounding of a monument. Unshaped stones were also used very often in combination with bricks for the construction of mortared faces of walls.

2.2.2 Bricks

In the Alluvial Valleys of Nile, Tigris and Euphrates rivers, mud, straw and reef have been used for almost ten millennia for the construction of simple buildings (mainly dwellings). Later, at about 8,000 years B.C. sun-dried bricks of high quality and regular size were in use (Figure 2.3). Reef mats incorporated at regular intervals in the horizontal joints permitted the construction of massive walls, pyramids and ziggurats of considerable height and of vaults and arches of several metres in span (Mainstone, 1975; Bei and Papayianni, 2003).

Later, fired bricks replaced the sun-dried ones to a certain degree. The fired bricks are of clayish consistency usually fired to a temperature in the range of 750°–850°C so that they become hard, durable and aesthetically attractive. The process of firing was developed about 6000 years ago by the people living in Mesopotamia, which is now a part of Iraq (Al-Kass, et al., 1989). Bricks used in areas around the Mediterranean Sea,

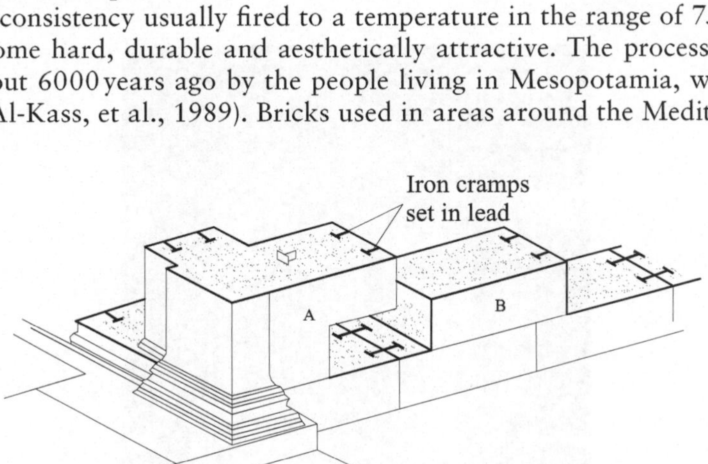

Figure 2.2 Unmortared cut-stone masonry. The north wall of Erehtheion, Acropolis, Athens (Stevens, G.P., 1927). (Courtesy: Zambas, K. (1982) Articulated ancient Greek and Roman buildings. UNDP/UNIDO, PROJECT (RER/79/015) National Report of Greece.)

Table 2.1 Mechanical properties of stones

	Aggregation properties			*Thermal properties*			*Mechanical properties*			
Rock type	ρ *(g/cm³)*	φ *(%)*	μ^a *(−log d)*	α *(1/10⁶)*	K $\frac{[\text{mcal}]}{\text{cm/s}\,°\text{C}}$	*k* *(10⁻²cm³/s)*	*H*	*E (Mb)*	*S (kb)*	*R(kb)*
Granite	2.5–2.7	0.1–4	9–6	5–11	3–10	0.5–3	5–7	0.3–0.6	0.8–3.3	0.1–0.7
Gabbro	2.8–3.1	0.3–3	7–5	4–7	4–6	1–2	5–6.5	0.5–1.1	1.1–3.0	0.1–0.7
Rhyolite–andesite	2.2–2.5	4–15	8–2	5–9	2–9	0.4–3	5–6.5	0.6–0.7	0.6–2.2	0.01–0.7
Basalt	2.7–3.1	0.1–5	5–1	4–6	2–5	0.4–1.5	4–6.5	0.5–1.0	0.5–2.9	0.1–0.9
Quartzite	2.5–2.7	0.3–3	7–4	10–12	8–16	2–8	4–7	0.6–1.0	1.1–3.6	0.1–1.0
Marble	2.4–2.8	0.4–5	6–3	5–9	3–7	05–1.5	2–4	0.2–0.7	0.4–1.9	0.04–0.3
Slate	2.6–2.9	0.1–5	11–8	8–10	3–9	0.5–3	3–5	0.3–0.9	0.5–3.1	0.05–1.0
Sandstone	2.0–2.6	1–30	3-0	8–12	2–12	0.4–5	2–7	0.03–0.8	0.2–2.5	0.01–0.4
Limestone	1.8–2.7	0.3–30	9-2	4–12	2–6	0.4–1.5	2–3	0.1–0.7	0.2–2.4	0.1–0.5
Shale	2.0–2.5	2–30	9-5	9–15	1–8	0.3–2	2–3	0.1–0.4	0.3–1.3	0.02–0.5
Soapstone	2.5–2.8	0.5–5	6-4	8–12	2–7	0.4–1.5	1	0.01–0.1	0.1–0.4	0.01–0.1
Travertine	2.0–2.7	0.5–5	5-2	6–10	2–5	0.4–1	2–3	0.1–0.6	0.1–1.5	0.02–0.1
Serpentinite[b]	2.2–2.7	1–15	7-3	5–12	3–9	0.5–3	2–5	0.1–0.5	0.7–1.9	0.05–0.1

Source: Adapted from Robertson, E. (1982). Physical properties of building stone. The National Academies Press, Washington, D.C. with permission of National Academies Press.

The symbols used as column headings are as follows: ρ – bulk density in *d*; *a* – linear thermal expansion of millionths of reciprocal degrees Celsius; K – thermal conductivity in millicalories per centimetre-second-degree Celsius; *k* – diffusivity in hundredths of square centimetres per second, *H* – hardness by Mohs's scale, *E* – Young's modulus of elasticity in megabars, *S* – uniaxial compressive strength in kilobars and *R* – modulus of rupture in kilobars.

[a] Permeability units are negative logarithms; thus 9 means 10^{-9} darcies.

[b] Serpentinite is much like verde antique stone.

Figure 2.3 Contemporary unfired bricks built with mud mortar. (Courtesy: Papayianni, J. and Batsouka, C. (2017). Unfired mud brick masonries. MSc Thesis of Batsouka C. supervised by Prof. Papayianni. J.)

Figure 2.4 A brick from the Roman phase of Rotunda of Thessaloniki. (From Penelis, G. et al., 1989.)

for thousands of years, exhibit a high level of standardisation and high strength. Thus, the fired Babylonian bricks (600 B.C.) have an average dimension of 32 × 32 × 8 cm and 32 × 18 × 8 cm and a compressive strength of 91 kp/cm^2 (9.1 N/mm^2). The bricks of Rotunda, Thessaloniki, so much in the Roman (300 A.D.) as in the Christian phase (450 A.D.) have an average dimension of 32 × 42 × 5cm. The Roman ones have an average strength of about 130 kp/cm^2 (13.0 N/mm^2) while the Christian ones 100 kp/cm^2 (10.0 N/mm^2). At Hagios Andreas of Peristera, Greece, the strength of the bricks exhibits an average value of (10.0 N/mm^2) 100 kp/cm^2. The bricks of defence walls of Thessaloniki have the following dimensions 0.40 × 0.40 × 0.05, 0.40 × 0.35 × 0.05 and 0.40 × 0.30 × 0.0 5m, even though their construction covers a very long period of time (300 B.C. to 1400 A.D.) (Penelis, G., 1982). Bricks bear usually on their surface engraved a seal which, regardless of the reason of its existence helps in the bonding of mortar and bricks (Figure 2.4) (Penelis, G., et al., 1989).

In Table 2.2, information is given for fired bricks of various important monuments. It should be noted that the values of compressive strength of bricks given above should be considered as mean values.

2.2.3 Mortars

2.2.3.1 General

Mortars in masonry monuments and historic buildings play a vital role in the functioning of the structure and its outer skins. They serve, in the first place, to fill the gaps between units (bricks–stones) and thereby provide a uniform bearing without the need for accurate cutting or shaping of the stones. Many mortars have been used in the past, ranging from the mud mortars used to build walls from unfired mud bricks to the strong

Table 2.2 Mechanical properties of various types of bricks from 600 B.C. to 1400 A.D.

Monument	*Century*	*Dimensions (cm)*	*Compressive Strength (MPa)*	*Modulus of Elasticity (GPa)*	*Consistency*
Babylon	600 B.C.	32 × 32 × 8 32 × 18 × 8	9	Unknown	Clay and Sand
Rotunda of Thessaloniki Roman Phase	–300 A.D.	32 × 42 × 5	9.3	11.28	Clay and Sand
Rotunda of Thessaloniki Christian Phase	450 A.D.	32 × 42 × 5	10	6.34	Clay and Sand
Defense Walls of Thessaloniki	300 B.C – 1400 A.D.	40 × 35 × 5 40 × 30 × 5 40 × 40 × 5	Variable	Variable	Clay and Sand
Ruins of Galerian Palace of Thessaloniki	300 A.D.	32, .5 × 47, 0 × 4.5 20 × 41.0, 0 × 3.5	9.0–13.0	Variable	Clay and Sand
Hagia Sophia of Thessaloniki	618 A.D.	32 × 40 × 4–4.5 45 × 40 × 4–4.5	13.2	Unknown	Clay and Sand

pozzolanic mortars, used to make Roman concrete – *opus caementicium* – and similar latter hydraulic mortars that incorporated pulverised underfired brick in place of a natural pozzolana.

Of course, nobody can overlook either the strong influence of the geopolitical space, on construction materials, or the developing trends in technology, which appeared in the centre of the Roman Empire and then spread over the entire geopolitical space. A characteristic example is the pozzolanic concrete. This was used in the beginning in ancient Greece for a waterproof covering of the walls of cisterns – e.g. the cistern of Kameiros, Rodos (500 B.C. Malinowski, 1981). Then, it was developed as *opus caementicium* in Rome – 1st century B.C. to 1st century A.D. – and later, it was spread all over the Roman Empire as Roman concrete with many alternatives due to the local conditions. Besides the monuments of ancient Rome, many other examples can be found all over the Empire, e.g. the Fortress of Babylon Camp at Cairo, 1st century A.D. (Penelis, G and Penelis, Gr., 2018), the Rotunda of Thessaloniki (300–450 A.D.) (Penelis, G., 1982), the Galerian Palace of Thessaloniki (300 A.D.) (Athanassiou, et al., 2015), Hagia Sophia of Istanbul (532 A.D.) (Mainstone, 1975), Hagia Sophia of Thessaloniki (620 A.D.) (Theocharidou, 1992) etc.

It is well known that the materials which constitute the mortars are the following four:

- Binder
- Aggregate
- Water
- Additives.

Traditional mortars, based on their binder, are characterised as

- Mud mortars
- Lime mortars
- Roman mortars (pozzolanic mortars).

2.2.3.2 *Mud mortars*

Soil with 10%–20% content of clay is mixed with water until it is in a plastic (workable) state. Once applied, a mud mortar sets quickly on drying without the need for any curing procedure. The compressive strength of mud mortar ranges from 1.3 to 3.3 MPa depending on the consistency and the moisture content (CSIR India, 2017; Bei and Papayianni, 2003). Presence of moisture reduces the compressive strength drastically. This mortar has been used in the past not only for unfired brick masonry but also for construction of unshaped stone masonry walls. In the Balkan region and in Asia Minor, it had been used even up to the end of the 19th century A.D., while in India, it is used extensively in housing even today (Figure 2.5).

The biggest disadvantage of mud mortar is its low and unstable strength due to moisture, the serious shrinkage and erosion due also to moisture. For diminishing shrinkage effects, organic fibres (e.g. straw fibres) have been used since early times.

2.2.3.3 *Lime mortars*

Lime mortar is composed of lime and an aggregate like sand or sand and gravel mixed with water. It was used first in ancient Egypt about 6,000 years ago. It was used extensively in the past all over the world and continues to be used even today. The binder is primarily composed of *calcium hydroxide* $(Ca(OH)_2)$. Lime is produced first by heating pure *calcium carbonate* $(CaCO_3)$, namely limestone, at the grade of 954°C–1,066°C driving off *carbon dioxide* (CO_2) producing *calcium oxide* (CaO) called 'quick lime'. This procedure is carried out in a lime kiln. The quick lime is then slacked; in other words, it is mixed with enough water with which it reacts to form hydrated lime $(Ca(OH)_2)$ in the form of a slurry. This slurry is transformed in due time to 'lime putty'. This lime putty mixed with aggregate and water gives the lime mortar, which sets and hardens in due time.

During this procedure, hydrated lime $(Ca(OH)_2)$ turns naturally back into calcium carbonate $(Ca\ CO_3)$ by capturing carbon dioxide (CO_2) *from the atmosphere*. In other words, it obtains again its lithic texture and, therefore, sufficient strength. The usual composition of lime mortar has a ratio of 1:2 or 1:3 lime to aggregate.

Figure 2.5 Stone masonry with mud mortar: Ruins after the earthquake of 8 May 1995, Kozani, Greece.

The main properties of lime mortars are the following:

- The compressive strength of the mortar ranges between 0.5 and 1.0 MPa. The adhesive strength of mortar to bricks is less than 0.14 MPa, while the tensile strength is about 0.1 and 0.30 MPa. The above are calculated for a ratio of 1:3 lime to sand.
- The lime mortar has a low modulus of elasticity and, therefore, can sustain small deformations without cracking.

It should be noted that it is necessary for lime mortar to be exposed to the air so that the chemical reaction of capturing CO_2 and, consequently, its hardening can be facilitated. Having the above property in mind, it is concluded that *this mortar is not adequate either for piers with large cross section* and *very thick masonry or for piers and foundations in water*, since in both cases the atmospheric air cannot enter deeply into the mortar to activate the chemical reaction of transforming the hydrated lime *($Ca(OH)_2$)* into limestone *($CaCO_3$)* again (Swallow and Coerrington 1995, 1996).

2.2.3.4 Pozzolanic mortars

As it was noted before, pozzolanic mortars had been known in Greece since 500 B.C., but they had limited use for waterproofing of works, like cisterns, aqueducts etc. Romans, in their effort to overcome the inadequacies of lime mortars for buildings with impressive dimensions or for harbours and foundations in the water, introduced pozzolanic mortars in a large scale. Instead of using as aggregate the usual sand, they added volcanic ashes of a ratio 1:3 lime to volcanic ash for building constructions and 1:2 for harbour works, according to Vitruvius, 1st century B.C. (Vitruvius, 1998). Particularly in Rome, they used volcanic ashes from the beds of *Pozzuoli* near Naples called by Vitruvius *pulvis puteolanus* (in Latin). However, in reality this *pit sand* was not an aggregate, *but it reacted chemically* with hydrated lime and water giving a *hydraulic* binder which set and hardened *without the presence of atmospheric air.* At the same time, cementing matrix consisting of calcium-aluminium-silicate-hydrate (C-A-S-H) was reinforced with platy crystals of Al-tobermorite (Figure 2.6) which did not allow the deterioration of the cement matrix in case that salt water was present. The above chemical reactions had given to pozzolanic mortar the following basic properties:

- High compressive strength (up to 5.0 MPa)
- Ability of setting and hardening in water
- High durability even in sea water for almost 2,000 years.

In areas of the Empire, where pozzolanic ashes were not available, crushed and pulverised bricks were used. These bricks, as it is concluded from contemporary chemical and physical analyses (Table 2.3, Penelis, et al., 1982), were kilned at low temperatures (500°–600°) so that crystallic ceramic compounds could not be formed. In this respect, the amorphous ceramic material could react with the hydrated lime and the water in the same way as the pozzolanic ashes. From the same chemical analyses mentioned above, it has been observed that the pulverised bricks were of the same consistency with the construction bricks, only that these were inactivated hydraulically because of the existence of crystallic ceramic compounds which suggests that they were kilned at higher temperature. Our estimate is that the pulverised bricks belonged to the outer covering of the kilns and since they were

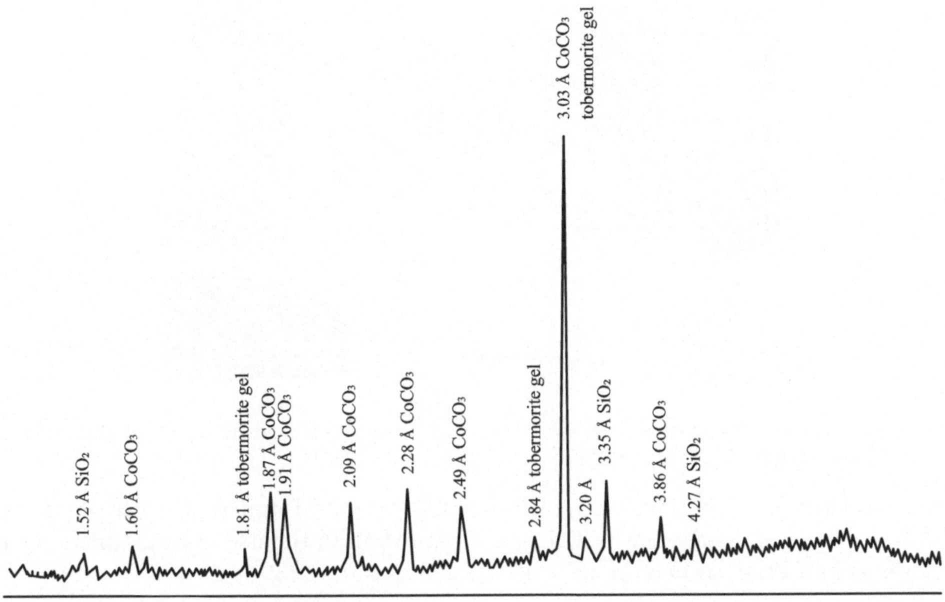

Figure 2.6 X-ray diffraction analysis of the mortar of Rotunda Thessaloniki. (From Penelis, G., 1982.)

Table 2.3 Rotunda of Thessaloniki. Chemical components of the mortar of Roman phase (Penelis, G., et al., 1982a)

	First analysis including coarse ceramic material			*Second analysis without coarse ceramic material*		
	Defined quantities			*Defined quantities*		
Composition	*(gr)*	*%*	*%*	*(gr)*	*%*	*%*
Loss on ignition	2.672	12.78		4.3045	21.52	
Insoluble	13.852	66.30		8.989	44.94	
Soluble	(4.372)	(20.92)		(6.71)	(33.54)	
SiO_2	0.540	2.58	12.35	0.640	3.20	9.53
R_2O_3	0.673	3.22	15.39	0.775	3.87	11.55
CaO	2.988	14.31	68.36	5.10	25.48	76.00
MgO	0.107	0.51	2.44	0.1175	0.59	1.75
SO_3	0.029	0.14	0.67	0.029	0.14	0.44
Na_2O	-	-	-	0.0060	0.03	0.09
K_2O				0.0400	0.20	0.59
Undefined material	0.035	0.16	0.79	0.0025	0.03	0.05
Total	20.896	100.00	100.00	20.003	100.00	100.00

not well fired, they chipped off. These chipped bricks due to their bad furnacing were chemically drastic and therefore proper for an efficient hydraulic component in the mortar (Karaveziroglou, et al., 1995; Moropoulou, et al., 1997; Papayianni 1997).

Consequently, the hydraulic reaction of the binder in the Roman mortars was achieved either with the use of volcanic ashes or pulverised underburnt bricks, depending on the local conditions of the work under construction. So, in the eastern regions of the Roman

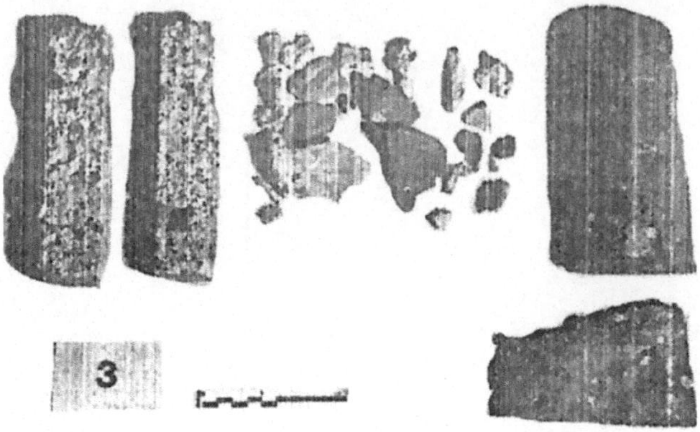

Figure 2.7 Rotunda of Thessaloniki mortar cores taken from the Roman phase of the monument. (From Penelis, G., et al., 1989.)

Empire and later in the regions of Byzantine and Ottoman Empires, the use of pulverised bricks was preferred, as there was a very long tradition in the use of brick masonry from Babylonians and Persians (Figure 2.7) (Penelis, G., et al., 1989).

2.2.3.5 *Aggregates*

The lime and pozzolanic mortars in monuments and historical buildings had as a basic component aggregates to a ratio of 1:2.0–1:3.0 (binder to aggregates). Sand, gravel and sometimes crushed brick constitute the aggregate conglomeration. Binder, as it happens with contemporary mortars, had the main role of filling the gaps among the aggregates and to connect them in a compound strong enough to withstand compressive and shear stresses.

The historic mortars, in general, do not fit modern standards, as very often, their aggregates contain not only a considerable number of fine components but also relatively high proportions of coarse grains >4 mm. Some aggregate characteristics of old mortars studied in the Laboratory of Reinforced Concrete and Masonry Structures of Aristotle University of Thessaloniki (AUTH) are given in Table 2.4 (Karaveziroglou, et al., 1995). The determination of proportion of binder to aggregate was based on the treatment of mortar samples with hydrochloric acid. The grading curves of aggregate are drawn in Figure 2.8. In the same figure, the lower and upper boundaries for grading curves valid in Greece for aggregates of contemporary concrete mixtures are also marked (Theocharidou, 1992, Papayianni, J. and Karaveziroglou, M., 1993a).

Taking into account the relatively high strength of these mortars, their hydraulic properties and the grading curves of their aggregates having a maximum grain size of 16 and sometimes even 32 mm, they may be characterised as *Roman concrete*, the ancestor of the contemporary one.

2.2.3.6 *Additives*

The excellent properties of ancient mortars are attributed by many historians to the additional use of organic additives in the old mixtures, such as egg whites, blood, milk of figs etc. However, this information is based on old texts' testimonies and is of low reliability. In fact, the analytical procedures of chemical and physical methods in use do not allow the verification of the above information in the lab.

Table 2.4 Mechanical properties of mortars of some important monuments of Thessaloniki

Monuments	*Century*	*Strength (N/mm^2)*	*Aggregate max size (mm)*	*Proportion of mortar to aggregate*	*Fines <0.075(%)*
Rotunda Roman	4th A.D.	2.3	16	1:3.1	28.25
Rotunda Christian	4–5th A.D.	3.7	25	1:2.5	35.40
Acheiropoietos	7th A.D.	-	16	-	46.69
Hagia Sophia	8th A.D.	4.5	19	-	24.03
Panagia Chalkeon	11th A.D.	-	25	-	20.30
St. Panteleimon	12th A.D.	-	25	-	29.63
Bey Hamami	15th A.D.	1.18	6	1:2.5	45.70
Minaret of Rotunda	17th A.D.	1.25	30	1:2.84	38.50

Source: Adapted from Karaveziroglou, M., Papayianni, I., and Penelis, G. (1995). Mortars and grouts in restoration of Roman and Byzantine monuments. *Proceedings of the European Seminar on Ancient and Traditional Mortar Technology and Their Importance in Conservation Projects*, Brussels.

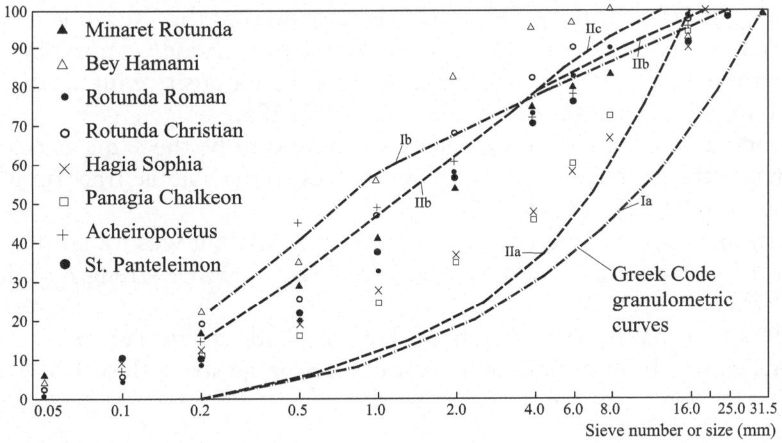

Figure 2.8 Granulometric curves of aggregates of some important monuments of Thessaloniki (Papayianni, J. and Karaveziroglou, M., 1993a). (Adapted from Penelis, G. (1995). Mortars and grouts in restoration of roman and byzantine monuments. *European seminar on Ancient and traditional mortar technology and their importance in conservation projects*. Rethymnon, Creta.)

2.2.3.7 *Case studies*

It is interesting to transfer here the authors experience from the analysis of the mortars, of some very important monuments of Thessaloniki, which cover a period of over one thousand years. This analysis was carried out after the earthquake of June 20, 1978, that caused a lot of damage to the monuments and so accelerated the restoration action.

- *In the Roman part of Rotunda, Thessaloniki* (300 A.D.), it has been found from chemical and X-ray analysis that the hydraulic action of the mortar is due to *pulverised brick material*. These bricks must have been fired at low temperatures (500°C–600°C), to prevent the formation of crystallic ceramic compounds. Besides, from the same analysis, it has been observed that the bricks of the masonry were of the same material of the pulverised bricks, only that these were inactivated in respect to hydraulic action, because of the existence of crystallic ceramic compounds, which suggested that they were fired at higher temperatures (750°C–850°C). It is also worth mentioning that the aggregate of the mortar was not only sand but a mixture of sand and gravel up to 16 mm. In other words, the mortar with which the Rotunda was built was a kind of fine concrete with a binder having hydraulic activity. It is quite characteristic that the mortar's average compressive strength is about 3.2 MPa.

The proportion of binder to aggregate was found to be as follows.	
1:2.5	by volume
1:4	by weight

- At the same time, the proportion of the complex silicates to the soluble fraction of the binder was found to be 9.53%.
- Concerning the Christian part of the same monument (end of 4th century A.D., beginning of 5th century A.D.), it has been found that the hydraulic action appears increased compared to the Roman part, and this is due to pozzolanic material. The increased quantity of pozzolanic material had a positive influence on the average compressive strength which is of about 3.7 MPa.
- The proportion of binder to aggregate was found to be the same as that of Roman part, while the proportion of complex silicates to the soluble fraction of the binder was found to be 22% (Table 2.5).
- Concerning Hagia Sophia of Thessaloniki (783 A.D.), it was found that the mortar has strong pozzolanic properties. The average compressive strength of the mortar is in the range of 4.5 MPa.
- On the other hand, the church of Hagios Andreas in Peristera, Thessaloniki (870 A.D.), was built with lime mortar of an average strength of 1.3 MPa.

Table 2.5 Rotunda of Thessaloniki, chemical components of the mortar of the Christian phase (Penelis, G., et al., 1982a)

	First analysis including coarse ceramic material			*Second analysis without coarse ceramic material*		
	Defined quantities			*Defined quantities*		
Composition	*(gr)*	*%*	*%*	*(gr)*	*%*	*%*
Loss on ignition	2.82	14.10		3.4155	17.08	
Insoluble	11.42	55.21		10.5645	52.85	
Soluble	(6.147)	(30.69)		(6.011)	(30.07)	
SiO_2	1.452	7.26	23.62	1.38	6.90	22.95
R_2O_3	1.38	6.90	22.45	0.9925	4.96	16.54
CaO	2.645	13.24	43.03	2.875	14.38	47.83
MgO	0.471	2.36	7.69	0.703	3.52	11.69
SO_3	0.021	0.11	0.35	0.037	0.19	0.62
Na_2O	0.037	0.18	0.61			
K_2O	0.06	0.30	0.98			
Undefined material	0.063	0.35	1.29	0.0235	0.12	0.37
Total	20.008	100.0	100.0	19.991	100.0	100.0

- Finally, the church of Vlatades Monastery in Thessaloniki (12th century A.D.) is built with a lime mortar containing a small quantity of chemically active pulverised bricks. The compressive strength of this mortal was found to be 1.9 MPa.

2.2.4 Timber

Until the end of the 18th century, timber was the only material that was available in quantity and in suitable sizes for the construction of roofs, decks, wall trusses, arch ties and other similar structural elements subjected to axial tension or bending. So, it was used very widely in co-operation with articulated and masonry structures. Timber, being a product taken from trees, is subjected to significant changes in its internal structure after its transformation to a building material. Therefore, it should be controlled carefully and continuously. In particular, it is subjected to considerable loss of moisture which results in pronounced shrinkage, splitting and serious distortion. The timber also becomes more vulnerable than trees to fire and to various forms of attack and decay.

The wooden parts of a building are the most vulnerable ones in comparison to stone or masonry parts. So, for the structural part of the building, timber of high durability was used and more particularly timber of oak, chestnut, beech, or pine trees, depending on the importance of the building and the timber availability in the region. Mechanical properties of some important timber qualities are given in Table 2.6 (Mühler, 1976).

The elastic properties of the above species are given below:

- Needle bearing trees: E(parallel to timber fibers) = 10.0 GPa
- Oak and beech trees: E(parallel to timber fibers) = 12.5 GPa.

The above must be considered as mean values.

Sensitive points of the timber constructions must be considered the joints of the members (Figure 2.9). There, due to the weakening of the cross sections and the joint formulations through cutting and drilling, the wood becomes more vulnerable. These points together with those exposed to high moisture and high fluctuation of temperature are the points where the timber constructions of old buildings present indications of decay (Mainstone, 1975; UNDP/UNIDO, 1984; Ulrich, 2006; Vitruvius, 1998).

2.2.5 Iron, cast-iron and steel

Iron is found in nature only in chemically combined forms, usually as one of its oxides. In order to extract it, it is necessary to heat the ore in the presence of carbon, which reacts as a reducing agent removing the oxygen out of the ore. From the antiquity (2000 B.C.) till

Table 2.6 Mechanical properties of some important timber qualities

Timber species	*Compression (MPa)*	*Tension (MPa)*	*Bending (MPa)*	*Shear (MPa)*
Spruce	45	90	65	6.7
Fir	40	84	62	5.1
Pine	47	104	87	10
Beech	53	135	105	8
Oak	55	90	94	11

Source: Adapted from Mühler, K. (1976). Holzbau, Beton Kalender 1976 Teil II, Verlag von Wilhelm Erns & Sohn. Berlin. pp. 573–626.

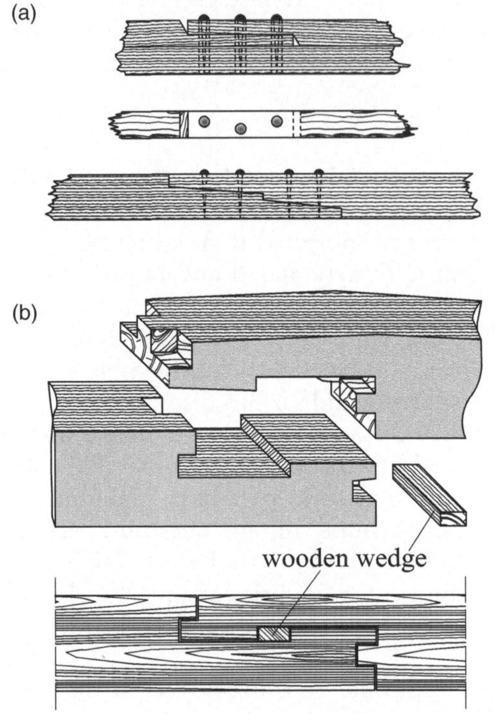

Figure 2.9 Ancient timber joints. Some examples. (a) Joints using nail connection; (b) joint using wooden wedge. (Adapted from Ulrich, R. (2006). *Roman woodworking*. Yale University Press, New Haven, USA.)

the Middle Ages, the temperature of the charcoal used for the *reaction of reduction* had been just sufficient to extract the iron but not sufficient to melt it. The products of this procedure was a spongy *bloom* and slugs, which were subsequently separated by *forging* to produce what was known as *wrought iron* (Mainstone, 1975; Drougas, 2009).

Due to the high cost and the production difficulties, wrought iron was used till 18th century A.D. mainly for high-value instruments like weapons and only secondarily in construction for the production of cramps, tie bars in masonry construction or nails, bolts and straps for jointing timber structural elements (Figures 2.10, 2.11 and 2.12) (Zambas, 1982; Drougas, 2009; Varoufakis, 2005; Harris, 1992).

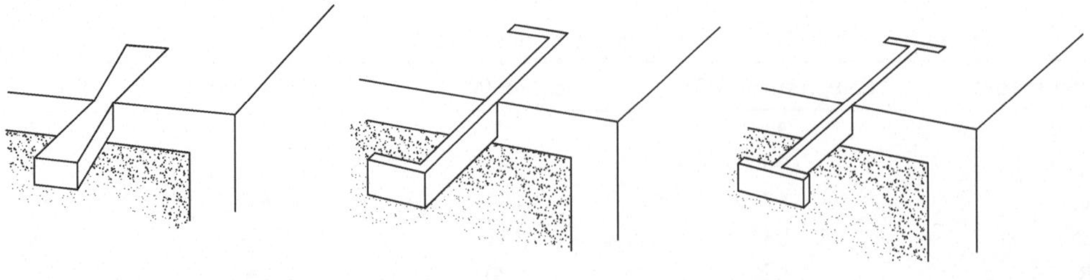

Figure 2.10 Cramps for cut-marble units connection, Acropolis of Athens. (Orlandos, A.K., 1955; Courtesy of Zambas, 1982.)

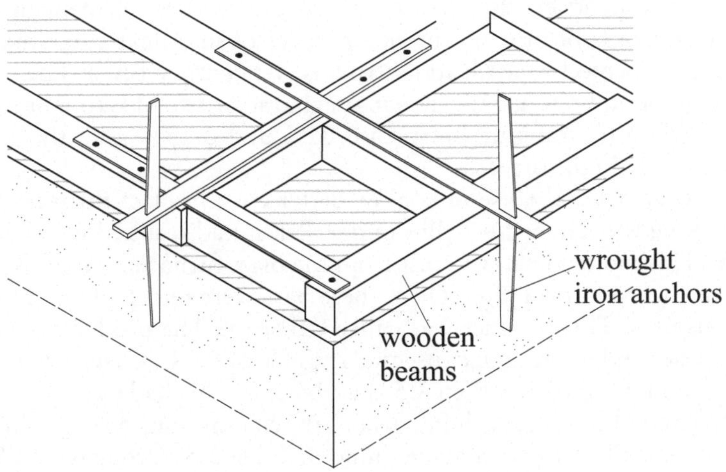

Figure 2.11 Anchor ties in masonry. (Adapted from Drougas, A. (2009). Investigation of the use of iron in construction from antiquity to the technical revolution. *Master's thesis Universität Politécnica de Catalunya*, Spain.)

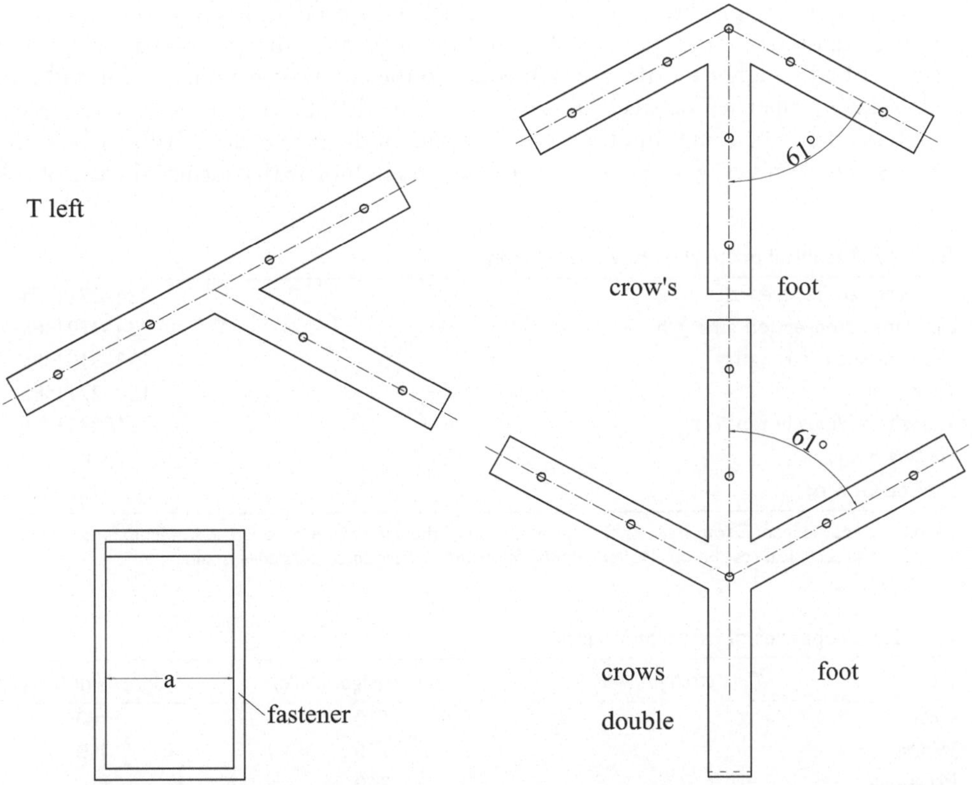

Figure 2.12 Wrought iron straps jointing timber trusses. (Adapted from Drougas, A. (2009). Investigation of the use of iron in construction from antiquity to the technical revolution. *Master's thesis Universität Politécnica de Catalunya*, Spain.)

Already, Vitruvius mentions the use of iron, as a structural element in the construction of walls, as a means of bracing the outer leaves of multileaf stone masonry, by the use of iron clamps covered by cast lead (Vitruvius, 1st century B.C.). The main technical properties may be found in Table 2.7. The main characteristic of wrought iron is the very low content of carbon (C) ranging between 0.05% and 0.25% and, therefore, the high ductility and relatively high softness.

Since 15th century A.D., the use of coal or better coke in blast furnaces had given the capability of melting the reduced ore, but in parallel, it had resulted in carbonising of the produced 'iron'. This molten iron was cast from the blast furnace in moulds in the form of bars which were known as *iron pigs*. Then, these bars were remelted and recast in desired forms for various uses. This product was called *cast iron*. The production process of cast iron was much easier and faster and, consequently, of lower cost. Cast iron contained a high percentage of carbon, ranging between 3.5% and 4.5%. This had to be, as a consequence, very hard, brittle and of low strength in tension, in relation to its strength in compression. Some basic properties of cast iron may be found in Table 2.8 (Drougas, 2009).

Between the second half of 18th and the first half of the 19th century, cast iron, due to its mechanical properties, was extensively used in building construction in the place of piers, columns, arches, corbels etc., in other words, in place of structural elements mainly under compression (Figure 2.13).

In 1858, the invention of *Bessemer converter* in England made possible the decarbonising of cast iron; so, a new product was born, *the steel*. This product had a carbon percentage ranging between 0.07% and 1.3% and, therefore, it was less hard then cast iron. At the same time, it had a higher tensile strength equal to the compressive one, and it had a high ductility. Main properties of steel are given in Table 2.9. Steel was perfect for tension, for compression and bending. In other words, it had all the properties of timber but almost at ten times higher values. Therefore, from the second half of 19th century, the use of rolled

Table 2.7 Technical properties of wrought iron

Ultimate tensile strength	234–371 MPa
Ultimate compression strength	234–372 MPa
Ultimate shear strength	193–310 MPa
Yield point	159–221 MPa
Young's modulus in tension	193 GPa
Melting point	1,540°C
Specific Gravity	7.8

Source: Adapted from Drougas, A. (2009). Investigation of the use of iron in construction from antiquity to the technical revolution. *Master's thesis Universität Politécnica de Catalunya*, Spain.

Table 2.8 Properties of cast iron varieties

Type	*Yield strength (MPa)*	*Tensile strength (MPa)*	*Brinell hardness*
Grey	-	170	180
White	-	170	450
Malleable	220	360	130
Ductile	365	480	170

Source: Adapted from Drougas, A. (2009). Investigation of the use of iron in construction from antiquity to the technical revolution. *Master's thesis Universität Politécnica de Catalunya*, Spain.

Figure 2.13 Typical corbels of cast iron. Allatini mansion 1863–1873, Thessaloniki. (Courtesy: Pasvantis, St. (2019) Photo files of Thessaloniki.)

Table 2.9 Main properties of hot rolled structural steel (EC3–2005, Table 3.1)

	Nominal thickness of the element t (mm)			
	t ≤ 40 mm		*40 mm < t ≤ 80 mm*	
Standard and steel grade	f_y *(N/mm²)*	f_u *(N/mm²)*	f_y *(N/mm²)*	f_u *(N/mm²)*
EN 10025-2				
S 235	235	360	215	360
S 275	275	430	255	410
S 355	355	510	335	470
S 450	440	550	410	550

Source: Adapted from EC3–2005.

steel sections in the form of decks, trusses, arches etc. (Figure 2.14) is found in all important masonry structures.

The main disadvantage of iron and steel members was their susceptibility to corrosion and fire. Additionally, due to the high strength and the subsequent reduced cross section, problems of buckling emerged in slender elements under compression.

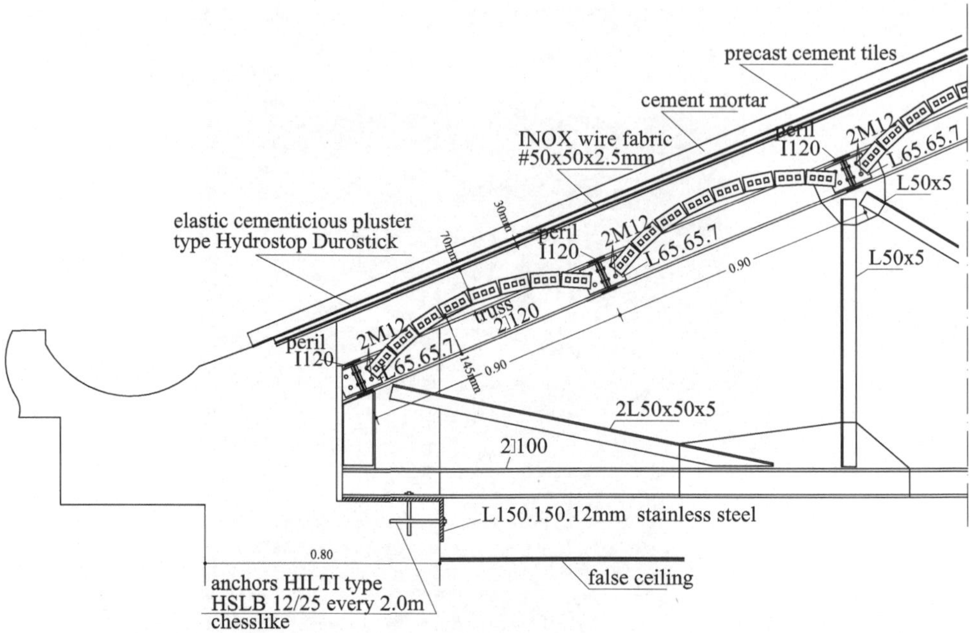

Figure 2.14 Hot rolled structural steel trusses, National Library of Greece, Athens, end of 19th century A.D. (From Penelis S.A., Archives, 2001a.)

2.3 CONSTRUCTION TECHNIQUES

2.3.1 Walls and piers of mortared stone and brick

The usual method of construction of walls, in the Roman period and during the Byzantine and Ottoman ones, was the following: First, the two outer faces were built with rectangular shape-like stones at a height of one metre using mortar at the bed joints. Simultaneously, the empty space between the external leaves was filled with rubble and with rich mortar of the same consistency as the joints. Next, a zone of bricks was built which usually consisted of four to five parallel layers which covered the entire thickness of the wall. Then the same procedure was repeated up to the upper level of the wall (Figures 2.15, 2.16 and 2.17) (Penelis, G., 1982; Choisy, 1873, 1883; Palladio, 1965).

Thus, bricks acted as ties between the two external leaves of the wall, diminishing the risk of buckling outward of the outer leaves. In parallel, they contributed to more uniform distribution of stresses on the cross section of the wall or the pier. Of course, there have been examples of walls built either solely with bricks or solely with stone or others with different relations between the layers of stone and bricks. Anyhow, the above-described wall texture is met in all big monumental buildings in the regions of Roman, Byzantine and Ottoman Empires, covering a period of more than two thousand years (Figures 2.18 and 2.19).

As far as the mortar in use, it was basically a pozzolanic one, based on volcanic ashes or on pulverised underburnt bricks as explained before. So, walls of this type, known as *opus caementicium,* were in reality a type of concrete known nowadays as *Roman concrete* with considerable compressive strength and durability for millennia. Of course, when the thickness of the wall was small (e.g. 0.7–1.0 m), the use of lime mortar was also in effect.

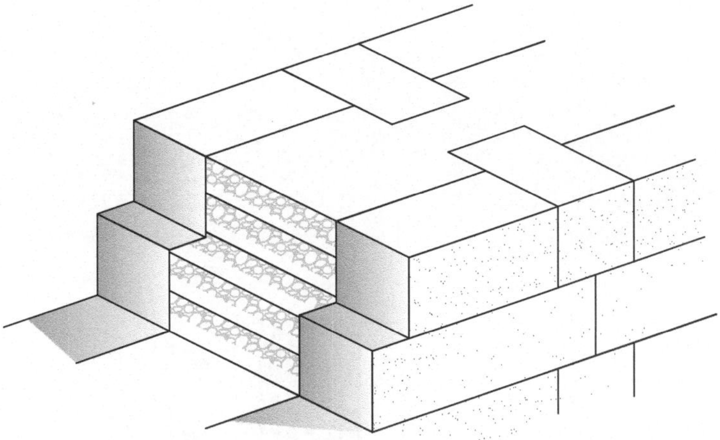

Figure 2.15 Roman concrete. Three-leaf masonry wall. (Adapted from Penelis, G. (1982). *The Masonry Roman and Byzantine Monuments in Greece. UNDP/UNIDO, PROJECT (RER/79/015) National Report of Greece.*)

Figure 2.16 Masonry of Hagia Sophia, Thessaloniki. (From Penelis, G., 1982.)

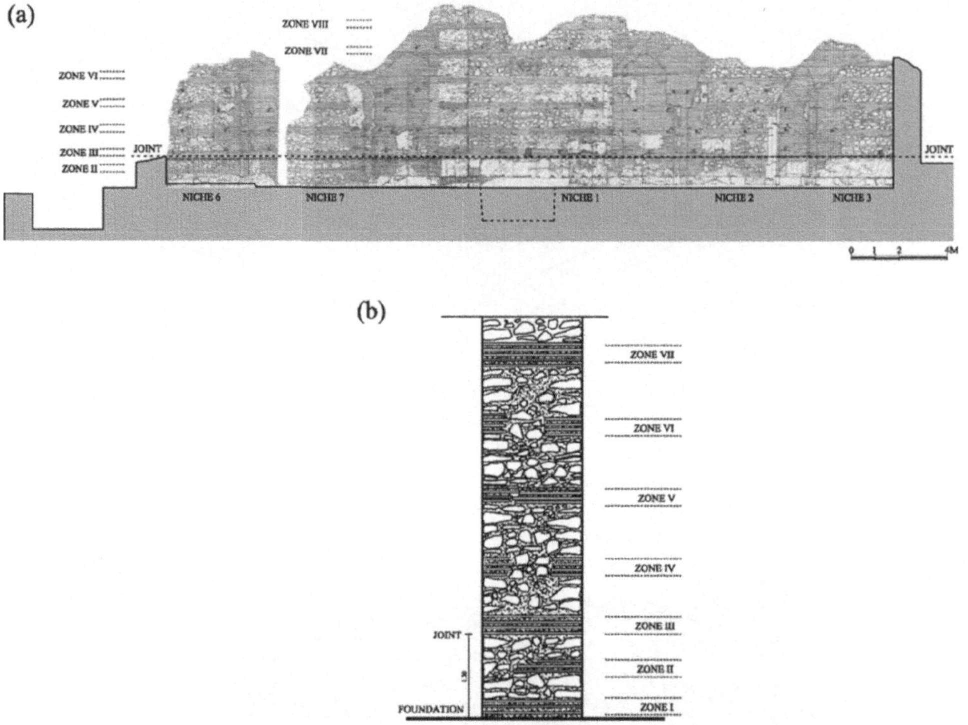

Figure 2.17 Galerian Palace, Ruins of Octagon, Thessaloniki. (a) Wall view; (b) cross section. (Adapted from Athanassiou, Ph., Malama, V., Misa, M., and Sarandidou, M. (2015). The restoration of the ruins of the Galerian complex (1994–2014). *Documentation and Interventions, Thessaloniki.* Ephorate of Thessaloniki Antiquities. Thessaloniki, Greece. Courtesy: Ephorate of Antiquities of Thessaloniki.)

Figure 2.18 Hagia Irene, Istanbul, external views. (From Penelis S.A., Archives, 2017.)

Figure 2.19 Mar Girgis, Old Cairo (Trajan Tower), external view. (From Penelis S.A., Archives, 2014a.)

In the Roman period, joints of mortar between the bricks were of smaller thickness than the bricks. Later, about the 4th century A.D., this relation was 1:1. Finally in the 6th century A.D., this relation was 1:1.5 (brick to joint thickness). These thick joints had as a result, due to shrinkage and creep of mortar, the appearance of visible deformations (settlements–drifts) without the expected cracks, which usually accompany large deformations in modern buildings.

The archaeologists attribute these thick joints to an attempt of sparing bricks which was a material of higher cost. It is the authors' opinion that probably, besides the cost, engineering reasons contributed to this tendency; the mortar shrinkage and creep attenuated the consequences of eventual foundation settlements on the expected cracking. The wall responded for years as a plastic body following the soil settlements without cracks.

In case of piers, the external faces usually were constructed exclusively with bricks or cut stone (ashlar) (Figure 2.20).

2.3.2 Timber-framed walls

For small structures like houses, small churches or small mosques, a complete braced frame of timber has probably provided in the past the best protection to seismic actions, although these elements increased the risk of fire destruction.

The timber elements were incorporated in the masonry in horizontal zones, one below the subsill plate of windows, one above the top header of windows, and one at the support level of the floor or the roof. Each element consisted of two horizontal beams at the two faces of the wall connected with transverse wooden ties forming with the horizontal

Figure 2.20 Mortared brick masonry of Hagia Sophia Thessaloniki. (From Penelis, G., 1982.)

beams a kind of horizontally installed ladder in the body of the masonry wall. At the same time, very often, diagonal and perpendicular struts were installed between the horizontal elements forming in this way trusses capable of resisting horizontal seismic actions (Figures 2.21 and 2.22) (UNDP/UNIDO, 1984).

The horizontal timber elements of each side of the building were connected with the corresponding timber elements of the transverse walls of the building. In this way, a sufficient protection for an out-of-plan failure of the walls was ensured in case of an earthquake.

Particularly, in case of buildings with more than two stories, the last storey was usually formed only with walls of wooden trusses cladded with narrow wooden boards at a space of about 2.0 cm from each other. The wall was then covered on both sides with plaster (Figure 2.23). At the lower storeys, the walls were built with stones or bricks and mud or lime mortars.

2.3.3 Timber and steel floors

Wooden beams (joists) have been used for spanning spaces for centuries. Wooden beams were extensively used in floor construction. The beams were installed in a distance of 30/40cm and were supported by the walls of the building, particularly on the timber horizontal ties, described in the previous paragraph. The floor was formed with boards of about 30 mm thickness, while ceilings were formed with boards, decorated properly. It is evident that this construction was very susceptible to fire (Figure 2.24).

Since 19th century A.D., in case of important works with big spans, floors had been formed with cast iron or steel beams in double T form, arranged with an in-between distance of almost 0.80 m. The span was bridged by flat vaults of hollow bricks; the deck was finalised with a concrete screed and marble plates, tiles or wooden boards (Figure 2.25).

Figure 2.21 Timber-framed walls, house in Halkidiki, Greece. (Courtesy: Didaskalou, Gr. (2018). Geoter Diduskalou Sp. (Archives) Thessaloniki.)

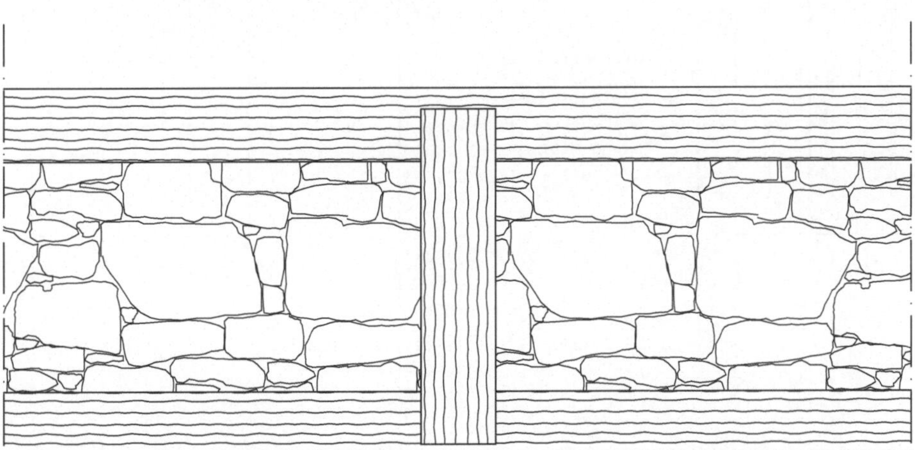

Figure 2.22 Horizontal grid of longitudinal beams combined with transverse ties.

Figure 2.23 Two-storey house in Greece. Timber structure of the second floor. (Courtesy: Tsintziras, G. (2018). Ten Brinke Group (Archives) Unknown, Thessaloniki.)

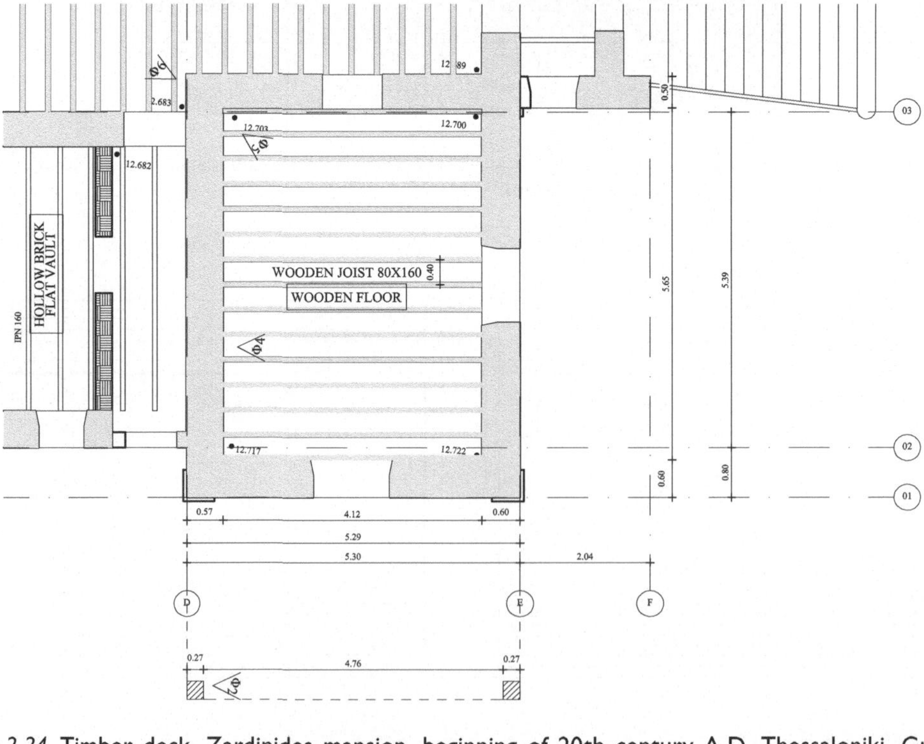

Figure 2.24 Timber deck, Zardinides mansion, beginning of 20th century A.D. Thessaloniki, Greece. (From Penelis S.A., Archives, 2014b.)

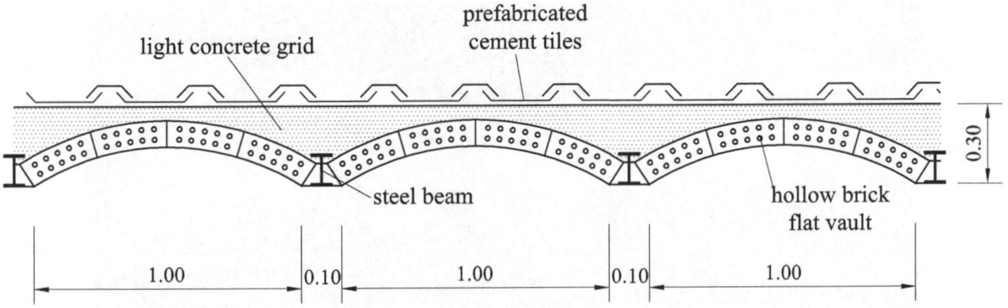

Figure 2.25 Structural steel beams, forming a deck of the National Library of Greece Athens. (From Penelis S.A., Archives, 2001a.)

2.3.4 Timber and steel roofs

The timber roof with trusses was the basic structural system for roofing buildings of small- and medium-size spans for thousands of years. The connections of the principal tie to the rafters were the crucial point for the roof trusses, since the ties safeguarded the supports from thrust that otherwise would develop even for dead loads. The roof was finalised with purlins, common rafters boarding and tiles of various types. This type of roofing was very appropriate for rain and snow drainage. At the same time, it was a very effective tool in the hands of architects for proper formation of the buildings style (Figure 2.26).

Since 19th century A.D., in case of important buildings with big spans, the timber structures have been replaced by cast iron or steel trusses (Figure 2.27).

2.3.5 Arches and barrel vaults

Arches of bricks or stone were one of the usual spanning elements in the past, and they were used almost invariably from Roman times onwards whenever large loads had to be carried. The intrados of arches was usually semicircular. However, the extrados was circular up to an angle of about 60° from the keystone of the arch being then widened by an external backing which formed a kind of a pier for the rest of the arch up to 90° (Figure 2.28).

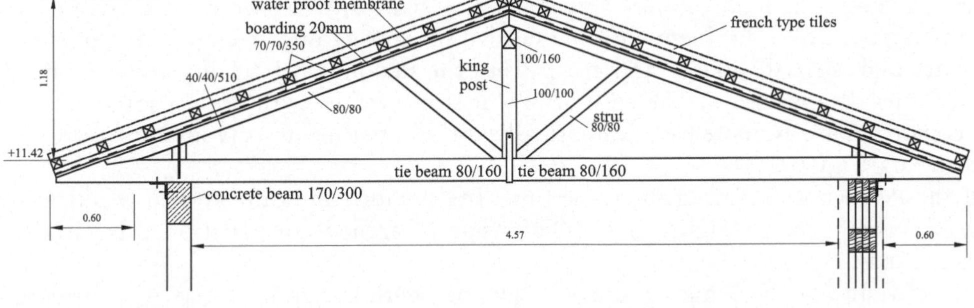

Figure 2.26 A characteristic type of timber roof at Zardinides Mansion, beginning of 20th century A.D. (From Penelis S.A., Archives, 2014b.)

Figure 2.27 Steel truss roof of the National Library of Greece, Athens (end 19th century). (From Penelis S.A., Archives, 2001a.)

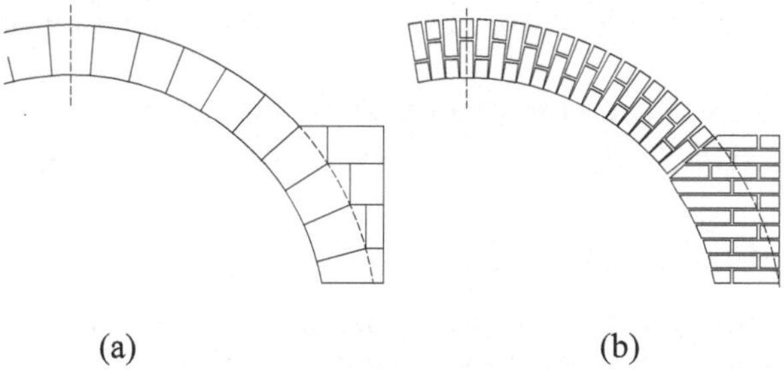

Figure 2.28 Formation of the springers of semicircular arches (a) unmortared cut stone (voussoirs); (b) mortared brick masonry arch. (Adapted from Hart, F. (1965). Kunst and Technik der Wolbung. Verlag George Callwey, München, Germany. Courtesy: Verlag Georg P.W. Callwey, München.)

Bricks were arranged radially in one, two or even three layers depending on the span of the arch and the loads above (Figure 2.29). Usually, arches were built using centring means (Figure 2.30). Problems of time needed for setting and strengthening of the joint mortars and their shrinkage and creep were the major issues during construction. All these issues disappeared in the case of precut stone arches with no wet joints, but then, the construction cost increased dramatically as the cut-stone units (voussoirs) were labour consuming (Figure 2.31).

It should be noted that, at the same time, the ecphoric building system was developed but for smaller spans (Figure 2.35). This type of arched construction was rather rare (Hart, 1965).

Barrel vault has the same structural behaviour with the arch. It may be considered as a series of arches placed one by the other. Accordingly, the same construction technique was therefore followed as in the case of arches.

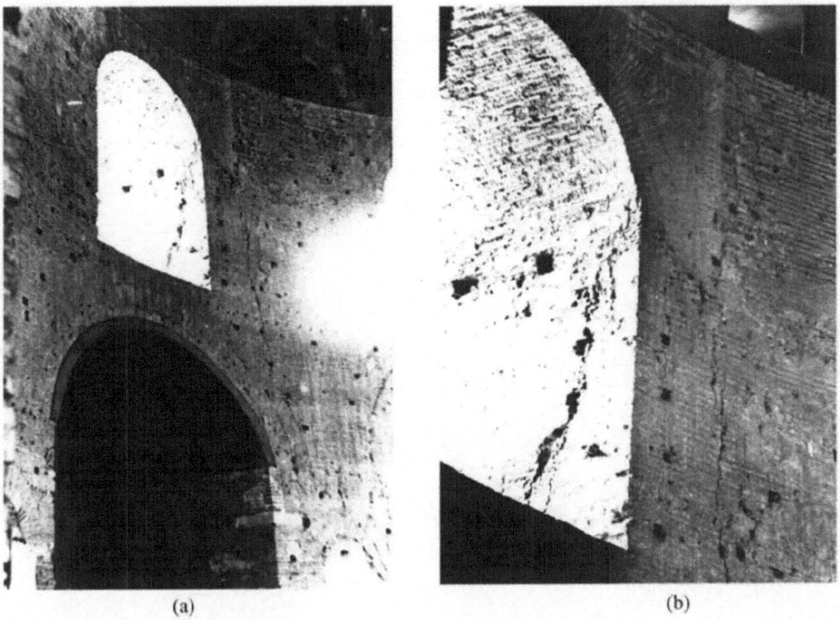

Figure 2.29 Barrel vault above a niche of Rotunda of Thessaloniki. (a) General view; (b) detail. (Courtesy: Ephorate of Antiquities of Thessaloniki (1978) Archives of Acheiropoietos church, Thessaloniki.)

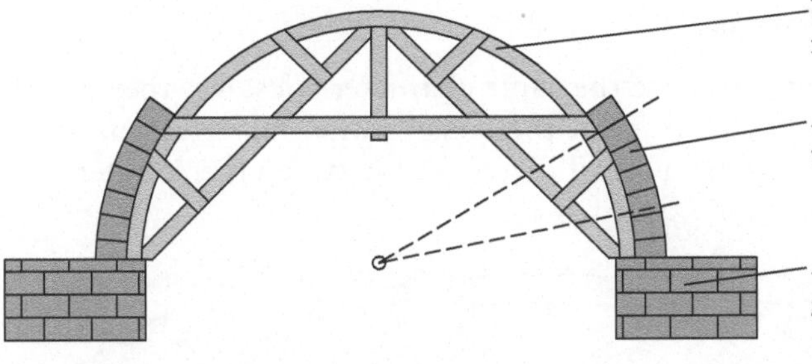

Figure 2.30 Temporary timber framework (centring) which is removed once the arch is closed. (Adapted from Choisy, A. (1873). Lárt de batir chez les Romans. Ducher etc i.e Paris.)

2.3.6 Groined and domical vaults

Groined vaults were built with the same technique as the barrel vaults up to their intersection, where two cross ribs were formed. It is the authors' opinion that the cross ribs were the first elements which were constructed on a framework and then the construction of the vaults followed, keeping the radial symmetry of the construction at the successive steps, so that horizontal forces of each member vault be balanced by the thrusts of the others. The same technique was followed for the construction of the domical vaults. Usually, these types of construction were accompanied by timber ties at the base of the groin, so that the thrust to the supporting piers can be minimised.

The above type of construction was in use very often for the formation of floors, at the galleries of churches or at the lower storeys of important buildings, in general.

Figure 2.31 Arch of Orange Theater. Voussoir arch. (Adapted from Hart, F. (1965). *Kunst and Technik der Wolbung*. Verlag George Callwey, München, Germany. Courtesy: Verlag Georg, D.W. Callwey, München.)

In this case, the space above the groined or domical vault was filled with light crushed pottery covered with a screed of Roman concrete and finishing of stone or marble plates above (Figures 2.32 and 2.33).

2.3.7 Domes and related forms

Domes, semidomes and pendentives all have the same characteristics, e.g. they are characterised by a double curvature. In this respect, the construction material could not be easily cut stone, since this would require a small divergence between the counterfaces of

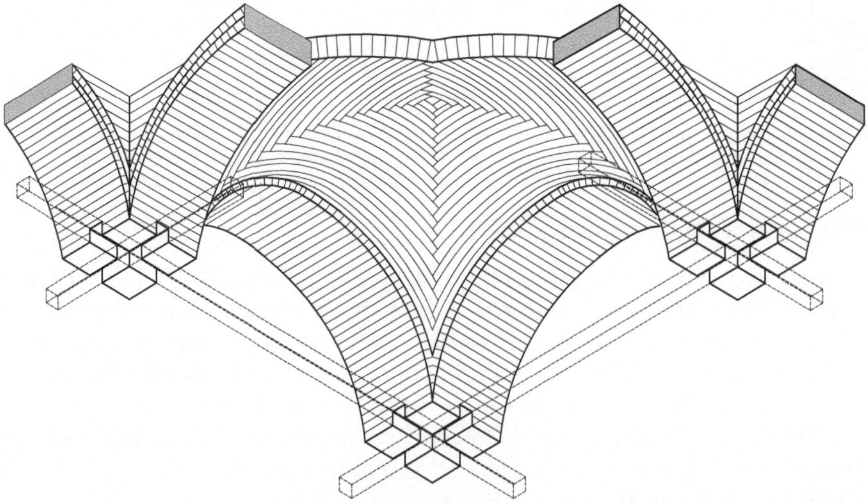

Figure 2.32 Domical vault. (Adapted from Choisy, A. (1873). *Lárt de batir chez les Romans*. Ducher etc i.e Paris.)

Figure 2.33 Groin vault, Mar Girgis, Cairo. (From Penelis S.A., Archives, 2014a.)

each stone in meridian as well as in ring directions so that the stones could adjust one by the other. An example of a stone dome is the Atreus Tomb of Mycenae with a diameter of 14.5 m (Figure 2.34). However, this dome was built following 'the ecphoric' system of construction (false dome) (Hart, 1965). The blocks were bedded more or less horizontally projecting beyond those below. In this way, the problem of inclined cuts of the stones was to a degree overpassed (Figure 2.35).

The development of the Roman concrete enabled the construction of structural elements of double curvature relatively easily, since the construction material of the joints could be poured *in situ* on a formwork or on a centring device. Pantheon of Rome is the most characteristic monument of the Roman period, with a dome of this type with a diameter of 43.44 m, being the biggest one up to Renaissance when St. Peter's Cathedral of Rome was constructed (Figure 2.36). Romans created at Pantheon dome a net of horizontal and meridian ribs using bricks and Roman concrete. On this net, after they had covered the spans between the ribs with moulds, they poured Roman concrete (Mainstone, 1975).

In the eastern part of the Empire and later, in Byzantine and Ottoman empires, domes were built with bricks and Roman mortar exclusively. The bricks were put in directions forming successive individual cyclic bands, so that the joints of the successive layers formed cones with their fisticius peak at the centre of the spherical dome (Figures 2.37 and 2.38). It is obvious that a proper centring was used although there are theories that Byzantines built it without centring using clever topographic techniques. The above procedure was also used for all other similar forms, like semidomes and pendentives.

2.3.8 Columns

Columns have been used as structural elements extensively in the past to create free space in important public buildings. Usually, they were foreseen in rows in the form of colonnades, in order to divide the space in smaller spans for bridging with wooden decks or roof trusses.

Figure 2.34 Atreus Tomb, Mycinae Greece (ecphoric building system). (From Penelis S.A., Archives, 2019b.)

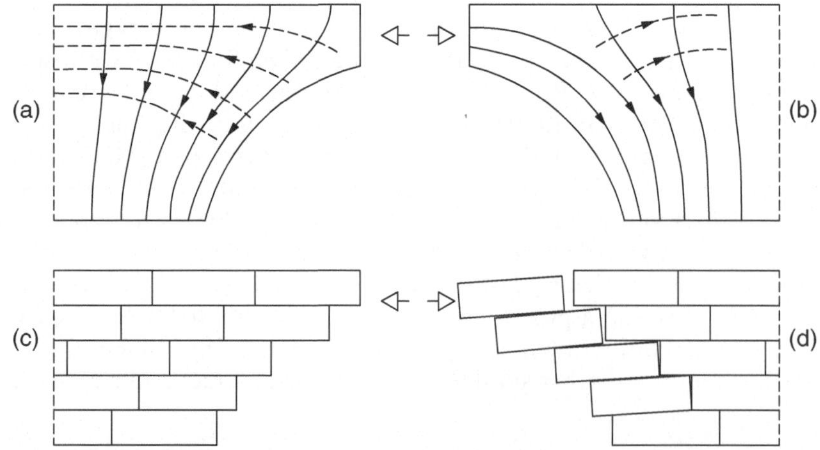

Figure 2.35 Internal static equilibrium of a false arch. Ecphoric building systems. (a) Half arch properly supported by its counterpoint; (b) half arch inadequately supported by its counterpoint; (c) structural behaviour in case (a); (d) structural behaviour in case (b). (Adapted from Mainstone, R. (1975). Developments in structural form. Penguin Books Ltd, England; Penguin Books.)

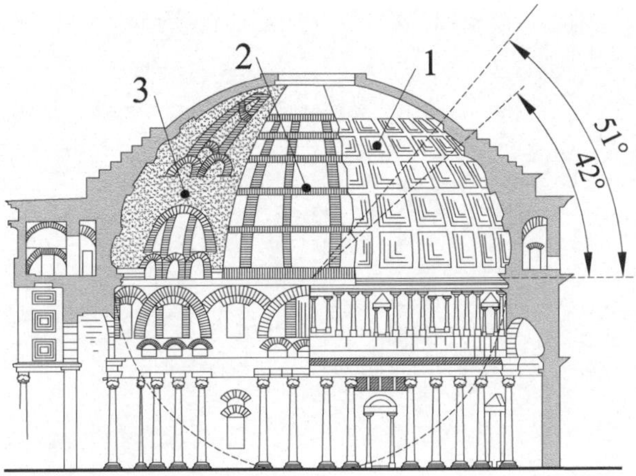

SECTION

1. Interior layer finished with lightening waffles
2. Structural brick ribs of rings and arches.
3. Outer layer of monolithic concrete reinforced with brick arches.

Figure 2.36 Section of Pantheon of Rome. (Courtesy: Andrade, Fr. (1995). La construccion romana (II). Revista de Edificación. RE. N°19 Teoriae historia de la arquitectura. p. 69.)

Figure 2.37 The Rotunda of Thessaloniki, general view. (From Penelis S.A., Archives, 2019d.)

The Christian church of basilica style is a typical example where arcades with colonnades have been used for separating the nave from the lateral aisles and the galleries (Figure 2.39).

For important buildings, the material in use was stone or marble, the column being in this way a main decorative element of the space. In this case, the column consisted of the

Figure 2.38 The Rotunda of Thessaloniki, the dome. (From Penelis, G., 1982.)

Figure 2.39 Acheiropoietos of Thessaloniki. Colonnades separating the nave from the aisles. (From Penelis S.A., Archives, 2019a.)

Figure 2.40 Acheiropoietos of Thessaloniki. Columns with Corinthian order capital. (From Penelis S.A., Archives, 2019a.)

capital, the column and the base. The capital and the base were of the shape of a cut pyramid in order to allow the spread of the loads in the areas out of the column. The column consisted usually of one stone element formed as a cylinder (Figure 2.40).

In churches, mosques or buildings of minor importance, columns were made of timber (posts). Either they were embedded in the soil or they were based on a plinth made of bricks, stones and mortar, like a small pier extended just above the floor of the space, so that the wooden column was protected from the moisture and rotting. Sometimes, the columns were formed with bricks and Roman mortar. After 19th century A.D., cast iron or steel columns have also been used.

2.3.9 Foundations

2.3.9.1 General

The foundation of a building or a bridge has been considered of high significance for thousands of years. The lower strength and stiffness of the soil in comparison to that of masonry had led to the conclusion that rock subgrade was the only safe ground for foundations. Even in the Bible, there are references to the significance of the rocky ground for foundations. So, the main task in the past for the foundation of big and important projects was for it to be built on bedrock. However, this was not an easy task. Big cities, where important buildings were constructed, were usually located in valleys or at coasts with diluvial or alluvial deposits. Therefore, bedrock could not be found at a reasonable depth below the ground level. So, the procedure described below was followed for foundations in almost all important buildings:

- The ground was excavated to take away the fill of recent deposits.
- In case that a rock stratum was found, masonry walls were built directly on this stratum, since stratum strength was considered reasonably to be higher than that of masonry.
- In case that a stiff layer of clay or sand or marl was found, foundations were layered at this layer. Depending on the strength and stiffness of this layer, being estimated qualitatively, the width and the depth of the foundation footings were decided upon.

- The material of the foundation was usually undressed stone with lime mortar and, later, lime pozzolanic mortar. In this way, the foundations might be characterised as Roman concrete foundations. These foundations could be built even under the water table.

In these buildings, soil deformations were usually of limited values influencing the serviceability of the building very little. However, there were cases of small buildings (churches, mosques etc.) which had been founded on fills. These buildings have presented serious settlements, accompanied by cracks, inclinations of the walls or even collapses.

Many times, when an acceptable soil for foundation could not be found at a reasonable depth, the loads were transferred deeper with the use of timber piles. Already Vitruvius 27B.C. (Vitruvius, 1998) makes reference to these types of deep foundations.

2.3.9.2 Case studies

Some of the most prominent monuments of Thessaloniki built during late Roman and Byzantine period were excavated locally up to their foundations during restoration works, after the earthquake of 1978. These monuments were the Rotunda (300 A.D.), the Octagon (in ruins) of the Galerian Palace (300 A.D.) and Hagia Sophia (620 A.D.). All these monuments were built on the stiff layer of clay of Thessaloniki which covers the lower part of the city up to the coast and beyond. Some details of these are given below:

- *Rotunda*: The foundation is a ring with an external diameter of 40.40m and a cross section 3.60 × 8.60m (depth to width). The foundation was built with Roman concrete (undressed stones and Roman mortar). The bottom of the foundation lies 5.0m below the ground level, while the water table is at −4.00m. It is interesting to note that the foundation ring is outwards, eccentric to the wall cylinder of the building, so that the eccentricity of the dead loads at the joint of the soil caused by the thrust of the dome is only 0.10m (Figure 2.41) (UNDP/UNIDO, 1984).
- *Octagon*: The foundation is also a ring with the external diameter of 35.0m and a cross section of 5.60 × 1.74m (width to depth) built with the same materials used for that of the Rotunda (Figure 2.42) (Athanassiou et al., 2015).

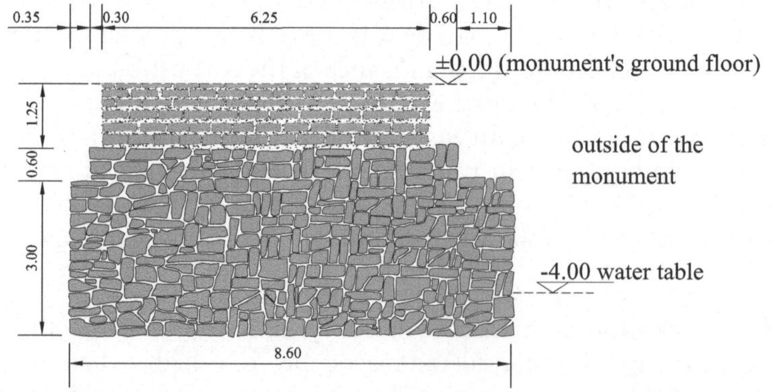

Figure 2.41 Cross section of the foundation ring of the Roman part of Rotunda Thessaloniki. (From Penelis, G., et al., 1980a.)

(a)

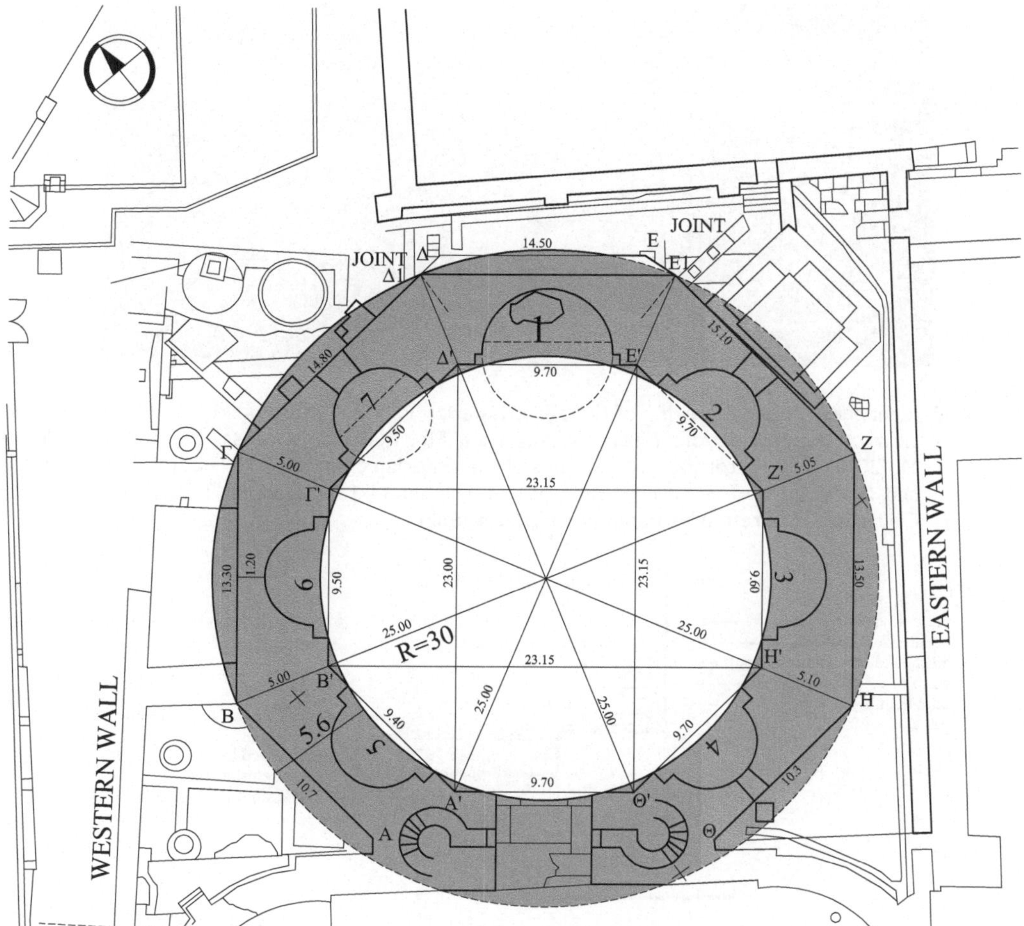

Figure 2.42 Foundation ring of Octagon, Galerian Palace, Thessaloniki. (a) Plan; (b) western view of the foundation. (From Athanassiou, Ph., Malama, V., Misa, M., and Sarandidou, M. (2015). The restoration of the ruins of the Galerian complex (1994–2014). *Documentation and Interventions, Thessaloniki.* Ephorate of Thessaloniki Antiquities. Thessaloniki, Greece. Courtesy: Ephorate of Antiquities of Thessaloniki.)

(*Continued*)

(b)

Figure 2.42 *(Continued)* Foundation ring of Octagon, Galerian Palace, Thessaloniki. (a) Plan; (b) western view of the foundation. (From Athanassiou, Ph., Malama, V., Misa, M., and Sarandidou, M. (2015). The restoration of the ruins of the Galerian complex (1994–2014). *Documentation and Interventions, Thessaloniki.* Ephorate of Thessaloniki Antiquities. Thessaloniki, Greece. Courtesy: Ephorate of Antiquities of Thessaloniki.)

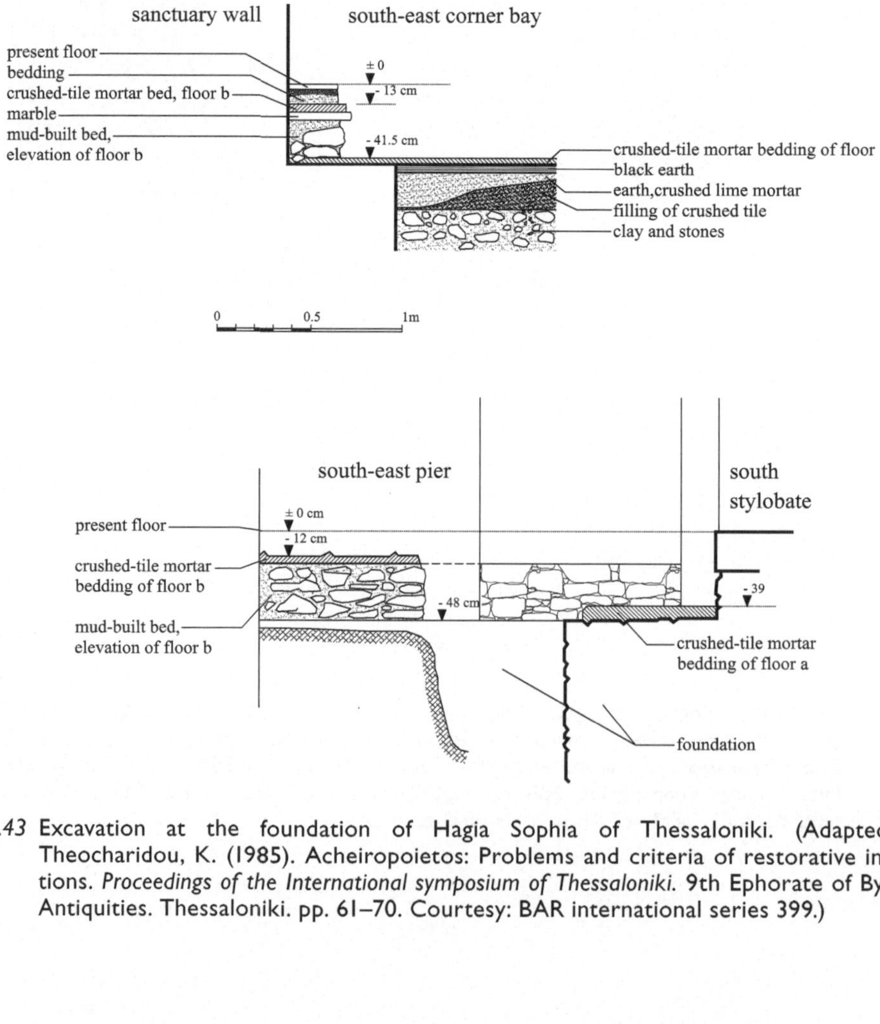

Figure 2.43 Excavation at the foundation of Hagia Sophia of Thessaloniki. (Adapted from Theocharidou, K. (1985). Acheiropoietos: Problems and criteria of restorative interventions. *Proceedings of the International symposium of Thessaloniki.* 9th Ephorate of Byzantine Antiquities. Thessaloniki. pp. 61–70. Courtesy: BAR international series 399.)

- *Hagia Sophia*: The foundations of the walls have cross section of 2.50 × 4.50 m (depth to width) being wider than the supported wall (2.0 m), so that the loads are distributed to the soil smoothly. The construction material is Roman concrete (Figure 2.43) (Theocharidou, 1988).

Similar findings were ascertained in Cairo (Babylon tower–Mar Girgis church (Figure 2.44)).

From all the above, it is concluded that for thousands of years, in case of important works, special attention was paid to the foundation and particularly to the soil layer on which the structure would rest.

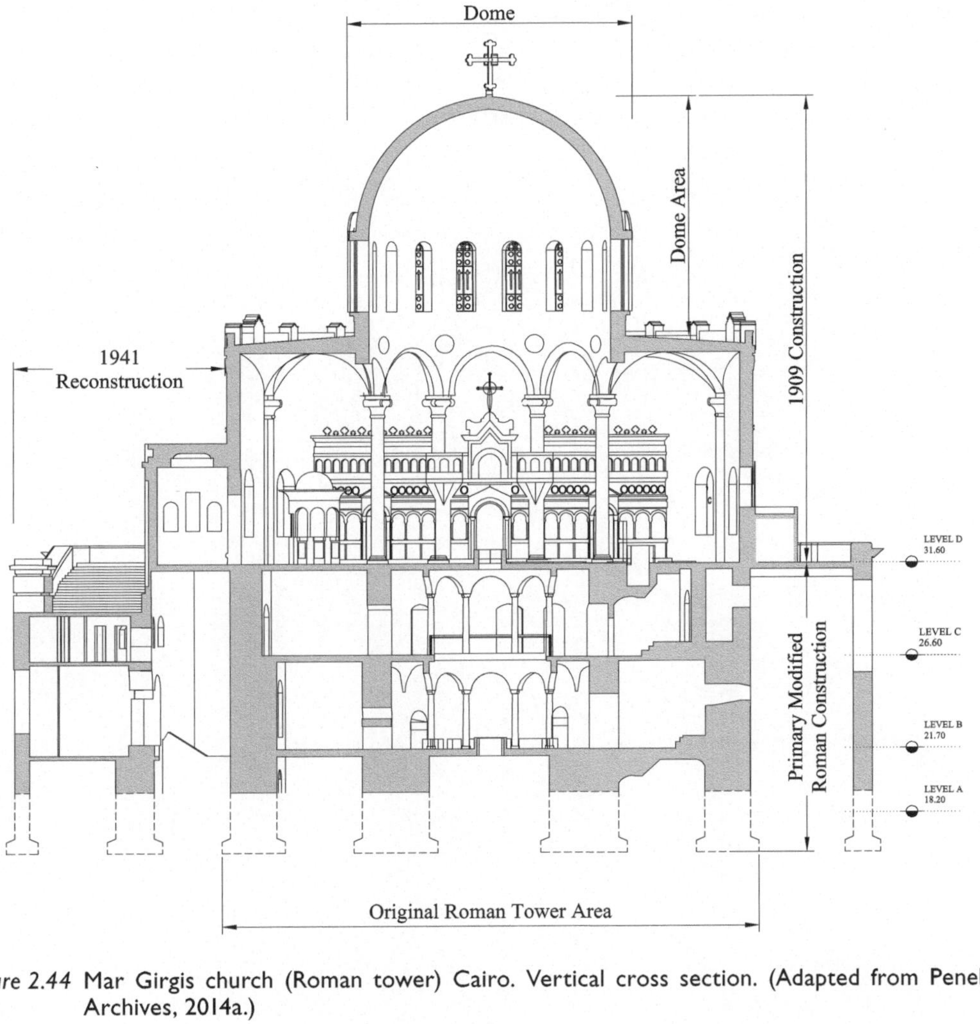

Figure 2.44 Mar Girgis church (Roman tower) Cairo. Vertical cross section. (Adapted from Penelis, Archives, 2014a.)

Chapter 3

Masonry mechanics

3.1 GENERAL

Brick or stone masonry has been the most common material in construction for thousands of years. However, its strength and deformation is very difficult to describe. The microstructure of masonry, i.e. the complex arrangement of horizontal and vertical joints, their thickness and the mechanical properties of the material of the units and mortar, influences the failure mechanism and strength anisotropy of the compound material.

Simplified approaches have been formulated in the past to analyse masonry structural behaviour under vertical and horizontal loading conditions. These methods had a semiempirical approach based, mainly, on the statistical evaluation of experimental results in connection with simplified analytical models for the formulation of closed formulas for mechanical properties of masonry. The main assumption in all these approaches is that the masonry functions as a homogeneous elastic material, although it presents in reality a strong nonelastic orthotropy, as far as its strength and stiffness are concerned, while its main characteristic is that it presents a *very low and unreliable tensile strength*.

Owing to the enormous advances of the computational mechanics in the last twenty years, a wide spread of numerical modelling techniques have been developed aiming at the simulation of an accurate structural behaviour of masonry. These models, starting from the simulating of bricks and joints with inelastic microelements, aim at the formulation of constitutional laws of inelastic orthotropic macroelements, proper for introduction of this output in FEM platforms for inelastic analysis of orthotropic masonry structures, taking into account early crack due to tension.

Closing, it should be noted that in the last decades, experimental and analytical research has been developed for an accurate approach to seismic behaviour of masonry, namely the response to permanent axial load combined with shear reversals aiming at the formation of the strength envelope (H–δ curve, envelope H–δ reversal curves), the M–θ curve and the corresponding ductility μ, which constitute the required values for seismic analysis and design.

In the following sections, a short presentation of these two approaches will be made, namely the simplified semiempirical one and the microelement analysis of masonry, both for gravity loads (static) and seismic recycling loading (shear loading).

3.2 MECHANICAL PROPERTIES OF MASONRY CONSIDERED AS A HOMOGENEOUS MATERIAL

3.2.1 Introduction

For many years and even now, for modern masonry structures, the structural systems were considered for the *analysis* as consisting of a linear elastic and homogeneous material. Then in *design*, based on the above analysis, the design strength of masonry in compression, shear, and out-of-plane bending should be taken into account *under the basic consideration that the in-plane tension capacity is zero*. Therefore, the determination of the mechanical properties of masonry as a homogeneous material, based on the properties of units (bricks or stones) and mortar joints, precedes the *analysis and design* of the structural system under consideration. Thereafter, the properties of the constituent materials are not used in analysis and design anymore. Under the assumption of a homogeneous compound, the following basic mechanical properties of masonry must be introduced in analysis and design of masonry structures:

- The compressive strength f_{wd}
- The shear strength f_{vd}
- The out-of-plane flexural strength f_{xd}
- The stress–strain (σ–ε) relationship
- The modulus of elasticity E_w
- The shear modulus G_w
- The Poisson ratio v_w.

A direct approach to the problem would be the determination of the above properties by extended laboratory tests using masonry units and mortar of the materials specified for the construction. Codes in effect specify the procedure for specimen preparation and testing. In Figure 3.1 (Tomazevic, 1999), the specimens specified by EN1052-1 (EN1052-1, 1999) are depicted. According to this standard, three identical specimens should be tested either in the form of small wallettes of at least 1.5 units length and 3 units height or

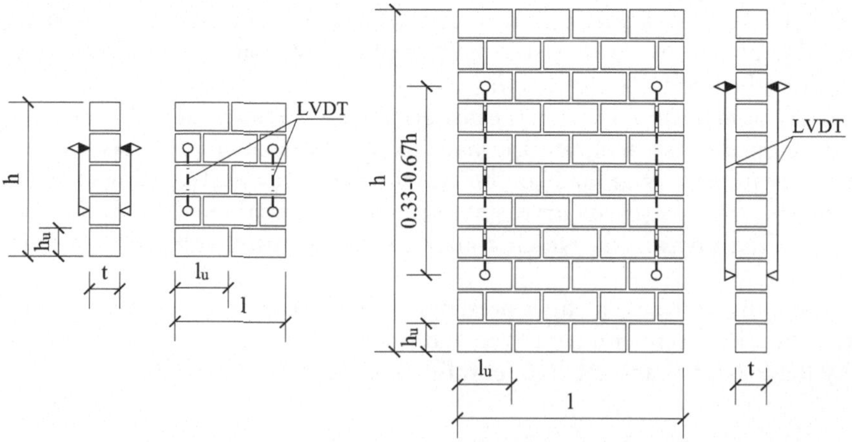

Figure 3.1 Masonry specimens for compression test. (Adapted from EN1052-1 (1999). Methods of test for masonry. Determination of compressive strength B.S.I., UK.)

walls of 1.0–1.80 m length and 2.4–2.7 height. The above procedure is time and cost consuming, even for conventional masonry buildings. Particularly for existing monumental buildings, this procedure is prohibitive, since it is not allowed to extract wall specimens for testing even in the dimensions of small wallettes. Therefore, since the early 20th century, extended investigations have been carried out for the determination of the main mechanical properties of masonry, using information based on the geometry of the masonry texture and the mechanical properties of the masonry units and of the mortar. These investigations have been based mainly on analytical models combined with statistical evaluation of laboratory tests.

3.2.2 Compressive strength and deformation

3.2.2.1 Failure theory based on elastic analysis

Consider the masonry specimen of Figure 3.2 subjected to axial compression P_y. The lateral stresses developing in the brick and the mortar, respectively, are also shown in this figure.

Under the assumption of linear behaviour of both materials, the corresponding lateral deformations for the brick ε_{bx} ε_{bz} and for the mortar ε_{mx}, ε_{mz} are given below (Hendry, 1981):

$$\varepsilon_{bx} = \frac{1}{E_b}\left[\sigma_{bx} + \nu_b\left(\sigma_y - \sigma_{bz}\right)\right], \tag{3.1a}$$

$$\varepsilon_{bz} = \frac{1}{E_b}\left[\sigma_{bz} + \nu_b\left(\sigma_y - \sigma_{bx}\right)\right], \tag{3.1b}$$

$$\varepsilon_{mx} = \frac{1}{E_m}\left[-\sigma_{mx} + \nu_m\left(\sigma_y - \sigma_{mz}\right)\right], \tag{3.2a}$$

$$\varepsilon_{mz} = \frac{1}{E_m}\left[-\sigma_{mz} + \nu_m\left(\sigma_y - \sigma_{mx}\right)\right], \tag{3.2b}$$

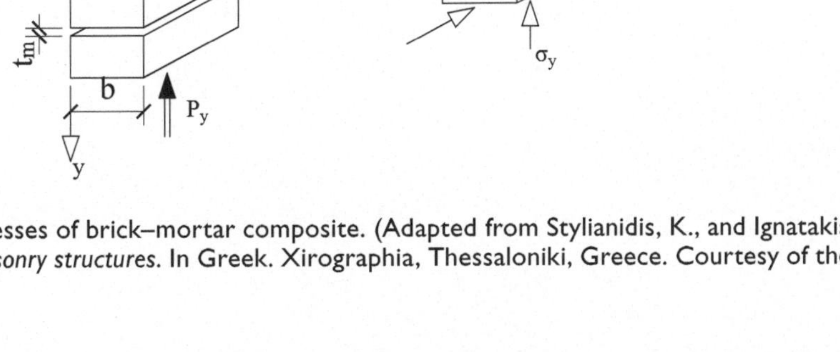

Figure 3.2 Stresses of brick–mortar composite. (Adapted from Stylianidis, K., and Ignatakis, Ch. (1997). *Masonry structures*. In Greek. Xirographia, Thessaloniki, Greece. Courtesy of the authors.)

where

E_b and E_m are the moduli of elasticity of brick and mortar, respectively, and ν_b and ν_m are the corresponding Poisson's ratios.

The lateral deformations of these two materials, due to bonding, are equal:

$$\varepsilon_{bx} = \varepsilon_{mx}, \tag{3.3a}$$

$$\varepsilon_{bz} = \varepsilon_{mz}. \tag{3.3b}$$

For equilibrium reasons, the lateral tensile force developing at the brick must be equal to the corresponding compressive lateral force of the mortar; in other words,

$$\sigma_{bx}.d.t_b = \sigma_{mx}.d.t_m, \tag{3.4a}$$

$$\sigma_{bz}.b.t_b = \sigma_{mz}.b.t_m, \tag{3.4b}$$

or

$$\sigma_{mx} = a.\sigma_{bx}, \tag{3.5a}$$

$$\sigma_{mz} = a.\sigma_{bz}, \tag{3.5b}$$

where

$$a = \frac{t_b}{t_m} \geq 1.0. \tag{3.6}$$

The introduction of equations (3.1) and (3.2) in equation (3.3) and then in equation (3.5) results in the following expression:

$$\sigma_{bx} = \sigma_{bz} = \frac{\beta v_m - v_b}{1 + a\beta - v_b - a\beta v_m}\sigma_y, \tag{3.7}$$

where

$$\beta = \frac{E_b}{E_m}. \tag{3.8}$$

Assuming a linear relationship for the envelope of the ultimate axial compressive stress and the corresponding lateral tensile stress for bricks (see Figure 3.3) results in the following relation:

$$\sigma_{bx} = \frac{1}{\varphi}\left(\sigma_{bc} - \sigma_y\right), \tag{3.9}$$

where

$$\varphi = \frac{f_{bc}}{\sigma_y}. \tag{3.10}$$

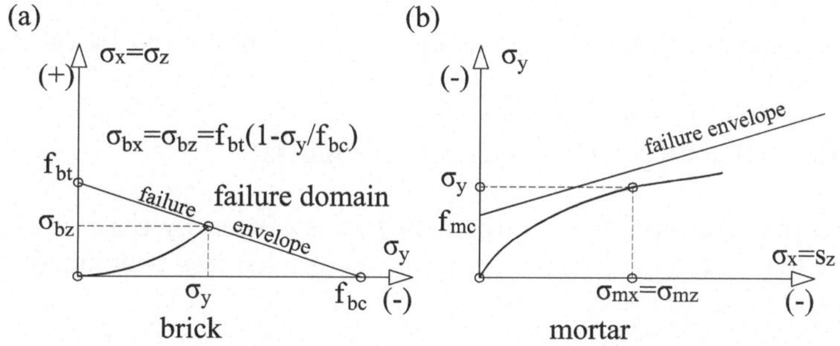

Figure 3.3 Assumed failure envelopes of constituent masonry materials under biaxial loading: (a) brick: compression–tension, (b) mortar: compression–compression. (Adapted from Stylianidis, K., and Ignatakis, Ch. (1997). *Masonry structures*. In Greek. Xirographia, Thessaloniki, Greece. Courtesy of the authors.)

Taking into account that at failure state

$$\sigma_y = f_{wc}, \tag{3.11}$$

where

f_{wc} the compressive strength of the wall,

and introducing equation (3.9) in equation (3.7), we get

$$\boxed{f_{wc} = \frac{1}{1+\dfrac{\varphi(\beta\nu_m - \nu_b)}{1-\nu_b + a\beta(1-\nu_m)}} f_{bc}.} \tag{3.12}$$

3.2.2.2 *Failure theory based on the strength of brick and mortar under multiaxial strength*

Hilsdorf (1969) has proposed a failure theory of brickwork under uniaxial compression, based on an assumed linear relationship between lateral biaxial tensile strength and axial compressive strength of the brick (Figures 3.2 and 3.3). This assumption was based on Hilsdorf's theory for confined concrete. According to this theory, the triaxial strength of mortar may be given by the following equation:

$$f_1' = f_c' + 4.1\sigma_2, \tag{3.13}$$

where

f_1' is the compressive strength of a laterally confined cylinder
f_c' is the uniaxial compressive strength of the cylinder
σ_2 is the lateral confinement stress of the cylinder.

This leads to a minimum *required* lateral confinement of the mortar joint of the order of

$$\sigma_{mx} = \frac{1}{4.1}\left(\sigma_y - f_{mc}\right), \tag{3.14}$$

where

σ_{mx} is the lateral compressive stress in the mortar joint caused by the confinement of the bricks
σ_y is the axial stress in y–y direction
f_{mc} is the uniaxial compressive strength of the mortar.

Having in mind that compressive confinement forces caused by the bricks to the mortar are equal to the tensile lateral forces, acting on the bricks, it is concluded that the lateral tensile forces on the bricks are equal to

$$\sigma_{bx} = \frac{t_m}{4.1t_b}\left(\sigma_y - f_{bc}\right). \tag{3.15a}$$

Introducing (3.15a) in (3.9), we get

$$\boxed{f_{wc} = \frac{f_{bt} + af_{mc}}{f_{bt} + af_{bc}} f_{bc},} \tag{3.15b}$$

where

$$a = \frac{t_m}{4.1t_b} \tag{3.16}$$

and

f_{wc} is the compressive strength of the wall.

The above value of f_{wc} in Equation (3.15b) must be reduced by a coefficient u_u characterised by Hilsdorf – the 'coefficient of nonuniformity', which is determined experimentally. This coefficient varies about $u_u \cong 1.30$.

3.2.2.3 *Semiempirical formulas for the compressive strength of masonry.*

In parallel, numerous experimental investigations have been carried out for the determination of the compressive strength of masonry wall. Some characteristic case studies are given below.

$$\boxed{f_{wc} = \frac{f_{bc}}{6} + \frac{\sqrt{f_{bc} \cdot f_{mc}}}{4} - \frac{f_{mc}}{20} + 1.40 \text{ [MPa]},} \tag{3.17}$$

(Tassios, 1992)

$$\boxed{f_{wc} = \left(\frac{2}{3}\sqrt{f_{bc}} - a\right) + \beta f_{mc}} \tag{3.18}$$

where

$\beta = 0.10$ for brickwork
$\beta = 0.50$ for stonework
$\alpha = 0.50$ for dressed stonework or brickwork

$\alpha = 2.50$ for raw stonework
(f_{bc}, f_{mc}, f_{wc} in MPa)

(Tassios and Chronopoulos, 1985)

$$\boxed{f_{wc} = 0.7\sqrt{f_{bc}}\sqrt[3]{f_{mc}},} \tag{3.19}$$

(Bröcker, 1961)

where f_{wc}, f_{bc}, f_{mc} are in MPa.

In the above equation (3.19), f_{bc} and f_{mc} refer to mean values of material qualities.

Most of the codes of practice, for strictly specified masonry construction schemes, include formulas or tables for the determination of the compressive strength of masonry. For example, EN1996-1-1/2005 specifies the following formula:

$$\boxed{f_{wc} = kf_{bc}^{07} \cdot f_{mc}^{0.3} \text{ (MPa)},} \tag{3.20}$$

where k is a coefficient dependent on the type of brick units and joint thickness, the last one varying between 6 mm and 15 mm. In this case, values of k range between 0.55 and 0.35.

It should be noted that the above formula (3.20) is not valid for thick joints which are the usual case in Roman, Byzantine and Islamic monuments. The results of an experimental research carried out in the Laboratory of Reinforced Concrete and Masonry Structures of Aristotle University of Thessaloniki (AUTH) (Papayianni, J. and Karaveziroglou, M., 1993b) may be used successfully for the extrapolation of the results of thin joints to thick ones (Figure 3.4).

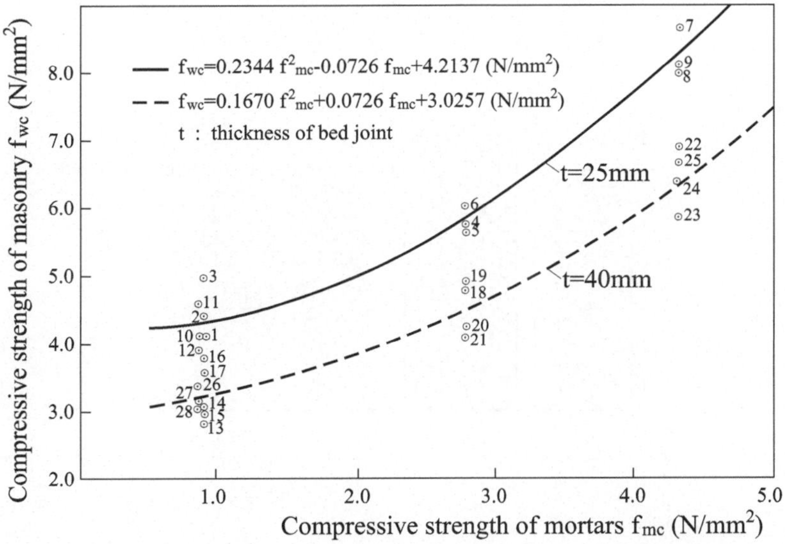

Figure 3.4 Compressive strength of masonry specimens in relation to the compressive strength of the mortar. (Papayianni, J. and Karaveziroglou, M., 1993b. Adapted from Penelis, G. (1995). Mortars and grouts in restoration of roman and byzantine monuments. *European seminar on Ancient and traditional mortar technology and their importance in conservation projects.* Rethymnon, Creta.)

Closing, it would be interesting to make here a special reference to a critical review of the various methods that have been presented so far on the estimation of unreinforced masonry compressive strength based on mortar and unit mechanical parameters (Marotta, et al., 2016).

3.2.2.4 *Compressive strength of three-leaf masonry walls*

Three-leaf masonry under compression is the most usual construction form in big masonry structures like fortification walls, walls or piers of monumental buildings of large dimensions etc.

In this type of construction, the compressive load is carried mainly by the external leaves (Figure 3.5) (Vintzileou, 2011 and Penelis, 1982). In fact, the compressive strength as well as the modulus of elasticity of the filling material is normally by an order of magnitude smaller than that of the external leaves. Therefore, as it was proven experimentally by Egermann (1993) and Binda, et al. (2006), the external leaves carry the bigger part of the applied load. Furthermore, due to the incompatibility of the deformations between the external and internal leaves, their bond may be broken; consequently, the external leaves are subject to out-of-plane deformations due to the higher dilatancy of the filling material. The latter, on the contrary, is confined by the resistance of the external leaves. The above description explains the crack configuration of three-leaf walls subjected in lab tests to axial compression (Figure 3.6).

According to various studies (Valluzzi, et al., 2004; Binda, et al., 1991; Egermann, 1993; Vintzileou and Tassios, 1995; Tassios, 2004), the compressive strength of a three-leaf wall can be expressed by a three-leaf model. Based on some simplified assumptions, the ultimate strength of the wall may be calculated using the compressive strength and the volumetric contribution of the external and internal layers:

$$\boxed{f_{wc,o} = \left(V_{ex}/V\right)\lambda_{ex} f_{ex,k} + \left(V_{inf}/V\right)\lambda_{inf} \cdot f_{inf,o},} \tag{3.21}$$

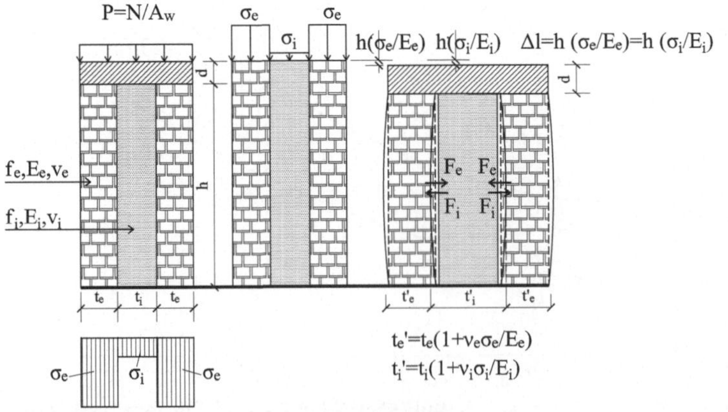

Figure 3.5 Stresses and deformations in external leaves and core in a three-leaf-masonry subjected to compression. (Adapted from Vintzileou, E. (2006). Grouting of three-leaf stone masonry: Types of grouts, mechanical properties of masonry before and after grouting, in Structural analysis of historical constructions, Lourenço, P.B., Roca, P., Modena, C., Agrawal, S. (eds.). New Delhi. pp. 41–58. Courtesy of the author.)

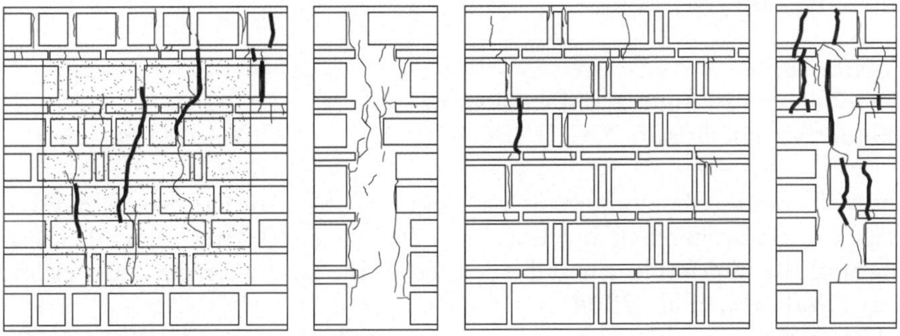

Figure 3.6 Typical failure mode of three-leaf-masonry in compression. (Adapted from Vintzileou, E. (2006). Grouting of three-leaf stone masonry: Types of grouts, mechanical properties of masonry before and after grouting, in *Structural analysis of historical constructions*, Lourenço, P.B., Roca, P., Modena, C., Agrawal, S. (eds.). New Delhi. pp. 41–58. Courtesy of the author.)

where

V_{ex}/V is the volumetric ratio of external layers (both ones)
V_{inf}/V is the volumetric ratio of infill
f_{wco} is the compressive strength of the three-leaf wall
$f_{ex,k}$ is the characteristic compressive strength of the external leaves
f_{info} is the compressive strength of the infill
$\lambda_{inf}, \lambda_{ex}$ are corrective factors, which take into account the influence of the mutual interaction between the external layers and the core; λ_{inf} must be higher than unity ($\lambda_{inf} > 1$) as the core is confined by the external leaves while λ_{ex} must be less than unity ($\lambda_{ex} > 1$) as the external leaves are subjected to thrust actions by the core due to its dilatancy.

The above considerations made by Egermann (1993) who has proposed the above formula (3.21) are qualitative, as far as λ_{ex} and λ_{inf} are concerned. Tassios (2004) has made an attempt to quantify λ_{ex} and λ_{inf} calibrating a series of experimental results (Vintzileou and Tassios, 1995; Toumbakari, 2002; Valluzzi, et al., 2001). The output of this attempt is given below:

$$\lambda_{inf} \cong 1.0, \quad (3.22a)$$

$$\lambda_{ex} \cong 1 - 0.06\ \zeta_e t_e h_w^{-4/3} \left(t_e, h_w \text{ in mm}\right), \quad (3.22b)$$

(Tassios, 2004)

where

$\zeta_e = E_{we}/f_{wc,e}$,

t_e is the external joint thickness
h_w is the distance between transverse connections
E_{we} is the modulus of elasticity of the external masonry leaves
f_{wce} is the compressive strength of the external leaves.

Mean values of ζ_e are given below:

$\zeta_e \cong 500$ for *rubble*
$\cong 1{,}000$ for clay units and partially cut stone
$\cong 1{,}500$ for well-cut stone.

In the above formulas, the compressive strength of the external leaves may be defined from flat jack tests (Gregorezyk and Lourenço 2000) (see Chapter 6), while the compressive strength of the internal leaf may be defined in the lab by testing cores extracts from the masonry (Valluzzi, et al., 2004).

In the last decades, analytical modelling of three-leaf masonry walls using finite element method was attempted by several researchers (Ramalho, et al., 2008; Binda, et al., 1994) with promising results.

From all the above, it is concluded that additional research is needed in the field.

3.2.2.5 *Compressive deformations*

The stress–strain curve (σ–ε) of masonry loaded perpendicularly to its bed joints may be simulated by a parabola (Figure 3.7) (Hendry, 1981):

$$\frac{\sigma}{f_{wc}} = 2.0\left(\frac{\varepsilon}{\varepsilon_u}\right) - \left(\frac{\varepsilon}{\varepsilon_u}\right)^2, \tag{3.23}$$

where

f_{wc} and ε_u are, respectively, the stress and the strain at the maximum ordinate of the curve. The initial tangent modulus is given by the expression

$$E_{wo} = 2.0\frac{f_{wc}}{\varepsilon_u}. \tag{3.24}$$

Modulus of elasticity E_w is defined as a secant modulus at service load condition i.e. at 1/3 of the axial strength.

The values of E_w present a high dispersion range:

$$\boxed{E_w = (400\text{–}1{,}000) f_{wc}.} \tag{3.25}$$

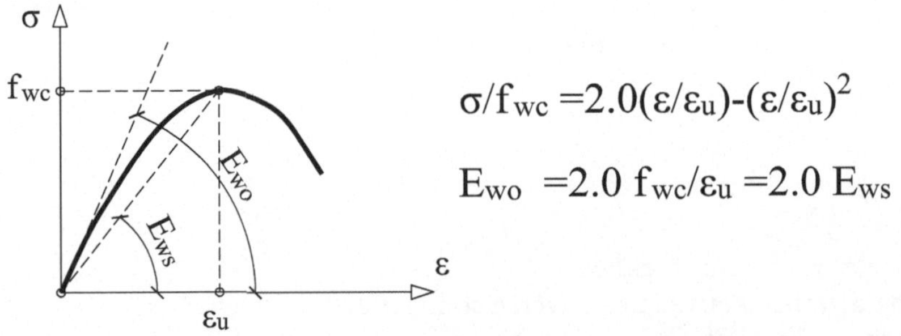

Figure 3.7 Stress–strain curve for brickwork.

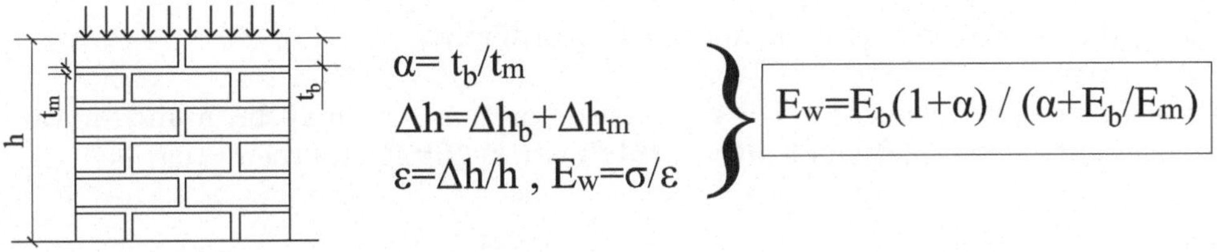

Figure 3.8 Analytical assessment of masonry modulus of elasticity.

In case of uncodified masonry (e.g. Roman masonry), E_w may be assessed analytically taking into account the mechanical properties of bricks and mortar (Figure 3.8) as follows (Penelis, et al., 1980a):

$$\boxed{E_w = E_b \frac{1+a}{aE_b/E_m},} \tag{3.26}$$

where

$$a = \frac{t_b}{t_m}$$

t_b is the thickness of the brick
t_m is the thickness of the mortar
E_b is the modulus of elasticity of bricks
E_m is the modulus of elasticity of mortar
E_w is the modulus of elasticity at service load condition.

Eurocode EN1996-1-1/2005 does not define a formula for σ–ε curve. However, for *design purposes* σ–ε diagram of concrete (Figure 3.9) is adopted, and the following value is specified for modulus of elasticity:

$$\boxed{E_w = 1.000\ f_{wc,k},} \tag{3.27}$$

while for long-term values of modulus of elasticity, the following expression is specified:

$$\boxed{E_w^{\text{longterm}} = \frac{E_w}{1+\varphi_\infty},} \tag{3.28}$$

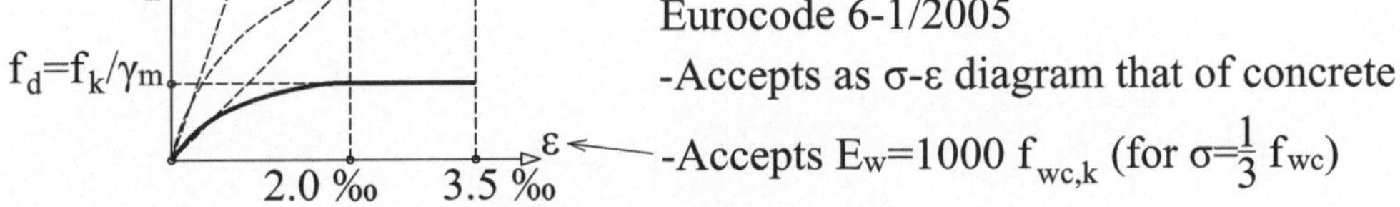

Figure 3.9 σ–ε diagram for masonry design at ultimate limit state.

where

φ_∞ is the final creep coefficient specified accordingly.

On the other hand, values of Poisson's ratio for masonry may be found in various references (EN 1996-1, 2005; Hendry, 1981; Tassios, 1992), and their range is

$$\boxed{\nu_w = 0.10\text{–}0.30.} \tag{3.29}$$

3.2.3 Tensile strength

Generally, the masonry tensile strength f_{wt} is a small percentage of the corresponding compressive one, and what is more important is it *has an unreliable value*. Therefore, in design as in concrete case, f_{wt} *is considered to be of zero value* (Figure 3.10) for in-plane loading.

As it has been found by correlating a large number of test results, the ratio between the tensile and compressive strength of any type of masonry varies between (Tomazevic, 1999)

$$\boxed{0.03\, f_{wc} \le f_{wt} \le 0.09\, f_{wc}.} \tag{3.30}$$

Failure mechanisms of masonry under tension are depicted in Figure 3.11.

For bending of walls out of plane, due to wind or earthquake, codes specify a *tensile strength* given in proper tables for each type of masonry unit and mortar.

The failure scheme in this case is depicted in Figure 3.12.

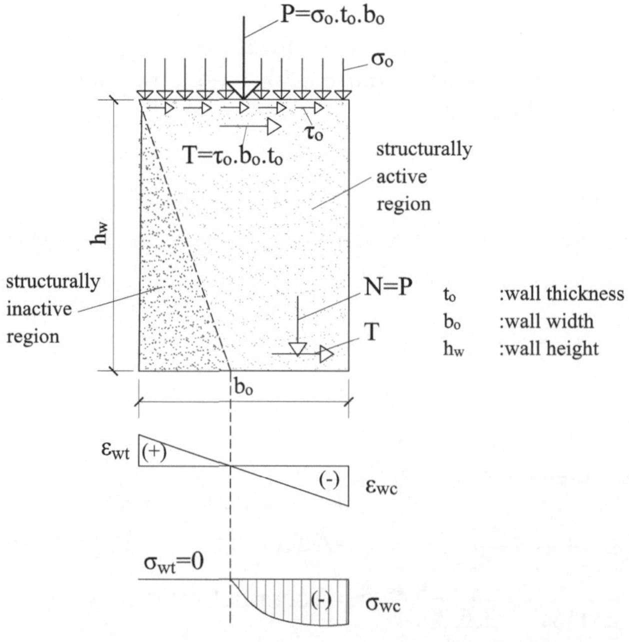

Figure 3.10 Conceptual design of a free-standing wall to an eccentric compressive axial load.

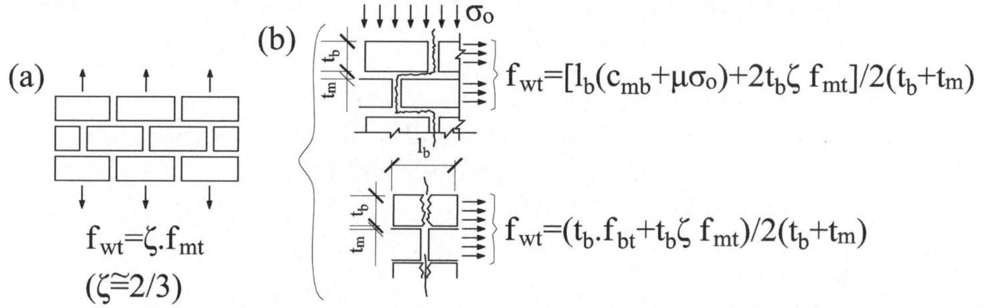

Figure 3.11 Failure mechanisms of masonry under tension. (a) Vertical tensile failure; (b) horizontal tensile failure. (Adapted from Tassios, Th. (1992). *Masonry mechanics*. In Greek. Athanassopoulos – Papadonis Ltd, Athens, Greece.)

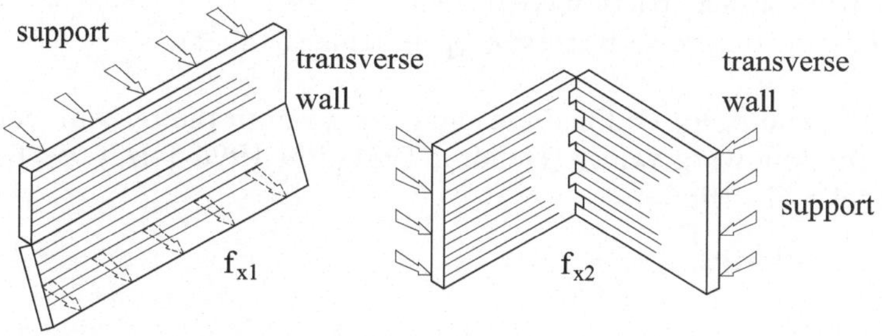

Figure 3.12 Wall failure parallel and vertical to bed joints. (Adapted from Tomazevic, M. (1999). Earthquake – resistant design of masonry buildings. Imperial College Press, London, UK.)

3.2.4 Shear strength and deformation

3.2.4.1 Shear strength

Brick masonry walls are frequently subjected to shear in addition to compression, particularly in seismic regions. The failure envelope of this type of loading may be found in Figure 3.13. This envelope may be divided into three basic regions, according

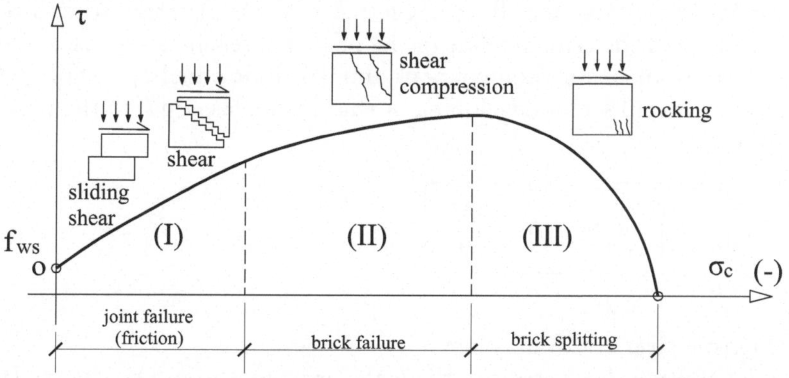

Figure 3.13 Typical failure envelope of masonry to compressive and shear loading. (Adapted from Stylianidis, K., and Ignatakis, Ch. (1997). *Masonry structures*. In Greek. Xirographia, Thessaloniki, Greece. Courtesy of the authors.)

to the relevant failure mechanism which corresponds in each of them (Stylianidis and Ignatakis, 1997).

- *Shear slide at the mid-height of the wall (region I).*
  At low precompression, failure cracks tend to develop stepwise along the brick mortar joints at an angle of approximately 45°. In this case, shear strength is given by the expression (Mohr–Coulomb theorem)

$$\boxed{f_v = f_{vo} + \mu_c \sigma_d,} \tag{3.31}$$

where
f_{vo} is the shear strength under zero compression stress
μ_c is a constant defining the contribution of compression stresses in developing friction resistance (friction coefficient)
σ_d is the design compression stress perpendicular to shear.

Values of f_{vo} and μ_c for codified masonry are given in the relevant codes and range between the following values (Hendry, 1981; EN 1996-1-1/2005; Tomazevic, M., 1999; Penelis, G., 1980a):

$-f_{vo}$: 0.15–0.30 MPa,

$-\mu_c$: 0.30–0.50.

The above equation (3.31) is valid for values of σ_d less than 2.0 MPa, and it has been also adopted by EN1996-1-1/2005.
In the same region, and particularly for very low axial stresses, the failure mode may have the form of a sliding plane at a bed joint.

- *Break of masonry units at the mid-height of the wall (region II)*
  As compression increases, the principal tensile stress causing the crack becomes inclined at an angle *greater than* 45° to the horizontal and passes vertically through the bricks as well as along the bed joints. This type of failure is characterised as *shear compression failure mechanism.*
  A number of investigators have suggested that the criterion of failure should be the exceedance of the critical value of the principal tensile stress. Assuming that the wall behaves as an elastic, homogenous and isotropic panel up to failure ($H = H_{max}$), the tensile principal stress developing at that instant is equal to the tensile strength of masonry:

$$\boxed{f_t = \sigma_1 = \sqrt{\left(\frac{\sigma_o}{2}\right)^2 + (\beta \tau_{bmax})^2} - \frac{\sigma_0}{2},} \tag{3.32}$$

where
f_t is the tensile strength of masonry
τ_{bmax} is the average shear stress in the wall for its maximum resistance H_{max}
β is the shear stress distribution factor equal to $\beta = 1.5$ for orthogonal wall cross section.
σ_o is the compressive stress.

The above equation (3.32) may be modified easily to the following expression (Tomazevic, 1999):

$$\boxed{f_v = \tau_{h\max} = \frac{f_t}{\beta}\sqrt{\frac{\sigma_d}{f_t}+1}.} \tag{3.33}$$

The principal plane orientation in this case is defined by the following expression:

$$\varphi_t = 0.50\arctan\frac{2f_v}{\sigma_0}. \tag{3.34}$$

The above procedure is more suitable in case of piers where the axial load is high.

A similar procedure was introduced by the first of the authors of this book in the case of Rotunda of Thessaloniki, for the assessment of the bearing capacity of the main piers with a cross section 6.25 × 6.25 m under axial loads and shear (Figure 3.14) (UNDP/UNIDO, 1984, Penelis, G., 1982).

- *Local compressive failure at the bottom corner of the wall (region III)*

  Finally, for high values of compression stresses σ_o, the failure of axial compression with bending prevails (Figure 3.15), presenting splitting mode of failure at the compressed corner.

  In this case, the failure moment M_{Rd} may be obtained from the following expressions (EN1996-1/2005 Greek National Annex).

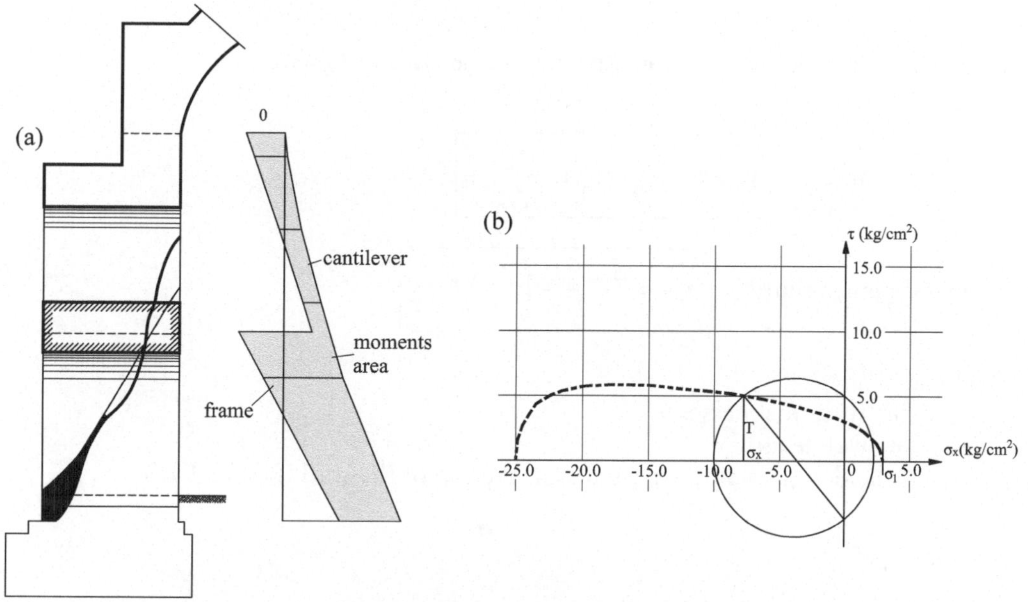

Figure 3.14 Rotunda. Shear–compression failure of piers P_2 and P_3. (a) Shear–compression crack of P_2 and P_3 in relation to the calculated slope of the crack, (b) σ–τ envelope of masonry based on experimental results for σ_{wd} (−25 kp/cm²) and τ_d (3.5 kp/cm²) and the $\sigma_1(+)/\sigma_2(-)$ experimental envelopes for low-strength concrete. (Kupfer, 1973) (Penelis G., 1992)

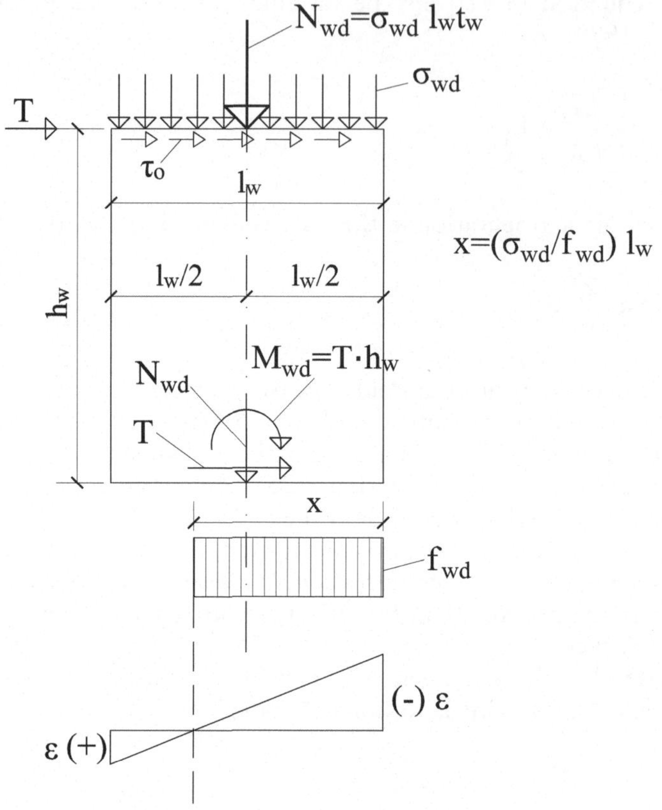

Figure 3.15 Determination of M_{Rd} for in-plane compression failure of a wall.

In-plane failure: $$\boxed{M_{Rd} = \frac{\sigma_{wd} t_w l_w^2}{2}\left(1 - \frac{\sigma_{wd}}{f_{wd}}\right)}, \quad (3.35)$$

Out-of-plane failure: $$\boxed{M_{Rd} = \frac{\sigma_{wd} t_w^2 l_w}{2}\left(1 - \frac{\sigma_{wd}}{f_{wd}}\right)}, \quad (3.36)$$

where
t_w is the wall thicknesses
l_w is the wall length
σ_{wd} is the design compressive stress at the top of the wall (Figure 3.15)

$$\sigma_{wd} = \frac{N_{wd}}{l_w \cdot t_w}$$

f_{wd} is the design compressive strength of masonry.

The above equations (3.35) and (3.36) have been obtained under the assumption of rectangular normal stress block diagram at failure.

3.2.4.2 Deformations

Shear deformations of masonry present strong nonlinearity τ–γ. Shear modulus G_w increases with axial compression.

Eurocode EN1996-1-1/2005 adopts a mean value

$$\boxed{G_w = 0.40\ E_w.} \tag{3.37}$$

Having in mind that there is the relation in theory of elasticity between G and E,

$$G_w = \frac{E_w}{2(1+\nu_w)}, \tag{3.38}$$

it is concluded that Eurocode EC-6 has adopted indirectly for ν_w a value

$$\boxed{\nu_w = 0.25.} \tag{3.39}$$

3.3 AXIAL COMPRESSIVE LOAD COMBINED WITH CYCLIC SHEAR LOADING

3.3.1 Introduction

The above loading scheme simulates with high precision the combination of gravity loads with seismic actions on the main structural elements of masonry constructions, namely, the walls which carry in-plane the above actions. Therefore, it is essential to make a short presentation of the response up to failure of the above typical structural elements (Figure 3.16).

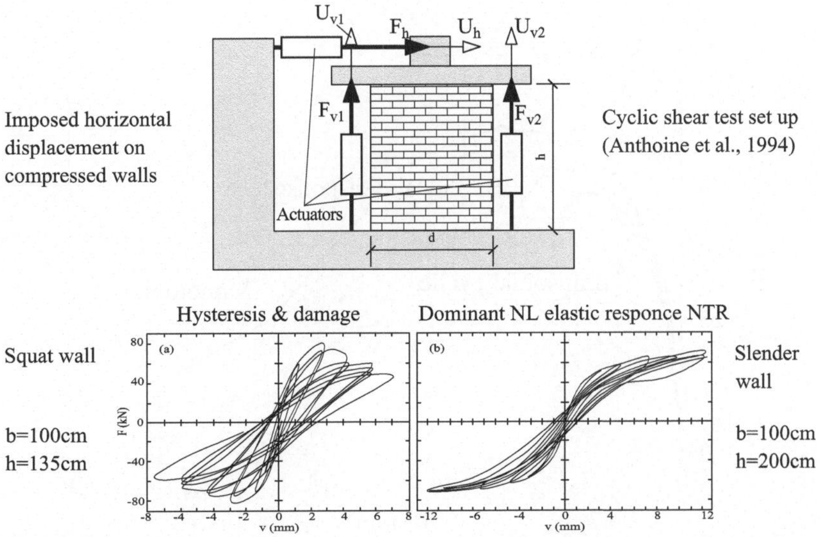

Figure 3.16 Imposed horizontal displacement on a compressed wall. (Adapted from Gambarotta, L. (2007). *Mechanics of masonry structures: arches, shear walls and vaults*. Seminari DICAT University of Genova, Italy.)

According to the results of earthquake damage analysis and subsequent experiments, three types of failure modes may appear which have already been described in detail in Subsection 3.2.4 (Figure 3.13).

It is not a simple task to model a heterogeneous material like masonry in a step-by-step procedure under cyclic loading up to failure.

A relatively good prediction of structural behaviour may be achieved, by FEM procedures, where masonry units and mortar joints are modelled as continuous elements with links between them (microelement models) (Ignatakis, et al., 1993). However, conclusions on hysteretic behaviour can be obtained with sound reliability only by experimental simulation of seismic behaviour of masonry walls (Tomazevic, 1999) (Figures 3.17 and 3.18).

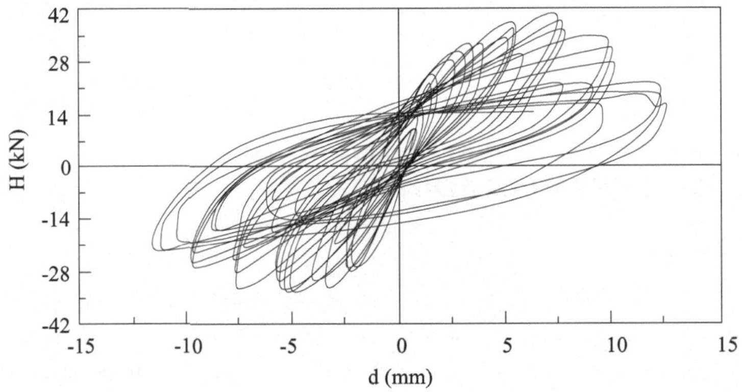

Figure 3.17 Typical lateral displacement–lateral resistance hysteresis loops, from a seismic resistance test on a masonry wall. (Adapted from Tomazevic, M. (1999). Earthquake – resistant design of masonry buildings. With permission of Imperial College Press, London, UK.)

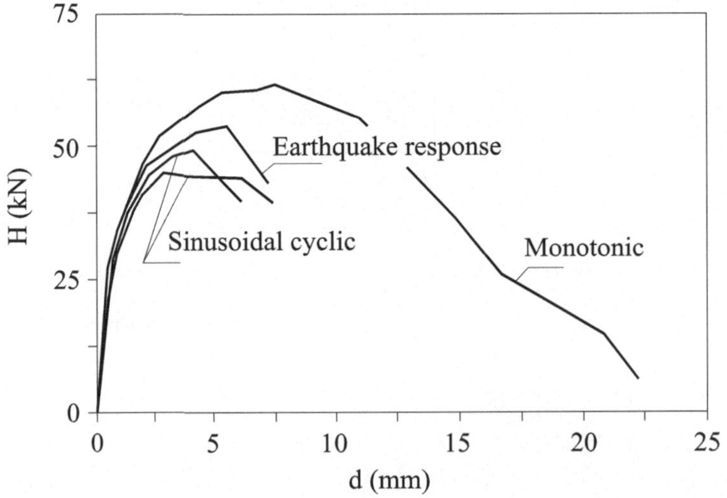

Figure 3.18 Hysteresis envelopes of the same type of wall with different displacement patterns. (Adapted from Tomazevic, M. (1999). *Earthquake – resistant design of masonry buildings*. Imperial College Press, London, UK. With permission of Imperial College Press.)

3.3.2 Mechanical properties

3.3.2.1 Strength

Strength issues related to compression and shear have been examined in detail in Subsection 3.2.4, and, therefore, no additional reference will be made here.

3.3.2.2 Displacements

The envelope of an experimental test of a masonry panel subjected to axial load combined with cyclic shear loading may be simulated by a curve of the form depicted in Figure 3.19. In this curve, three characteristic points (limit states) are defined (Tomazevic, 1999):

- *Crack limit* described by displacement d_{cr} and resistance H_{cr} at the formation of the first significant cracks (Point A). At this point, the slope of the envelope curve presents a characteristic modification.
- *Maximum resistance* with characteristic coordinates $H_{\max}$–$d_{H\max}$
- *Ultimate displacement* with characteristics coordinates $d_{\max}$–$H_{d\max}$

This curve, for simplification reasons of analysis may be idealised by a bilinear curve. The first branch of this curve passes through Point A.

Its slope expresses the *elastic stiffness* of the wall K_e:

$$K_e = \frac{H_{cr}}{d_{cr}}. \tag{3.40}$$

The second branch is a horizontal line with an ordinate H_u (ultimate resistance) which is defined by taking into account the equal energy dissipation capacity of the actual and the idealised wall.

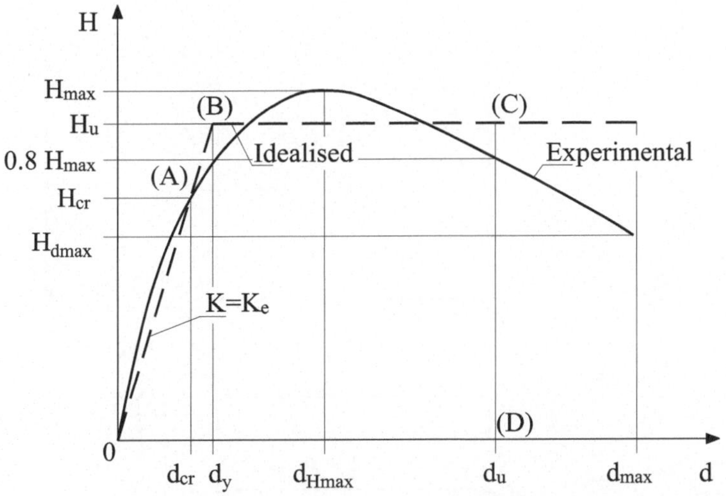

Figure 3.19 Idealisation of a *H*–d diagram with bilinear relationship. (Adapted from Tomazevic, M. (1999). *Earthquake – resistant design of masonry buildings*. Imperial College Press, London, UK. With permission of Imperial College Press.)

Based on statistical evaluation of a big number of tests (over 66) (Tomazevic, 1999), it could be possible to introduce for H_u a value

$$H_u = 0.9H_{\max}. \tag{3.41}$$

This line, which is the plastic branch of the constitutional law of the panel, intersects the initial slope line at Point B with coordinates *(d_y, H_u)* which is characterised as *the conventional yield point* of the panel under consideration.

Finally, the ultimate idealised displacement d_u is defined conventionally as the displacement value for which H is reduced (descending branch) to 80% of $H_{\max}$ (Point C).

Based on the above analysis, the idealised bilinear H–δ diagram (O, A, B, C, D) may be easily obtained in case of existing experimental results. *However, for safety assessment of existing buildings, such experimental results do not exist.* Therefore, the bilinear diagram in Figure 3.19 for masonry panels must be assessed by procedures based on statistical evaluation of extended experimental results. The procedure proposed by Tomazevic (1999) is presented below.

- The elastic stiffness of the wall panel K_e is estimated by the expression (Figure 3.20)

$$K_e = \frac{G_w A_w}{1.2h_w\left[1 + a'\frac{G_w}{E_w}\left(\frac{h_w}{l_w}\right)^2\right]}, \tag{3.42}$$

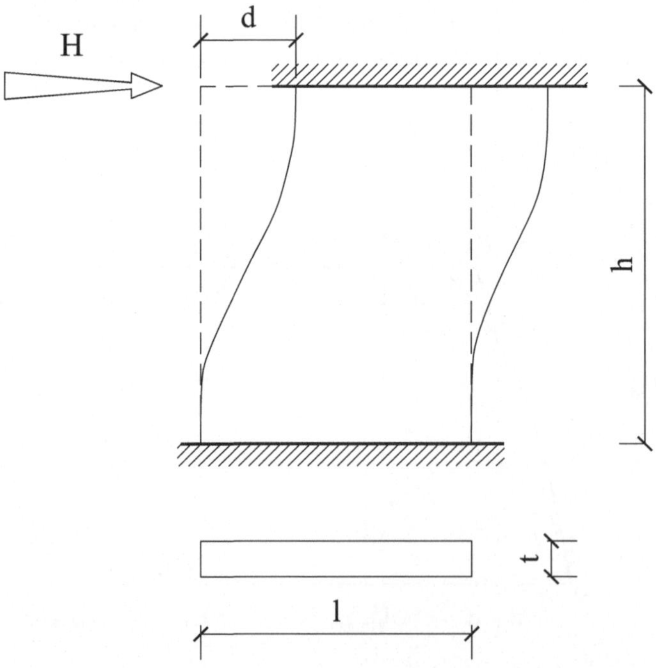

Figure 3.20 Deformation of a fixed-end wall subjected to lateral loading. (Adapted from Tomazevic, M. (1999). *Earthquake – resistant design of masonry buildings*. Imperial College Press, London, UK.)

where
h_w and l_w are the wall dimensions
t_w is the wall thickness
h_w/l_w is the aspect ratio
A_w is the area of the wall cross section $(A_w = l_w \cdot t_w)$
E_w, G_w are the elastic and shear moduli of masonry.
a' is a coefficient determining the position of the bending moment inflection point along the height of the wall ($a' = 0.83$ in case of a fixed ended and $a' = 3.33$ in case of a cantilever).

In the above slope, both bending and shear deformations have been taken into account.

- The ordinate of Point B on the above line of initial slope is estimated by the expression

$$H_u = 0.9\ H_{max},$$

where H_{max} is the shear strength of the wall resulting from equation 3.31 and Point B represents the yield point of the wall with coordinates

$$H_u = 0.9 H_{max}, \tag{3.43a}$$

$$d_y = 0.9 H_{max} / K_e. \tag{3.43b}$$

- Finally, the ultimate idealised displacement d_u (Point C) is estimated by the expression

$$d_u \cong (2.0 \div 3.0)\ d_y. \tag{3.44}$$

The above expression means that the estimated available ductility μ_w of an individual wall panel is

$$\boxed{\mu_w \cong 2.0\text{–}3.0.} \tag{3.45}$$

The above value of μ_w seems to be conservative. In fact, the statistical evaluation of the experimental results on this issue of Pavia and ISPRA, tests carried out by the second author (Penelis, Gr., 2006a) have resulted in higher values of μ_w:

$$\boxed{\mu_w \cong 3.0\text{–}4.0.}$$

The above ductility leads to a *behaviour factor q* of a masonry building to the level of

$$\boxed{q = 1.50\text{–}2.00.} \tag{3.46}$$

The above information for the idealised H–δ bilinear curve of individual URM panels is essential for the inelastic static design (pushover analysis) of masonry buildings. In this context, extended analytical attempts have been made for the analytical formulation of a H–δ curve, based on the geometrical dimensions of the panel wall and its support conditions, the compressive strength of masonry, the friction coefficient μ_c, the cohesion f_{uo} and the axial load N (equation 3.31) (Magenes and Calvi, 1997; Penelis, Gr., 2006a).

3.4 MECHANICAL PROPERTIES OF MASONRY UNDER BIAXIAL LOADING

3.4.1 Introduction

For new masonry buildings, the analysis and design for the time being is based on the assumption of a homogeneous material. All codes of practice in effect, are based on this assumption, and the mechanics of masonry presented in the previous subsections constitutes the basic analysis and design procedure for a new building.

On the other hand, existing masonry constructions are an important part of the historic and cultural heritage of many countries, and the analysis of their safety conditions as well as the required structural interventions on them is a challenging task. In fact, the requirements for preserving the authenticity of the building impose soft interventions. In this respect, a refined analysis and design should be implemented to take into account all safety reserves of the construction. So, in the last forty years, serious research has been carried out for the experimental investigation of masonry under biaxial loading, and, at the same time, numerical modelling techniques have been developed, thanks to the enormous advancement of the computational tools. A short overview of both procedures will be presented below.

3.4.2 Experimental results

Thanks to the extended experimental investigations (Page, W., 1982), the following failure envelopes have been obtained (Figures 3.21 and 3.22).

From Figures 3.21 and 3.22, the following conclusions may be drawn:

- The structural behaviour is strongly orthotropic and complicated.
- The joint direction in relation to the principal stresses is crucial for the form of the failure envelope.
- Experimental results for the ± region of a masonry strength envelope are depicted in detail in Figure 3.22. These diagrams are very important for the shear compression of walls.

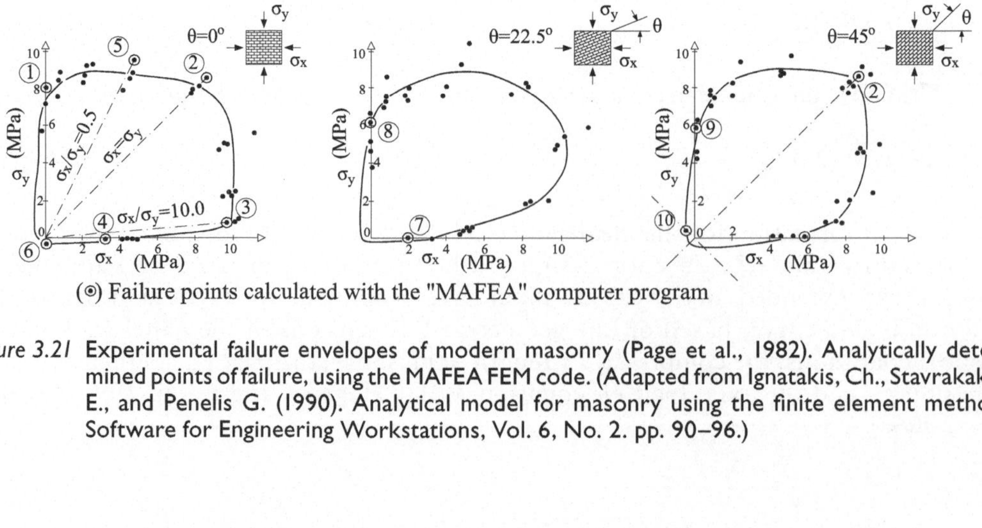

Figure 3.21 Experimental failure envelopes of modern masonry (Page et al., 1982). Analytically determined points of failure, using the MAFEA FEM code. (Adapted from Ignatakis, Ch., Stavrakakis, E., and Penelis G. (1990). Analytical model for masonry using the finite element method. Software for Engineering Workstations, Vol. 6, No. 2. pp. 90–96.)

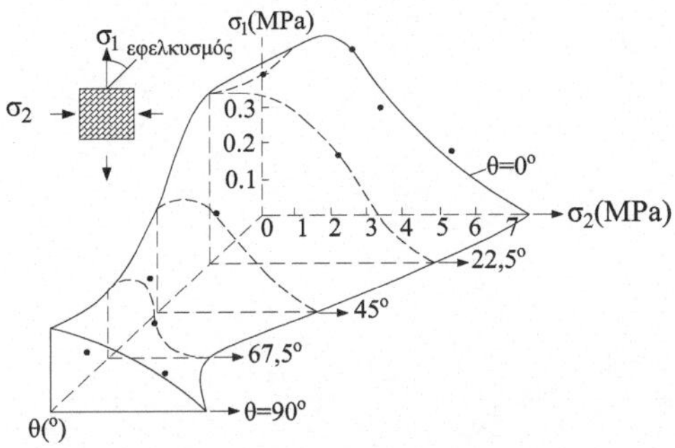

Figure 3.22 Experimental failure envelope of modern masonry (Page, et al., 1982) under compression and tension. (Adapted from Karantoni, F. (2004). *Masonry constructions.* In Greek. Papasotiriou, Athens. Courtesy of the author.)

3.4.3 Analytical results

Detailed *micromodelling* is probably the most accurate tool available to simulate analytically the real behaviour of masonry. It is particularly adequate to describe the response of masonry in the form of a panel. Each constituent material, namely the bricks, the mortar joints and the brick–mortar interface, is represented explicitly; that is, they are divided into finite elements, and they are characterised by their mechanical properties. Then the model is loaded step by step till failure. In Figures 3.23, 3.24 and 3.25, the failure envelopes are depicted from the analysis of a specimen of Roman masonry of Rotunda

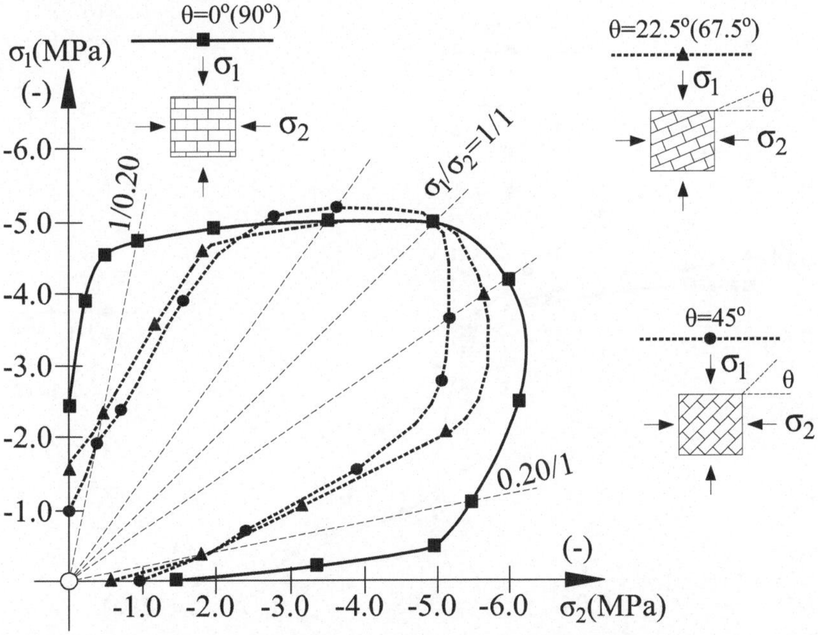

Figure 3.23 Failure envelopes of Rotunda Thessaloniki masonry at (–/–) region. (Adapted from Ignatakis, Ch., Stavrakakis, E., and Penelis G. (1990). Analytical model for masonry using the finite element method. Software for Engineering Workstations, Vol. 6, No. 2. pp. 90–96.)

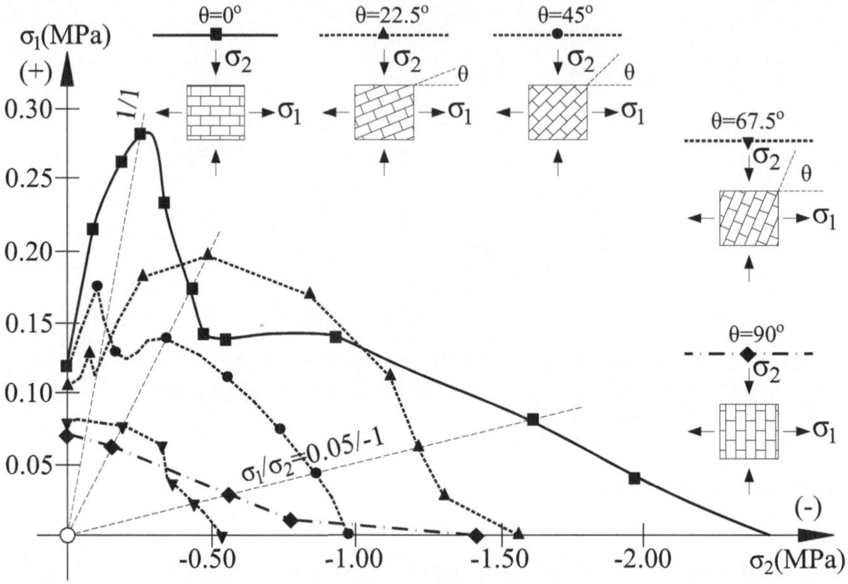

Figure 3.24 Failure envelopes of Rotunda Thessaloniki masonry at (∓,±) region. (Adapted from Ignatakis, Ch., Stavrakakis, E., and Penelis G. (1990). Analytical model for masonry using the finite element method. *Software for Engineering Workstations*, Vol. 6, No. 2. pp. 90–96.)

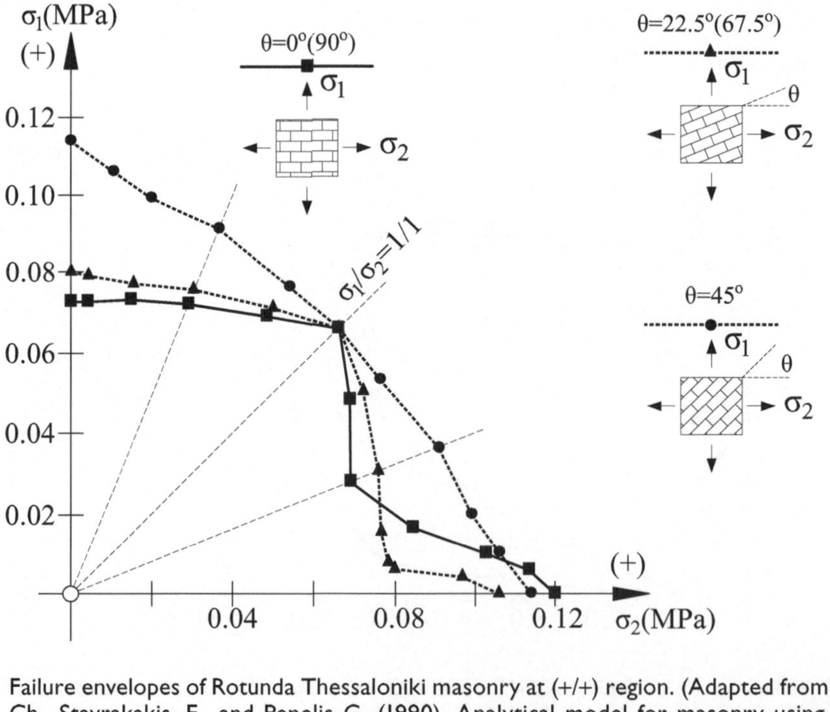

Figure 3.25 Failure envelopes of Rotunda Thessaloniki masonry at (+/+) region. (Adapted from Ignatakis, Ch., Stavrakakis, E., and Penelis G. (1990). Analytical model for masonry using the finite element method. *Software for Engineering Workstations*, Vol. 6, No. 2. pp. 90–96.)

Table 3.1 Mechanical properties of Roman masonry (Rotunda of Thessaloniki) constituents

Materials	*Dimensions*	f_c *(MPa)*	f_t *(MPa)*	E_0 *(MPa)*	v_o	ε_v *(‰)*
Bricks	30 × 40 × 5cm	$f_{bc} = 9.00$	$f_{bt} = 1.20$[b]	$E_{bo} = 10{,}700$	0.19[c]	2.50[c]
Mortar	$t_{bed} \approx 4$ cm	$f_{mc} = 1.20$[a]	$f_{mt} = 0.40$[b]	$E_{mo} = 1{,}350$	0.19[c]	2.50[c]
Joints	$t_{vert} \approx 3$ cm	Shear/tensile strength: $f_{jso}/f_{jt} = 0.17/0.075$[c] MPa Friction coefficient: $\mu = 0.75$[c]				

Source: Adapted from Penelis, G. karaveziroglou, M., Stylianidis, K., and Leohdaridis, D. (1980a). Restoration and strengthening of Rotunda of Thessaloniki. Archives of Ephorate of Antiquities of Thessaloniki.

[a] Prismatic strength from the strength of a short specimen $f_{mc}^{c} = 1.80$ MPa.

[b] Direct tensile strength from the flexural strength $f_{bt}^{fl} = 2.25, f_{mt}^{fl} = 0.65$ MPa.

[c] From the literature according to relative measurements.

Thessaloniki (thick mortar joints) with mechanical characteristics presented in Table 3.1. The analysis was carried out using the computer code MAFEA (masonry finite element analysis) which was developed in the Laboratory of Reinforced Concrete and Masonry Structures of Aristotle University of Thessaloniki (AUTH) (Ignatakis, et al., 1990).

The following table gives the required information on mechanical properties of the constituent materials.

	Masonry units		*Bed joints*
f_{bc}	Uniaxial compressive strength	f_{mc}	Uniaxial compressive strength
f_{bt}	Uniaxial tensile strength	f_{mt}	Uniaxial tensile strength
E_b	Initial modulus of elasticity	E_m	Initial modulus of elasticity
$\surd_b$	Poisson's ratio	$\surd_m$	Poisson's ratio
ε_{1u}	Ultimate compressive strain (uniaxial)	ε_{1u}	Ultimate compressive strain (uniaxial)

Bonding parameters of materials

f_{vo} is the shear strength under zero compression (cohesion)
μ_c is the friction coefficient of joints.

The relevant stress–strain σ–ε diagram is depicted in Figure 3.26. The failure envelopes at the (∓) and (±) regions are in very good qualitative agreement with the relative experimental results of Samara Singhe and Hendry (1982) and Page (1981).

Owing to the enormous advancement of the computational tools, the last decades have witnessed a wide spread of the numerical modelling techniques of heterogeneous material like masonry similar to the one presented in the preceding paragraph. It is noteworthy to make reference to procedures which replace mortar joints and bonding between mortar and bricks by proper inelastic unilateral springs diminishing in this way the computational cost (Figures 3.27 and 3.28) (Baraldi, et al., 2018).

It should be noted that micromodelling is a very useful tool for the determination of the mechanical properties of masonry of monumental constructions, as it is well known that it is almost impossible to be allowed by the authorities to extract big masonry samples for biaxial testing. Therefore, from the information for the properties of the constituent materials, conclusions must be drawn for the masonry mechanical properties under biaxial strain and, therefore, all necessary information for orthotropic behaviour.

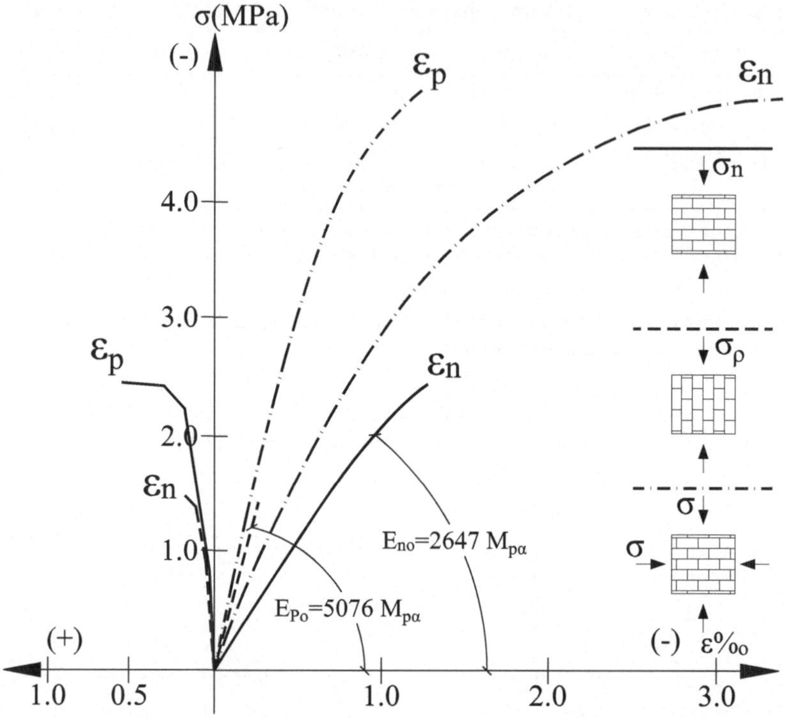

Figure 3.26 Stress–strain curves of Roman (Rotunda of Thessaloniki) masonry under uniaxial and biaxial compression. (Adapted from Ignatakis, Ch., Stavrakakis, E., and Penelis G. (1990). Analytical model for masonry using the finite element method. *Software for Engineering Workstations*, Vol. 6, No. 2. pp. 90–96.)

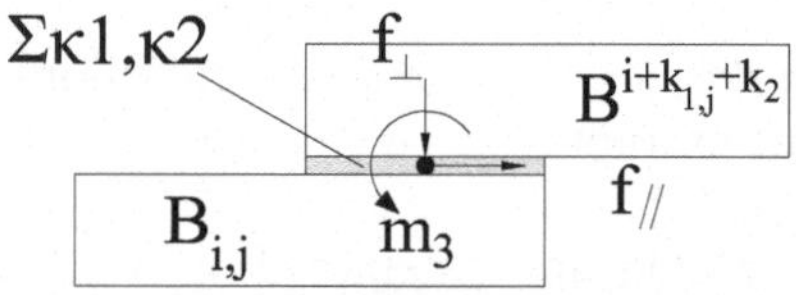

Figure 3.27 Interface between adjacent blocks and interface actions. (Adapted from Baraldi, D., Reccia, E., and Cecch, A. (2018). In-plane-loaded masonry walls: DEM and FEM/DEM models: A critical review. New trends in mechanics of masonry. Meccanica, Vol. 53, Springer Science and Business. pp. 1613–1628. Courtesy of Springer Science and Business.)

Closing, it is important to note that micromodelling of masonry is an indispensable preliminary step for the implementation of FEM methods of inelastic analysis using homogenisation strategies.

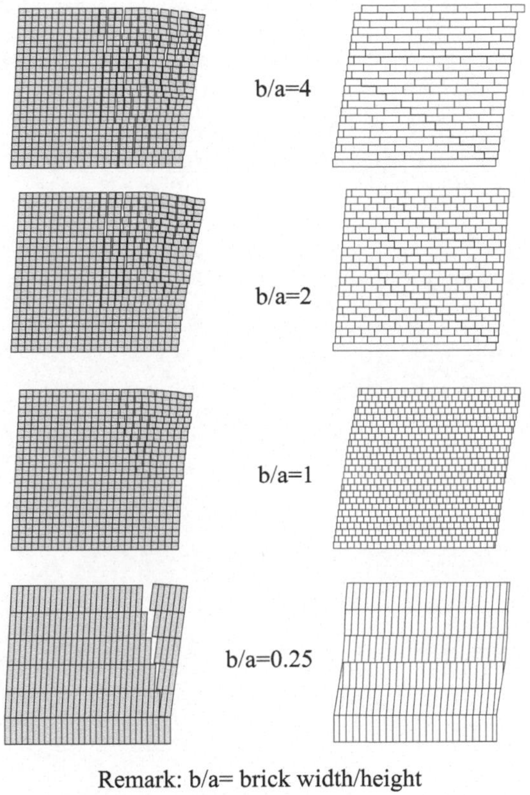

Figure 3.28 Collapse mechanism of masonry panels with h/l = 1 subject to self-weight and increasing lateral loads. (Adapted from Baraldi, D., Reccia, E., and Cecch, A. (2018). In-plane-loaded masonry walls: DEM and FEM/DEM models: A critical review. *New trends in mechanics of masonry. Meccanica*, Vol. 53, Springer Science and Business. pp. 1613–1628. Courtesy of Springer Science and Business.)

Chapter 4

Structural behaviour of masonry in monumental buildings

4.1 GENERAL

Masonry structures cover a very wide range of monumental works in space and time, covering a period of almost 6,000 years.

The examination of the bearing capacity of any structure and, consequently, of a masonry one, assumes that the following parameters must be taken into consideration:

- The strength of the materials of the structure
- The texture of the structure
- Its geometrical form and dimensions
- The existing loading and constraints
- The expected loadings at failure.

In this context, the quantitative evaluation of the load-bearing capacity based on the existing evidence is probably the most difficult problem for a structural engineer, much more difficult than the design of a new building. Extensive site investigations are required, first of all, for the damage survey and then for the geometry of the structural system, the quality of the construction materials, the gravity loadings and the quality of the foundation soil. Subsequently, extensive calculations are needed, using the above information in order to determine the residual resistance of the structure, first for gravity loads and second for the combination of gravity load and seismic action in areas of high seismicity like Mediterranean basin or other variable actions like wind, fire, flood etc. (Penelis, G. and Penelis, Gr., 2019). It is apparent that the above-described procedure constitutes the main issue of this book.

However, at the same time, it is important for the structural engineer, who is involved in structural restoration design, to know in depth qualitatively the basic forms and their structural behaviour of monumental buildings. In fact, the structural engineer in charge must understand from the beginning, qualitatively, the structural behaviour of a building under consideration to various basic actions, so that he can make a preliminary *diagnosis* of the *causes* that are responsible for the existing *damages* (cracks, drifts, settlements, splitting etc.). In this way, he will be in a position to focus on the detailed research, *in situ* and in laboratory, to the critical points of the building, and he will conduct the analysis and design using suitable models for a credible *safety evaluation* (see Chapter 8).

Therefore, an effort is made in this chapter to give a qualitative approach of the structural behaviour of the basic structural forms of masonry monumental and historical buildings, which are met in Europe and particularly in the South, where the consequences of the seismic action are traumatic every now and then.

4.2 CONCEPTUAL APPROACH TO THE STRUCTURAL BEHAVIOUR OF MASONRY ELEMENTS

As it was already analysed in previous chapters, masonry, like concrete, is a material of medium and reliable compressive strength but low and unreliable tensile one ($\frac{1}{10}$–$\frac{1}{7}$ of compressive strength). In this respect, unreinforced masonry is an inappropriate material for bending, e.g. for beams and slabs.

In fact, in the case of a simply supported beam (Figure 4.1a) made of masonry or concrete, *but without any reinforcement*, the ultimate load q_{ult} that the beam can carry may be calculated using the following formulas:

$$\frac{q_{1ult}l^2}{8\cdot\left(bh^2/6\right)} = \sigma_{tult} \leq\approx \frac{f_{cd}}{10} \tag{4.1a}$$

or

$$\boxed{q_{1ult} \cong\leq 0.133\frac{bh^2 f_{cd}}{l^2},} \tag{4.1b}$$

where

b, h are the cross-section dimensions of the beam
l is the length of the beam
f_{cd} is the compressive design strength of masonry or concrete.

The above value of q_{1ult} is very low, usually lower even of the self-weight of beams for spans longer than 1.5 m. In parallel, as it was pointed out before, this value is also unreliable.

Now, if the same beam is reinforced in its tensile region with steel rebars capable of balancing the compressive design strength of the upper zone (Figure 4.1b), the ultimate load that the beam can carry may be obtained from the following expressions:

$$\frac{q_{2ult}l^2}{8\left[0.85h(0.25\ bh)\right]} = \sigma_{cult} \leq\approx f_{cd} \tag{4.2a}$$

or

$$\boxed{q_{2ult} \cong 1.70\frac{bh^2}{l^2}f_{cd}.} \tag{4.2b}$$

This means that the invention of the reinforced concrete at the end of the 19th century *made possible the use of stone material (concrete) to carry bending monuments*, in other words, its use for beams, slabs etc. In fact, the ultimate load q_{2ult} in the second case is at about fifteen times bigger than q_{1ult}, and, most important, it has a stable and reliable value.

The above inadequacy of masonry to carry bending monuments had been recognised since the early stages of human civilisation. So, masonry has been used for centuries *only*

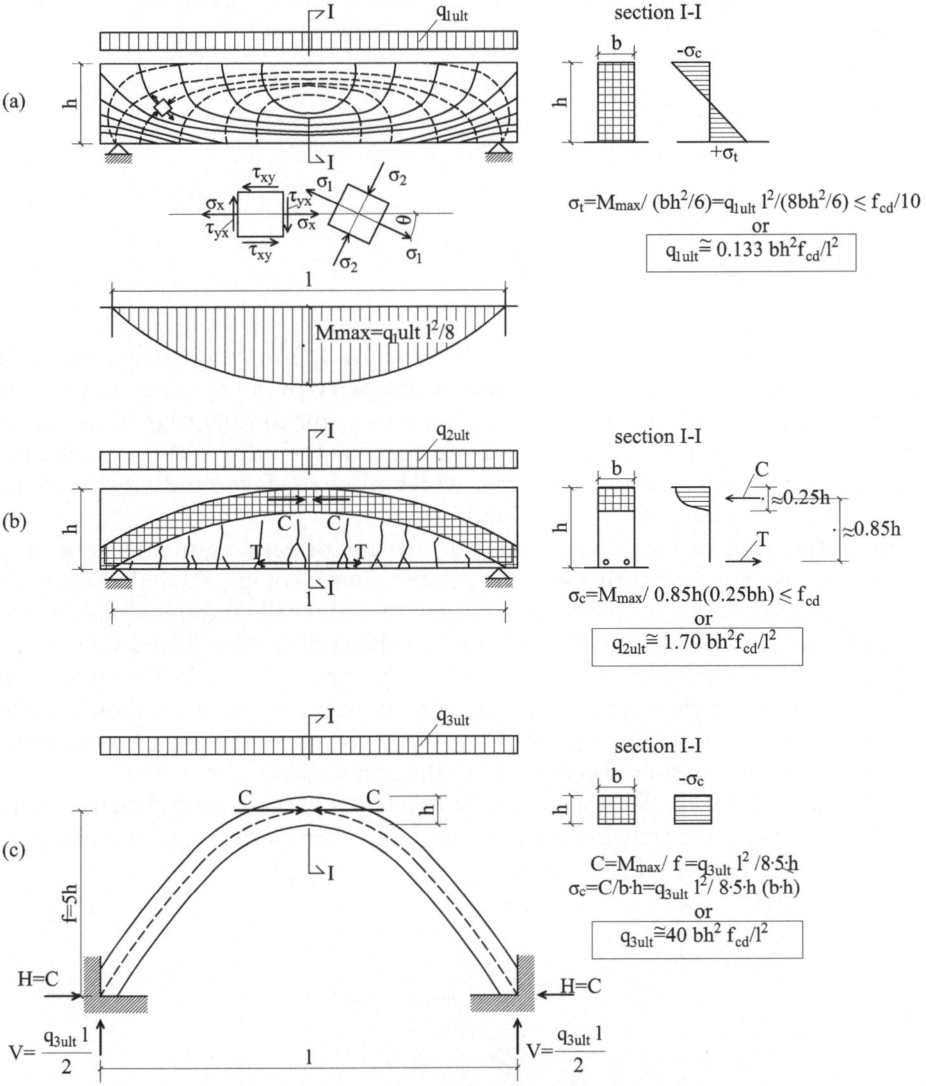

Figure 4.1 Comparison of the bearing capacity of a beam and an arch: (a) an unreinforced concrete beam; (b) a reinforced concrete beam; (c) an arch.

under compression or at maximum *under compression and shear*, in other words, for *walls, arches, vaults and domes*. Therefore, through centuries, *a static of the form* was developed step by step, namely a proper formulation of the structural system, so that the external loads could be transferred to the foundations only via compressive load paths. This was achieved by replacing *the beam* by *an arch* properly supported at its ends by a system capable of carrying *horizontal thrusts* (compression combined with shear).

It is interesting to stay a little bit more at the structural function of an arch. Consider the arch of Figure 4.1c. This arch, having a parabolic form of second order and an upward arrow $f = 5.0h$, can carry an ultimate load q_{3ult} which may be obtained by the following expression:

$$\frac{q_{3ult}\ l^2}{8f(hb)} = \sigma_{cult} \leq\approx f_{cd} \tag{4.3a}$$

or

$$\boxed{q_{3ult} \leq \frac{40bh^2 f_{cd}}{l^2}.} \tag{4.3b}$$

Comparing equation (4.3) with equations (4.1) and (4.2), it may be easily concluded that the arch has a significantly higher load-bearing capacity than the concrete beam, either unreinforced or even reinforced. However, it is important to note that in the case of an arch, while the bending monuments are transformed via the form of the structure to axial compressive forces at the axis of the arch, which are very favourable for the masonry, horizontal thrusts are developed at the supports.

It is interesting to note here that this transformation of bending moments of the beam to axial forces and horizontal thrusts of the arch is not a computational trick. In fact, if we examine carefully the load paths of compressive and tensile internal forces in a beam, unreinforced or reinforced, we will discover that the compressive forces *C–C* constitute within the height of the beam an arch closed by the tensile forces *T–T* so that no thrust reactions are needed at the supports. In case the material is not in position to carry the tensile forces *T–T* at the lower region of the beam, these forces must be introduced as thrust reactions from the supports to balance the compressive thrust *C–C*.

The techniques that were developed step by step in the past centuries had as a main task to invent proper paths for transferring the horizontal thrusts that appear at the supports to the foundations (Figure 4.2).

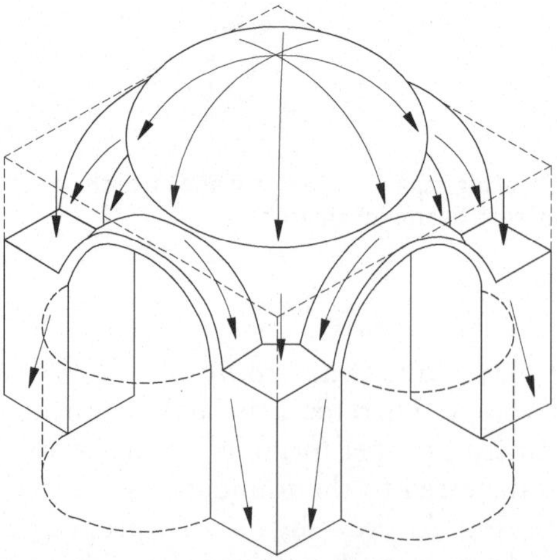

Figure 4.2 Load path at a central nucleum system. (Adapted from Mainstone, R. (1975). Developments in structural form. Penguin Books Ltd, England.)

4.3 MASONRY STRUCTURAL ELEMENTS

4.3.1 Masonry walls

The basic 'loading' patterns and the failure modes of a masonry wall are given below (Guerra, 1979; Penelis, G., 1998; Tomazevic, 1999).

4.3.1.1 Axial compressive loading

As it was noted before, the basic structural function of a masonry wall is to carry compressive loads, since masonry has a rather high and credible compressive strength in contrast to its low and incredible tensile strength.

The failure mode of a slender masonry wall under axial compression is given in Figure 4.3.

Usually in case of three-leaf walls due to the inadequate connections of the two external leaves (see Chapter 3) of the masonry, the failure appears in the form of vertical cracks at the outer leaves of the wall, due to the development of horizontal tensile stresses resulting from the dilatancy of the core and the transverse constraints to the external skins causing their buckling (Penelis, G., 1982).

4.3.1.2 Axial compressive load combined with horizontal out-of-plane seismic action

The out-of-plane loading appears in the case of free-standing walls or masonry systems without diaphragmatic action of the floors (one or multistorey buildings) as it will be explained later. Two failure modes may appear.

i. *High N–Low V* (Figure 4.4)
   In this case, the base of the wall is completely under compression. The more compressed side of the wall, due to simultaneous failure of the transverse constrains of the two leaves, results in local bucking and splitting of the outer leaf.
ii. *Low N–High V* (Figure 4.5)
   In this case, due to the high tensile stresses that develop at one side of the wall, horizontal cracks appear there, and the uplifting and overturning (rocking) of the wall is most probable.

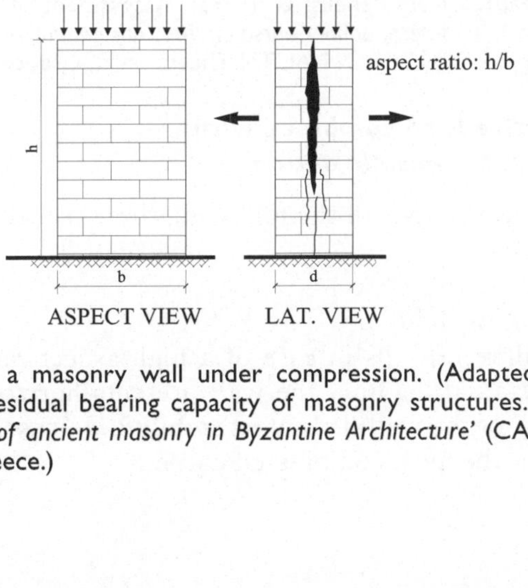

Figure 4.3 Failure mode of a masonry wall under compression. (Adapted from Penelis, G. (1998). Assessment of residual bearing capacity of masonry structures. *1st International Course on 'the conservation of ancient masonry in Byzantine Architecture'* (CAMBA-98) ICCROM-AUTH Thessaloniki, Greece.)

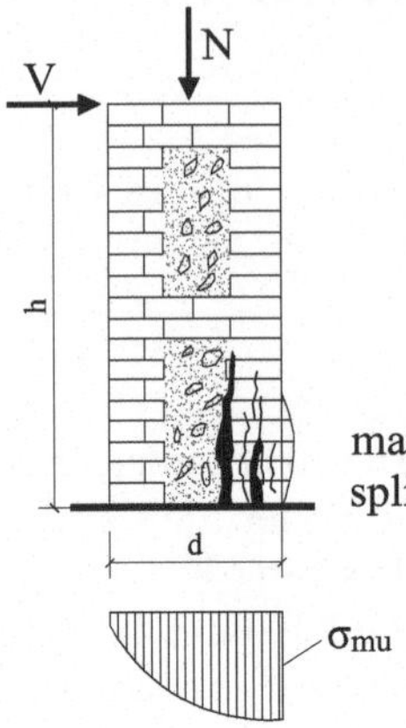

Figure 4.4 Splitting at the compressed edge of a masonry wall. (Adapted from Penelis, G. (1998). Assessment of residual bearing capacity of masonry structures. *Ist International Course on 'the conservation of ancient masonry in Byzantine Architecture'* (CAMBA-98) ICCROM-AUTH Thessaloniki, Greece.)

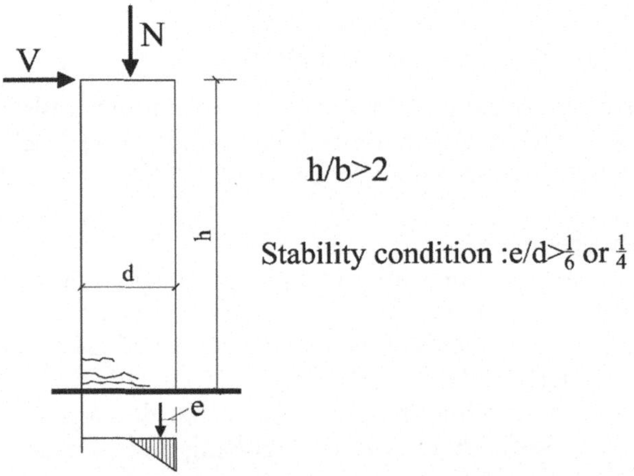

Figure 4.5 Horizontal tensile cracks (uplifting) and eventual overturning of the wall (out-of-plane overturning). (Adapted from Penelis, G. (1998). Assessment of residual bearing capacity of masonry structures. *Ist International Course on 'the conservation of ancient masonry in Byzantine Architecture'* (CAMBA-98) ICCROM-AUTH Thessaloniki, Greece.)

4.3.1.3 Axial compressive load combined with in-plane horizontal seismic action

The failure mode in this case is one of the following three (Penelis, G., 1998; Tomazevic, 1999):

i. *Bending failure* (Figure 4.6*a*)

This type of failure prevails in case of a high aspect ratio h/b (higher than 2). In the case of low axial loading, the wall presents horizontal tensile cracks and, finally, overturning of the wall (loss of its stability). In case of high axial loading, it presents splitting at the most compressed corner.

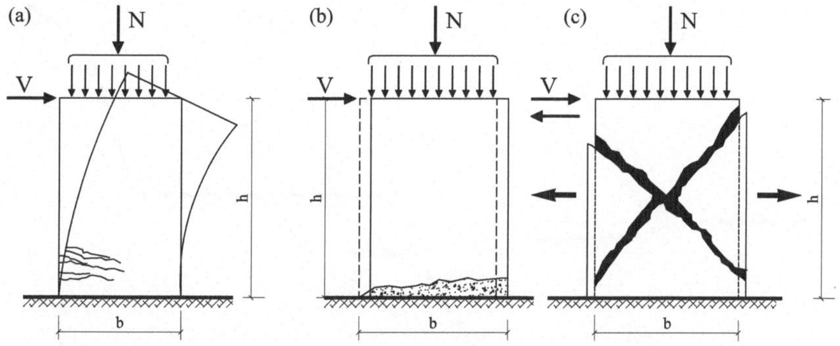

Figure 4.6 In-plane failure modes of masonry: (a) uplift failure; (b) sliding failure; (c) shear–compression failure. (Adapted from Penelis, G. (1998). Assessment of residual bearing capacity of masonry structures. *1st International Course on 'the conservation of ancient masonry in Byzantine Architecture'* (CAMBA-98) ICCROM-AUTH Thessaloniki, Greece.)

ii. *Sliding failure* (Figure 4.6*b*)

This type of failure prevails in case of low aspect ratio h/b (lower than 2) and when the axial loading is very low. This type of failure appears rather rarely.

iii. *Shear–compression failure*

This type of failure prevails in case of low aspect ratio as the previous one, and it is the most usual and critical failure mode, in masonry walls located between openings. It is caused by regional skew tension due to shear and axial stresses. It prevails for medium or high axial loading. It is a very dangerous failure mode, particularly in case of horizontal seismic action, because, after the formations of x-shaped cracks, the wall is unable to carry the gravity loads any longer, even when the horizontal seismic loads have stopped to act on the wall (Figure 4.6c).

4.3.1.4 Masonry walls subjected to foundation settlements

Four modes of failure may occur to a masonry element in case of foundation settlements (Penelis, G., 1998):

a. Inclined cracks due to settlements at one side (Figure 4.7a)
b. Vertical cracks at the base of the wall due to local sliding of the foundations (Figure 4.7b)
c. Vertical cracks at the bottom of the wall combined with inclined ones at both of its sides due to local settlements at the centre of the wall (Figure 4.7c)
d. Vertical cracks at the top of the wall combined with inclined ones at both sides, due to local settlements at the two sides of the wall (Figure 4.7d).

The above modes of failure are due to constraints caused to the wall by settlements of the foundations and their extent and seriousness depend on the size of these settlements.

4.3.2 Arches and barrel vaults

Arches and barrel vaults will be examined in common, since the behaviour of the barrel vault in Roman Byzantine and Islamic monuments usually coincides with that of the arch

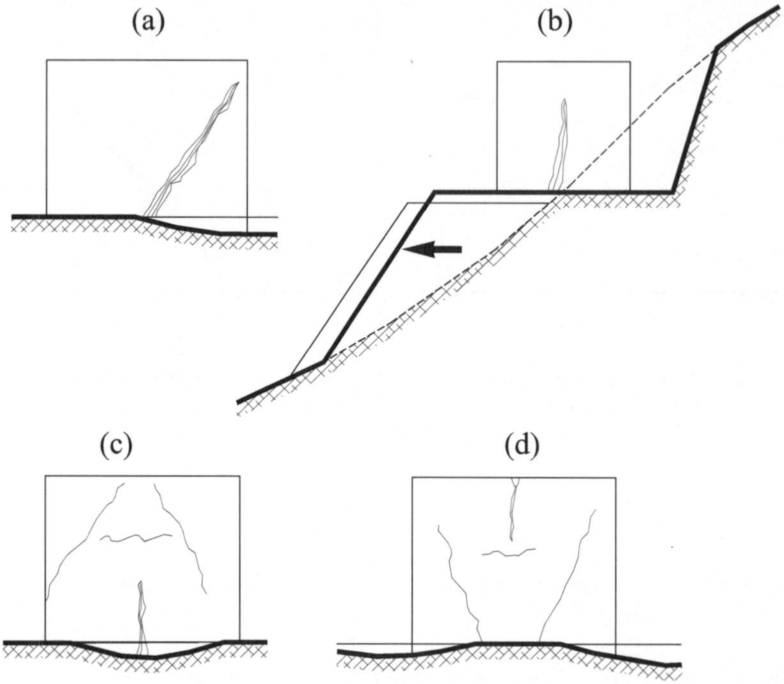

Figure 4.7 Cracks on masonry wall due to foundation settlements: (a) local settlement at corner; (b) local sliding; (c) local settlement at the centre of the wall; (d) local settlements at the corners of the wall. (Adapted from Penelis, G. (1998). Assessment of residual bearing capacity of masonry structures. *1st International Course on 'the conservation of ancient masonry in Byzantine Architecture'* (CAMBA-98) ICCROM-AUTH Thessaloniki, Greece.)

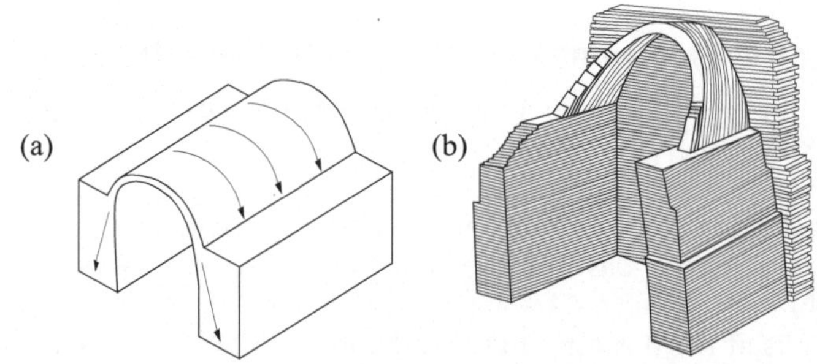

Figure 4.8 Structural behaviour of a continuous supported vault: (a) the behaviour of a vault supported by thick walls; (b) Ktesiphon Palace axisymmetric form of the main vault. (Adapted from Mainstone, R. (1975). Developments in structural form. Penguin Books Ltd, England; Hart, F. (1965). *Kunst and Technik der Wolbung.* Verlag George Callwey, München, Germany. Courtesy of Georg, D.W., Callwey Verlag.)

because of its continuous or narrow spaced supports (Figure 4.8). Therefore, barrel vaults of this type in a first approximation may be considered as a series of adjacent but independent arches, one by the other. Additionally, since *the circular arch* is the main form used in the above types of monuments, all references below will be to this form (Hart, 1965; Bouras, 1965).

In the *semicyclic* arches of buildings, where loading is permanent and uniformly distributed, the *load path method* (thrust line method) is the most suitable for an easy presentation of their behaviour (see Chapter 7). It is known from the theory of structures that an arch is safe *against crack* due to tension if the thrust line passes in the central third of each cross section along the axis of the arch, i.e. *when no tensile stresses develop*. Given that the thrust line for uniform loads is *parabolic* (see Section 4.2, Figure 4.1c), it is clear that, if the thickness t of the arch is less than a certain percentage of the radius a, there appear the five plastic hinges of Figure 4.9, which lead to collapse. The hinge B appears at an angle of $\beta \cong 58°49'$. From the diagram of Figure 4.10 Heyman (1977, 1986, 1989), it is observed that the ratio t/a of the thickness t towards the radius a in the case of a semicyclic arch should be greater than 0.33 (i.e. 3 times the value of the diagram corresponding to $\theta = 90°$), so that no tensile stresses exist at the body of the arch; that is, for an arch of $a = 5.0$ m radius (10.00 m span), a thickness of 1.65 m is required. Thus, for decreasing the thickness of the arch, already from the Roman period, *cracked areas were accepted* at the intrados of the crown of the arch.

At the same time, for the improvement of the form of the thrust line at point B (Figure 4.9), the arches were ended usually at an angle of about $\alpha = 60°$, while the remaining part had external backing and was formed as a pier so that the equilibrium condition

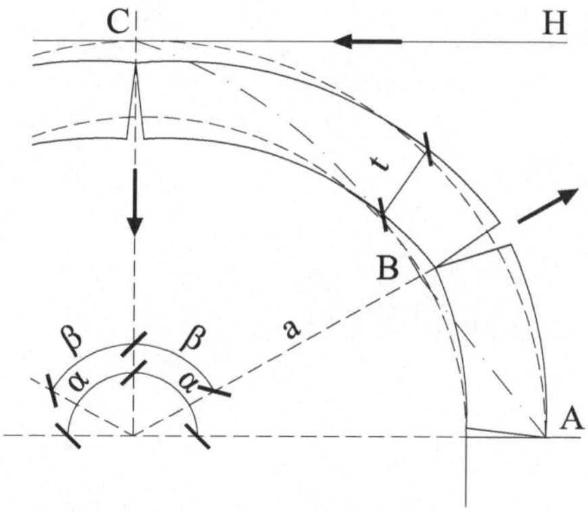

Figure 4.9 Plastic hinges in a semicircular arch or vault. (Adapted from Heyman, J. (1977). *Equilibrium of shell structures*. Clarendon Press, Oxford, UK.)

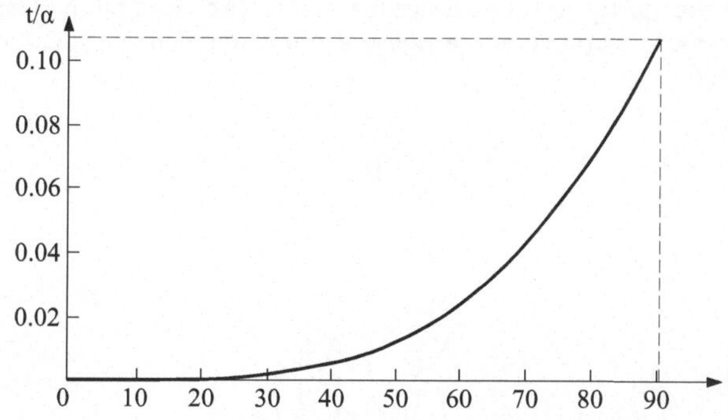

Figure 4.10 Determination of the ratio *t/a* (arch thickness to radius) in relation to the corner $\alpha°$ for the arch to comply with the no-collapse criterion. (Adapted from Heyman, J. (1977). Equilibrium of shell structures. Clarendon Press, Oxford, UK. Courtesy of Oxford University Press.)

at *A* appears improved because of the increase of the width of the section outwards (Figures 4.11 and 4.12) and the additional weight of the external backing.

The previous structural procedure, on which the diagram of Figure 4.10 is based, was presented 300 years ago (Couplet, 1730) and it is based on the following assumptions:

a. Unlimited compressive strength of the material
b. Zero tensile strength (acceptance of cracked sections)
c. No displacement of the supports
d. Symmetrical loading.

According to this method and the assumptions above, an arch with an angle of $\alpha \cong 58°$ and a ratio of $t/a = 0.02$ (i.e. an arch 0.60 m thick for 20 m span) would never fail (see Figure 4.10). However, the results in practice are less optimistic, because in practice, none of the above assumptions is completely valid.

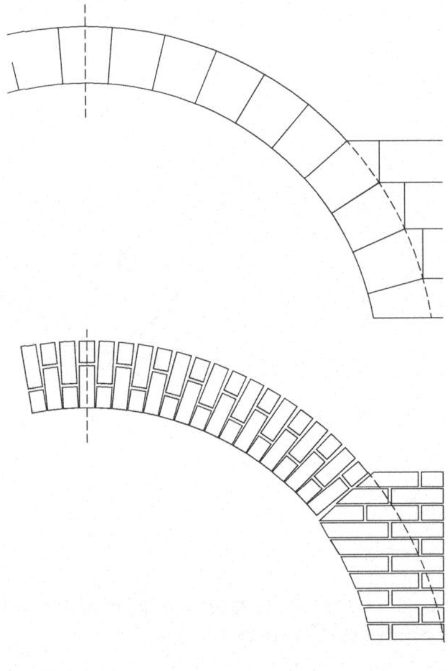

Figure 4.11 Formation of the springings of semicircular vaults. (Adapted from Hart, F. (1965). *Kunst and Technik der Wolbung*. Verlag George Callwey, München, Germany. Courtesy of Georg, D.W., Callwey Verlag.)

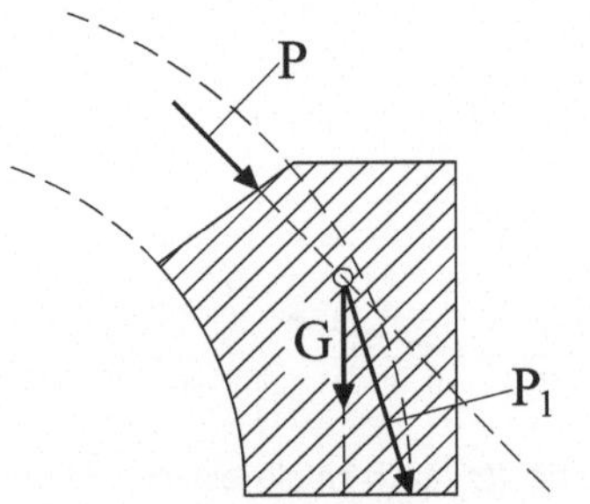

Figure 4.12 Improved static equilibrium at the broadened springing of a vault.

It is known that the basic problem in the arch constructions is the carrying of the horizontal thrusts by the supports. But while supports' stability and strength can be secured, it is not possible to diminish their horizontal displacements outwards and therefore, the widening of the cracks and the eventual collapse of the arch. On the other hand, the simultaneous loading with gravity and seismic loads destroys the symmetry of the loading pattern.

It is the authors' opinion that in order that an accurate estimation is made for the load-carrying capacity of the arches, they should be analysed not as individual elements but together with their supporting system up to the foundation. This should be done using an ultimate load step-by-step method based on more rational assumptions. Next, the method which was followed in the analysis of certain monuments of Thessaloniki (Christian phase of Rotunda, Hagios Andreas of Peristera, Vlatades monastery) will be outlined (UNDP/UNIDO, 1984; Penelis, G., 1982, 1992). It is noted that this method has been recently incorporated in a computational platform for an integrated inelastic static analysis for masonry structures of complicated monumental buildings (see Subsection 7.3.4) with 1D frame and shell elements (Penelis, G., et al., 2010).

At first, the interaction curve M–N (Figure 4.13c) is determined for the cross section of each structural element, based on the following assumptions:

a. Tensile strength is equal to the flexural strength of the material.
b. Constitutional law at the compressive zone is the same as that of concrete.
c. Compressive strength is equal to the prismatic one.

Next the structure is loaded with a small multiple, say 0.2, of the service loads. For this loading, the combinations of M–N at the critical sections are determined and then are plotted on their interaction M–N curve diagrams. There, using the analogy method, the position of the first plastic hinge is determined. Then, the same procedure is repeated for the new structural system with the first hinge, the new group M–N is again determined at the critical sections and so on, until a collapse mechanism is formulated and the load for which the collapse occurs is also calculated.

From this analysis and its computational development presented in Subsection 7.3.4, which was applied in many cases, the following conclusions can be drawn:

a. The hinge at the crown of the arch appears for a load smaller than the service one (Figure 4.14), at about 0.30–0.40 of the gravity load. As a result, cracks appearing at intrados at the crown of the arch are usually of no importance, as they are expected even for low gravity loads, and, therefore, they usually have no relation with eventual overloading of the structure from permanent or earthquake loads.
b. The collapse of an arch, so much for gravity loads as for earthquake loads, *is due to failure of the supporting system and not the arch itself.*
c. The failure of the supporting system is usually a *shear–compression* one and is due to the simultaneous action of the gravity loads, the arches' thrusts and the horizontal seismic shear forces (Figure 4.15).
d. At an advanced stage of horizontal displacements of the supports, the width of the cracks at the crown broadens, the compressive upper zone becomes very narrow, and so, it's not uncommon that splitting forces, which lead to horizontal cracks and splitting in the area, act on the arch's body at the crown (Figures 4.16 and 4.17).
e. To avoid these undesirable consequences of the thrust of the arch on the supports, the most effective way is the use of horizontal ties at the springing of the arch.

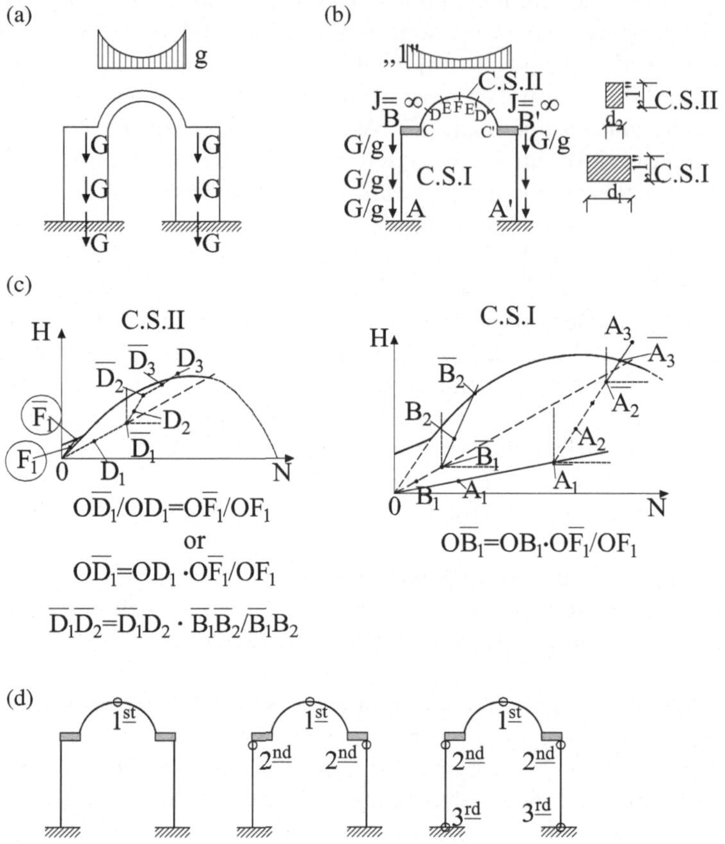

Figure 4.13 A step-by-step ultimate load method push-down analysis applied at the vaults and the arches of monuments of Thessaloniki damaged by earthquake (1978): (a) geometry of the vault and loading; (b) the first step of loading; (c) *M–N* envelop of a critical cross section; (d) successive steps of creation of plastic hinges. (Adapted from Penelis, G., Karaveziroglou, M., and Papayianni, I. (1989). Materials used in the restoration of historical buildings and monuments. *Seminario per Conservao Recuperaçäo e consolidacäo de monumentos e edificios histöricos*, Conselho Directivo National (ed) Lisbon; UNDP/UNIDO (1984). Repair and Strengthening of Historical Monuments and Buildings in Urban Nuclei, 6, Vienna.)

It is characteristic that in most Byzantine and Islamic arches, there existed timber ties anchored with metallic junctions in the masonry construction of the supports. These timber ties, either because they had rotted or for some reason of ignorance in latter centuries, have been removed leading to adverse results for these arched systems (Figure 4.18).

4.3.3 Groin vaults

The slicing notion used for the barrel vault also leads, in case of groin vaults, at once to some particularly simple solutions (Figure 4.19). For example, the cylinders of the vault may be sliced into independent circular arches not interacting with each other and

Figure 4.14 Hagia Sophia of Thessaloniki after the earthquake of 1978, cracks at the crown of the arches.

Figure 4.15 Hagia Irene of Istanbul, shear compression cracks at the main supporting walls of the main arch.

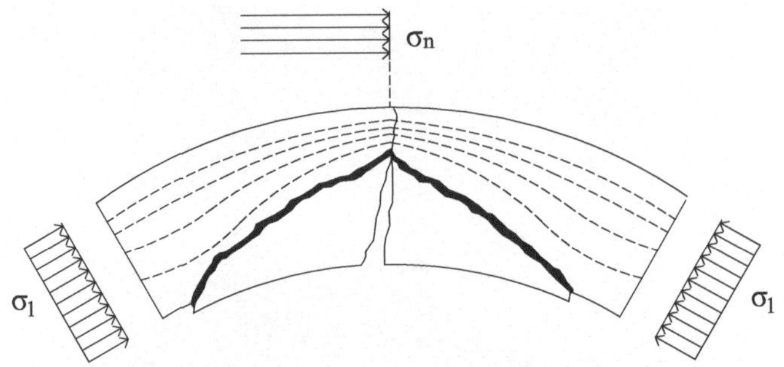

Figure 4.16 Splitting at the crown of an arch.

Figure 4.17 Bezesteni of Thessaloniki: crack pattern at the crown of an arch, compared to Figure 4.16. (From Penelis S.A., Archives, 1978.)

Figure 4.18 Vlatades monastery, Thessaloniki: the results of taking away of the ties of the arches.

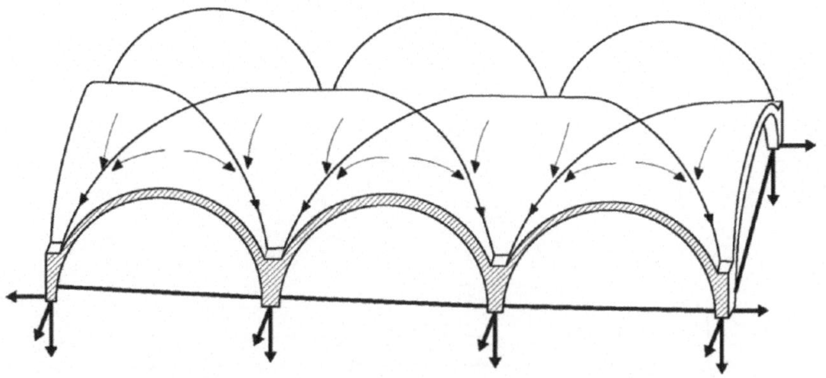

Figure 4.19 Load path at the groin vaults. (Adapted from Hart, F. (1965). *Kunst and Technik der Wolbung.* Verlag George Callwey, München, Germany. Courtesy of Georg, D.W., Callwey Verlag.)

supported, therefore, solely from the diagonal cross ribs reinforcing groins (Heyman, 1986). In this case, the thrust lines in the sliced arches are, of course, no longer circular. In fact, since, due to slicing, N_x and $N_{x\theta}$ are considered zero (membrane response is ignored), N_θ thrust line does not follow the axis of the slices. At the same time, the thrusts at the base of the sliced arches are not zero. The thrusts of the sliced arches lead to a pattern of forces which shows at once that 'flying buttress' supports or timber ties are required in this pattern. In fact, since the sliced arches carry their own weights to the diagonal ribs as vertical reactions and horizontal thrusts, evidently buttressing forces are required to stabilise the diagonal ribs.

The same kind of analysis may be carried out also for domical vaults (Figure 4.20). Most of the Roman, Byzantine and Islamic groined vaults are well buttressed or have timber ties at the base of the arches. In case these ties are deteriorated or generally destroyed, heavy damage occurs at the arches and the groin vaults themselves (Figure 4.21).

4.3.4 Domes

As known, in late Roman, Byzantine and Islamic monumental buildings, the dominant form used for roofing of big spaces was the spherical dome (Krautheimer, 1965; Penelis, G., 1992; Mainstone, 1975; Mango, 1975).

Spherical dome, as known from the shell theory (Heyman, 1977; Girkman, 1963; Penelis, G., 1965), behaves to its greatest extent as a membrane (Figure 4.22) and only near the supports exhibits bending moments in the form of disturbances. Even in the case of very thick domes, the deviations from the membrane theory are very small as it has resulted, at least, in Rotunda of Thessaloniki (Figure 4.23) according to a detailed analysis as a thick shell (solid-state model).

It is also known that the spherical shell (dome) for uniform self-weight exhibits compressive stress resultants as much in the direction of meridians N_φ as in the direction of the horizontal rings N_θ up to an angle from the perpendicular axis of $\varphi = 51.82°$ (Figure 4.24). The above fact, i.e. the membrane biaxial compressive state of stress, makes the spherical shell, at least up to an angle ($\alpha = 51.82°$), an ideal form for masonry construction, since there are no tensile stresses anywhere.

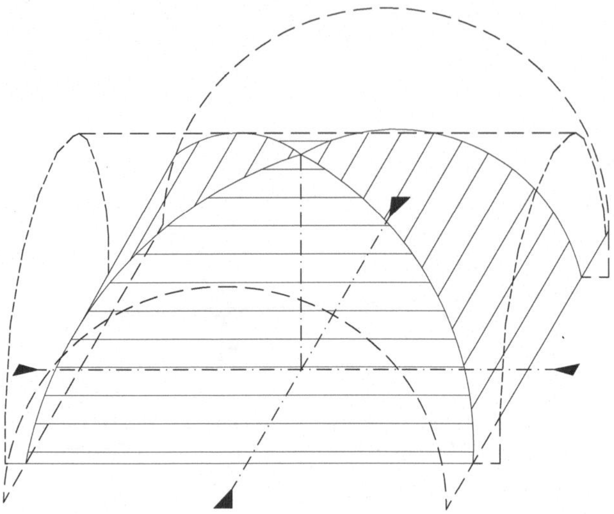

Figure 4.20 Domical vault. (Adapted from Hart, F. (1965). *Kunst and Technik der Wolbung*. Verlag George Callwey, München, Germany. Courtesy of Georg, D.W., Callwey Verlag.)

Figure 4.21 Hagia Anastasia monastery, Vasilika Thessaloniki. Splitting of the external piers of the convent building due to the deterioration of the wooden ties of the groin-vaulted deck of the first floor. (a) General view; (b) detail of a damaged pier. (From Penelis S.A., Archives, 2001b.)

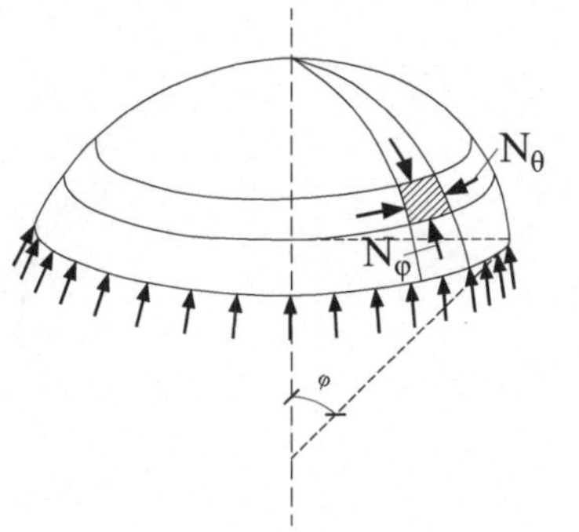

Figure 4.22 The membrane state of stress in shells in case of distributed gravity load.

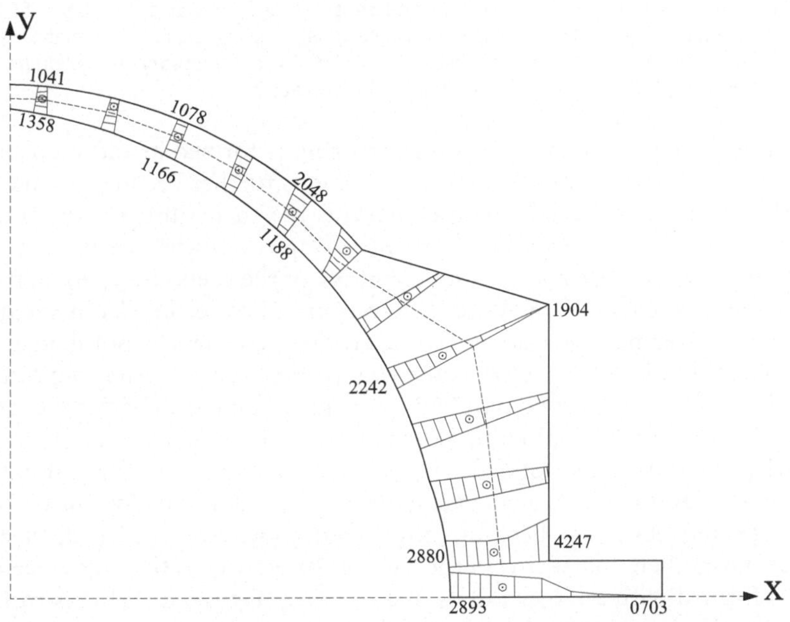

Figure 4.23 The state of stress (N_φ) in the dome of Rotunda of Thessaloniki for dead loads (FEM analysis). (Adapted from Penelis, G. (1998). Assessment of residual bearing capacity of masonry structures. *1st International Course on 'the conservation of ancient masonry in Byzantine Architecture'* (CAMBA-98) ICCROM-AUTH Thessaloniki, Greece.)

Even so, the spherical dome's problem begins exactly at the region where the dome ends, namely, at its support. Beyond any doubt, the Romans, the Byzantines and the Ottomans were aware of the fact that the spherical shell is under compressive ring forces for angles smaller than 51.82°. Thus, almost all spherical domes end at smaller angles. From this angle downward, a supporting system is formed, i.e. the *ring* and the *drum*, which beyond any doubt 'was designed' taking into account that it was not in a position, because of the construction material, to carry tensile ring forces. Thus, it was formed under the assumption that it acted as a series of independent buttresses which cannot develop tensile ring forces between them.

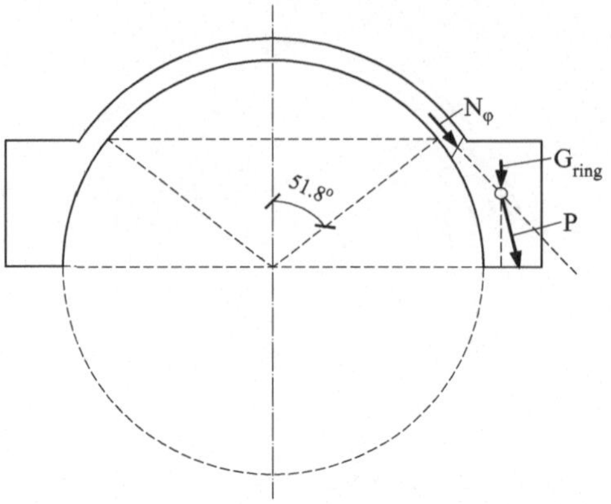

Figure 4.24 The boundary between compressive and tensile ring forces at a dome $\varphi = 51.8°$. (Adapted from Penelis, G. (1998). Assessment of residual bearing capacity of masonry structures. *1st International Course on 'the conservation of ancient masonry in Byzantine Architecture'* (CAMBA-98) ICCROM-AUTH Thessaloniki, Greece.)

Indeed, while in the intrados of the dome the ring is formed as the supplement of the hemisphere (Figure 4.24), from the outside, the ring is further broadened and assisted by loading with additional material which constrains the thrust line to stay in the central third of the base of the ring. In other words, the ring is nothing else but a series of buttresses that can carry the horizontal component N_φ of the shell safely, no matter if tensile ring forces cannot be carried any more. It is very usual to see cracks in the direction of meridians in the lower part of a dome. These cracks are generally not dangerous for the monument. They should be expected *as the ring cannot carry tensile ring forces*. On the other hand, it is quite usual to find windows piercing at regular intervals the lower part of the ring (i.e. Rotunda of Thessaloniki, Hagia Sophia, Dafni Monastery etc.).

Particularly, the formation of the ring of the Rotunda in Thessaloniki exhibits great interest mainly for the latter development of the dome in Byzantine monuments (Penelis, G., 1980a). As it has been already stated above, the shell ends at an angle of about 51.82°. From there, outwards in the case of the Rotunda, the inner face of the zone of the ring is formed with a larger radius of curvature (Figure 4.25). This formation had two advantages. First, the inner diameter of the building increased by 4.5 m, and second, the thrust line passed from the base of the ring to the support drum with an eccentricity of only 0.50 m. This formation is nothing else but a continuous pendentive like those formed by Anthemius and Isidorus in Hagia Sophia 250 years later, in order to cover a squared area with a circular dome and which, afterwards, was repeated in all of the cross-domed nucleum churches (Figure 4.57) (Penelis, G., 1982, 1992).

Finally, it is of some purpose to analyse the way of transmission of loads from the base of the ring to the base of the drum, in case, of course, a drum exists. In circular buildings (rotundas), which consist of the dome, the ring and the drum only, the last is foreseen with such a thickness, so that the thrust line always passes through the third of the cross section under the assumption of zero ring forces. A characteristic building of this type in Greece is the Rotunda of Thessaloniki, where it has been proven that the thrust line passes through the base of the drum almost through its centre line (Figure 4.26). This was accomplished by choosing for the drum a ratio of height to thickness equal to 2.8; that is, for a height of the drum of 18.0 m, its thickness was designed 6.25 m. In a cross-domed

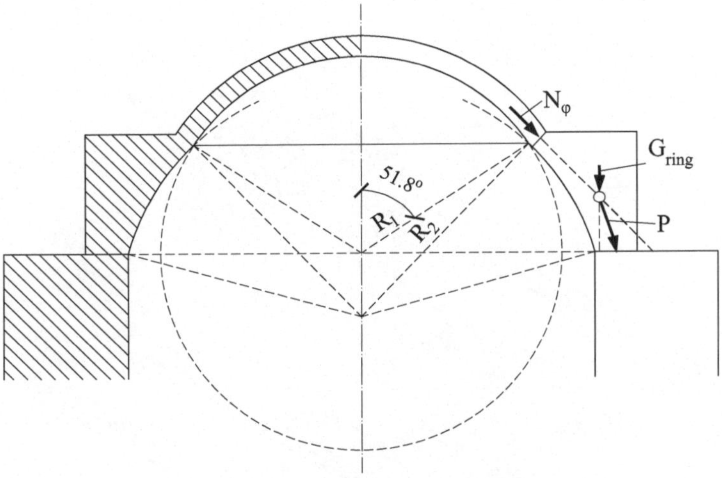

Figure 4.25 The formation of the dome of Rotunda, Thessaloniki, using two different radii of curvature. (Adapted from Penelis, G. (1998). Assessment of residual bearing capacity of masonry structures. *1st International Course on 'the conservation of ancient masonry in Byzantine Architecture'* (CAMBA-98) ICCROM-AUTH Thessaloniki, Greece.)

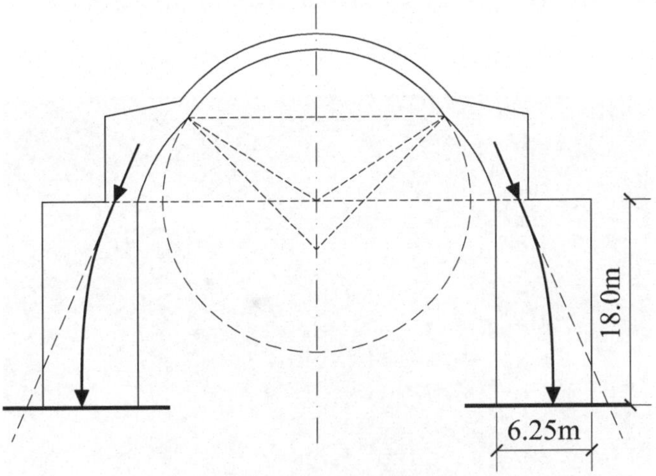

Figure 4.26 The load path (thrust line) at Rotunda, Thessaloniki, passing almost all along the height of the monument through the centres of the cross sections of the ring and of the cylindrical drum. (Adapted from Penelis, G. (1998). Assessment of residual bearing capacity of masonry structures. *1st International Course on 'the conservation of ancient masonry in Byzantine Architecture'* (CAMBA-98) ICCROM-AUTH Thessaloniki, Greece.)

church, where the dome is usually very large, the drum is kept very low and coincides with the ring (Figure 4.27). Openings are provided all around the ring for lighting the building as the tensile forces at the ring are not necessary for the equilibrium of the system.

In the Greek cross-octagon type (Krautheimer, 1965; Mango, 1975), there is a medium-sized dome with a medium-sized drum in height (Figure 4.28).

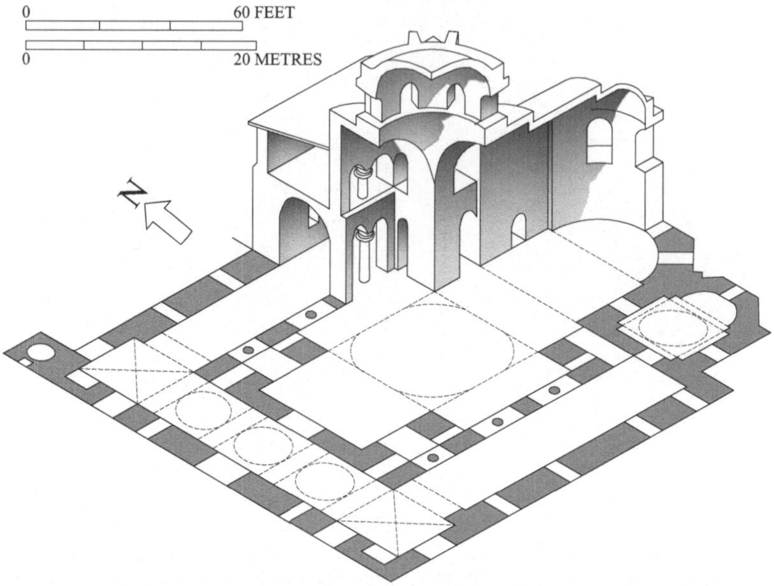

Figure 4.27 Hagia Sophia Thessaloniki: combination of a big dome with a low drum. (Adapted from Theocharidou, K. (1988). The architecture of Hagia Sophia, Thessaloniki. *B.A.R International Series 399*, Great Britain. Courtesy of BAR international series.)

Figure 4.28 Dafni monastery, Athens: medium-sized dome, medium-height drum.

In the Quincunx type (Krautheimer, 1965; Mango, 1975), the dome is small and the drum high. On the other hand, horizontal ring timber ties are foreseen in the base of the dome just after the ring at the height of the windows of the drum (Figure 4.29).

In the above three types of domes on a drum, a systematic effort has been made for aesthetic reasons to keep the drum thickness smaller than that in Roman circular buildings,

Figure 4.29 Agios Panteleimon, Thessaloniki, small dome, high drum formed in eight piers. (From Penelis, S.A., 2019c.)

as Pantheon or Rotunda. If the drum is low in height, this can easily be achieved even for large domes. As the drums grow in height, the domes become smaller in diameter, while the drums finally take the form of eight individual piers with skilfully designed cross sections to accomplish structural as well as aesthetic requirements (Figure 4.30). In case of high-rise drums, timber ties have been foreseen at the base of the dome.

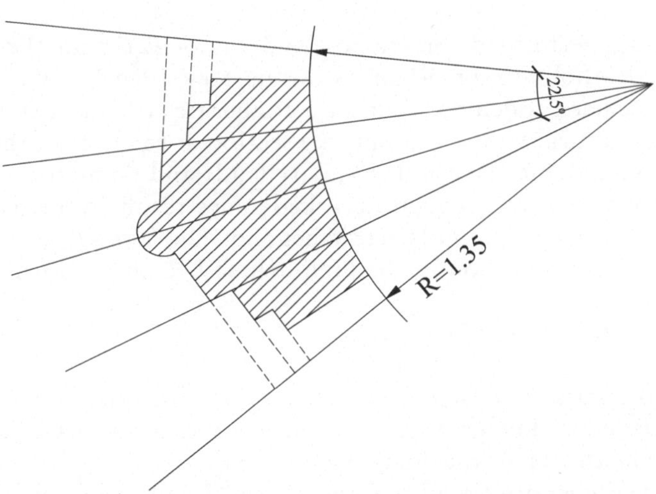

Figure 4.30 Vlatades monastery Thessaloniki, cross section of the drum piers. (Adapted from Penelis, G., Karaveziroglou, M., Stylianidis, K., and Leohdaridis, D. (1980c). Restoration and Strengthening of Vlatades Monastery. Archives of the Ephorate of Antiquities of Thessaloniki.)

Figure 4.31 Rotunda, Thessaloniki, meridian cracks at the dome.

From an analysis performed for Hagios Andreas of Peristera (Penelis, G., et al., 1980b) and for Katholikon of Vlatades Monastery (Penelis, G., et al., 1980c), Thessaloniki, it has been concluded that if the timber ties which exist at the base of the dome are taken structurally into account, the section at the base of the drum is found to be completely under compression for gravity loads and dome thrusts. In case these timber ties are not taken into account, cracked zones result at the base of the drum, but they lie within allowable boundaries. When, now, horizontal seismic action is also taken into account, the situation in the high drums becomes critical at their base in case the ties do not exist.

In conclusion, the following can be said:

a. In Roman, Byzantine and Islamic domes, cracks in direction of meridians is a usual phenomenon and is of no great concern when it does not surpass the angle $\alpha = 51.8°$ measured from the top of the dome (Figure 4.31).
b. Drums of the domes are sensitive at their base. Usually they exhibit inner horizontal cracking especially if the ring-like timber ties existing at the base of the dome have deteriorated or been removed. These cracks are dangerous and can lead to the dome's collapse (Figure 4.32).

Figure 4.32 Hagios Andreas, Peristera: horizontal cracks at the base of the drum, internal view.

4.4 MASONRY STRUCTURAL SYSTEMS – VULNERABILITY

4.4.1 General

The architectural forms of the Roman, Byzantine and Islamic monumental buildings around the Mediterranean basin till the end of the 19th century may be grouped, according to their structural behaviour, into a limited number of *structural types* (Penelis, G., 1998). It is important to note that these monuments are exposed to strong earthquakes from time to time. These structural types may be classified as follows:

i. *Free masonry walls*
   Under this category are classified mainly fortifications, for example, the defensive walls of Istanbul, or of Thessaloniki, and walls of ruined constructions staying free as the palace of Paleologos at Mystras, Peloponnese.
ii. *Timber-roofed basilicas*
   Under this type are classified all public halls of large dimensions. Their main characteristic is that they don't exhibit horizontal thrusts for gravity loads. This does not mean that they lack certain secondary arch elements, such as arcades and apses.
iii. *Domed circular buildings*
   This group includes a limited number of imposing monumental buildings, belonging basically to the Roman period, like the Pantheon of Rome, the Rotunda of Thessaloniki, the Octagon of Galerius in Thessaloniki (ruins) etc.
iv. *Vaulted and domed basilicas*
   The following architectural forms can be classified under this structural system (Krautheimer, 1965; Mango, 1975):
   a. Basilicas with vaulted wings and a central dome
   b. Mixed form of a basilica and a quincunx
   c. Mixed form of a basilica with a cross-domed church.

v. *Cross-domed central nucleum*
The following architectural forms can be classified under this structural system (Krautheimer, 1965; Mango, 1975):
a. The cross-domed church
b. The octagon-domed church
c. The Greek cross-octagon plan
d. The quincunx
e. The quincunx enveloped by a three-side narthex
f. Islamic mosques.

vi. *Secular buildings*
The following architectural forms may be classified under this structural system:
a. Dwellings
b. Monastery buildings
c. Governmental buildings
d. Castles.

vii. *Towers*
The following architectural forms may be classified in this structural system:
a. Bell towers
b. Minarets.

In the above structural systems, (i) to (vii), the structural elements that have been presented in the previous section (Section 4.3) are combined to a degree, in order to formulate the structural system of a monumental building.

4.4.2 Free masonry walls

The individual wall is the basic element of fortification works (Penelis, G., 1998; UNDP/UNIDO, 1984; Penelis, G., 1982). It is characterised by its great thickness in relation to its height (thickness to height ratio about 1:2.5–3.0). It behaves for vertical loads as a disk with, theoretically, uniform distribution of stresses. Because of its great thickness and its uniform mass distribution (by height), it exhibits very small natural period of about 0.1 sec. As a result, in case of an earthquake, the ground acceleration is not amplified by the structure. This has as a result that the risk for overturning, even for ground accelerations of about 0.30–0.33 g (Figure 4.33), is very small. However, the ramparts, which behave as setbacks, are very vulnerable to earthquake actions; therefore they usually present extended damages in case of a strong earthquake.

The most serious issues for these walls arise either from differential settlements which cause diagonal shear cracks or from the texture of their construction in relation to earthquake actions.

The first issue appears to be of lower significance because of the use of bed joints of high thickness. The creep of the thick joints during the construction phase contributed to the adjustment of the wall to the long-term ground settlements without serious self-constraint on the wall.

The second problem is due to the lack of homogeneity in the wall body, due to their structural texture consisting of three leaves, even though connections exist at intervals of about 1.0 m between the external leaves consisting of a number of layers of bricks (4–5 layers) throughout the wall body (Binda, et al., 2000, 2003; Binda, 2008). This leads to concentration of stresses on the two external leaves of the wall, particularly in case of seismic

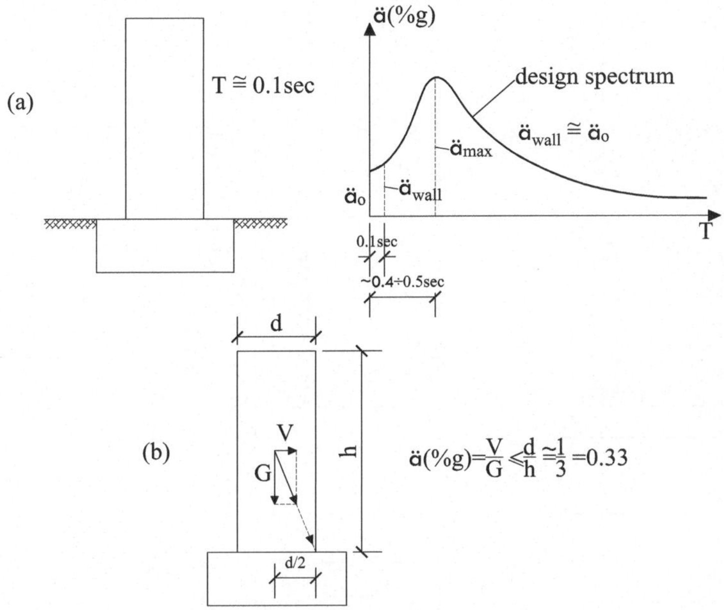

Figure 4.33 Seismic base shear of fortification walls: (a) explanation of the low seismic amplification; (b) ultimate base shear of fortification walls.

action resulting in their separation from the core of the wall. So, these leaves buckle and collapse (Figures 4.34 and 4.35) after the interruption of the transversal brick connections.

4.4.3 Timber-roofed basilicas

Timber-roofed basilicas consist of an external masonry rectangle and two or four internal colonnades with arcades carried by columns which form the nave and two or four lateral aisles. Usually, above the aisles and on the arcades rest the galleries (Theocharidou, 1985; Mavropoulou, 1986; Krautheimer, 1965; Mango, 1975). The timber roof (truss) over the nave is raised usually to some height above the gallery roofs to leave space for clerestory lighting between the two upper parts of the walls of the nave. It should be noted that this form of buildings pre-existed the Christian basilicas as Greek and Roman public buildings and continued to be used for secular uses up to the end of the 19th century. After the industrial revolution, the timber roofs and decks were replaced by steel trusses and beams (Penelis, G. and Penelis, Gr., 2001) (Figures 4.36 and 4.37).

It is clear enough that each structural element in this case, because of the lack of horizontal stiffness in the roof and the floors of the galleries, functions independently, so much for vertical as for horizontal seismic loads.

For permanent vertical (gravity) loads, all the elements (outer walls–colonnades with arcades) are free of horizontal thrusts. This has as a result the uniaxial compression of these elements. Only constructional eccentricities produce bending moments in the structure. The only elements under bending are the timber floors of the galleries and sometimes the timber joists of their roofs, when they are not trussed. Even the roof of the nave, as it is trussed, is affected mainly only by axial forces.

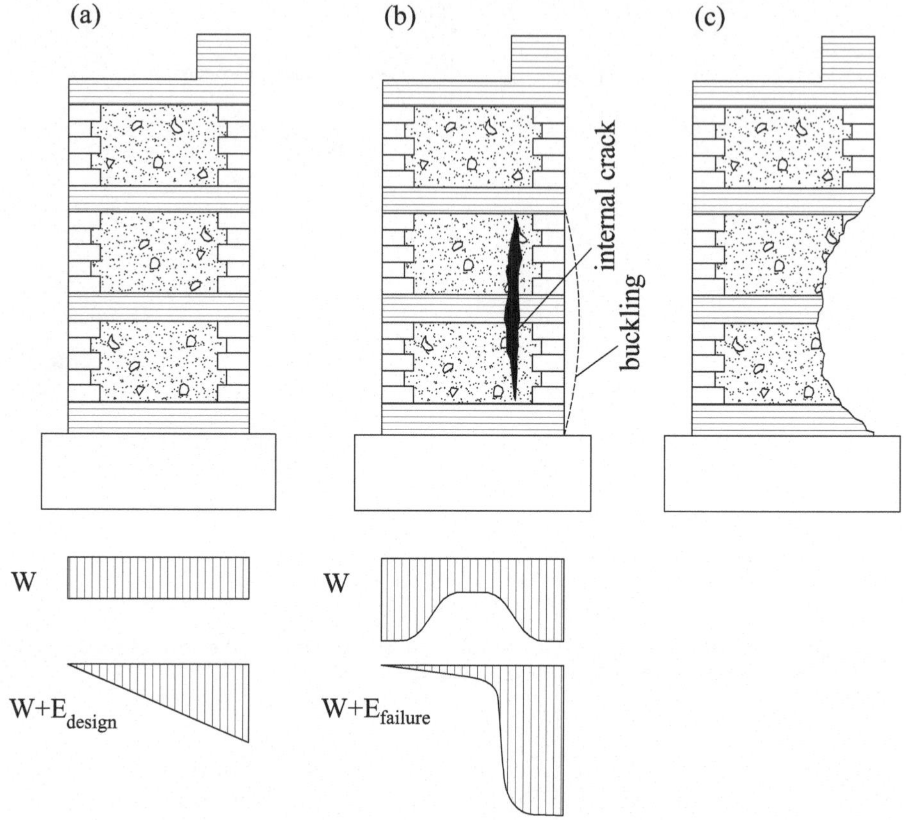

Figure 4.34 Structural behaviour of fortification walls: (a) elastic design for dead load (w) + seismic design action; (b) inelastic design for dead load (w) + seismic action at failure; (c) response to a strong earthquake, local collapse.

However, for horizontal seismic action and sometimes for wind, the whole system is very vulnerable. The absence of horizontal diaphragms, makes unavoidable in case of an earthquake, the vibrations of each masonry wall and each colonnade in an independent way, transversally to their main direction. Under these circumstances, it is clear that, if it is taken into consideration that the ratio of thickness to height in the walls of a large basilica is of about 1/15, the seismic coefficient (base shear to vertical loads) cannot surpass a maximum value of $\varepsilon = 0.05–0.06$ because, otherwise, the masonry will be overturned (Figure 4.38). Even so, many basilicas have survived for centuries even though they have passed through a series of earthquake actions with a peak ground acceleration, more than double of a value $\varepsilon_o = 0.05–0.06$. This is due to the fact that the structural elements (masonry walls, colonnades with arcades) are very flexible to their transversal direction and as a result they have a long natural period of about $T = 1.0$ s or more. Thus, they are found mostly out of phase of the predominant period of ground acceleration, and so, they have not collapsed (Figure 4.38).

This does not mean that they are not vulnerable to seismic action. So, even though the basilicas are, mostly, out of phase during an earthquake excitation, their structural

Figure 4.35 Damage in the body of the fortification walls of Thessaloniki.

Figure 4.36 National Library of Greece, Athens: general view.

elements (walls, columns), due to their low stiffness, undergo large displacements at their top independently from each other. Consequently, they present a series of quite significant damages (vertical cracks), mainly at the joints of the walls. Thus, usual damages resulting from an earthquake are (Figure 4.39) (Penelis, G. and Stylianidis, K., 2001)

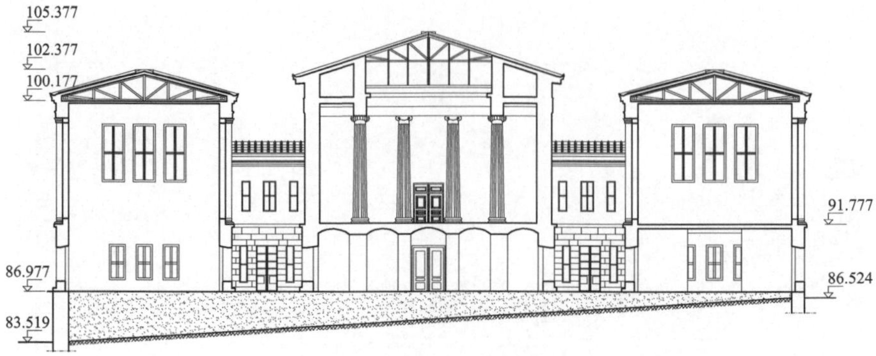

Figure 4.37 National Library of Greece, Athens: cross section.

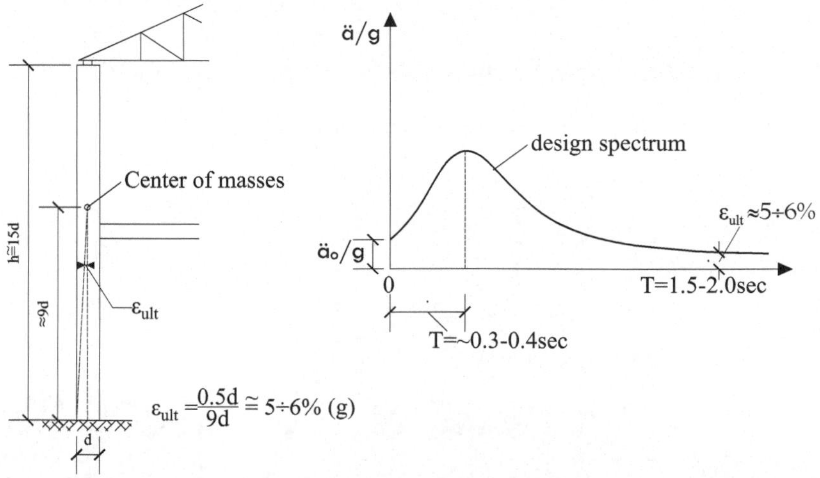

Figure 4.38 **Structural behaviour of timber-roofed basilicas towards seismic action.**

a. Permanent drifts of masonry walls and colonnades out-of-plane, because of unilateral settlement of the foundations during the earthquake excitation (Figure 4.40).
b. Dislocations of masonry walls, perpendicular to each other, at their joints (Figure 4.41).
c. Dislocations of the colonnades from the walls transverse to them at their vertical joints. Damages appear as a crack of the last arch which connects to colonnade with the walls (Figure 4.42).
d. Cracks at the crowns of the arches of the arcades and the apse (Figure 4.43).

It is obvious that, in a structural system like this, any structural intervention usually aims towards damage repairs of a local character and not to general improvement of the structural system given that not only the nature of these weaknesses, but also the nature of system's advantages (high flexibility, long natural period) are inherent in its form.

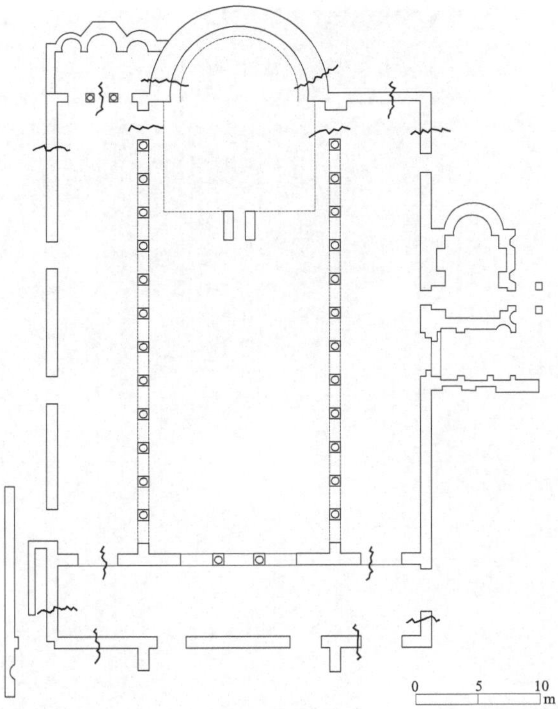

Figure 4.39 Acheiropoietos of Thessaloniki, cracks survey after the earthquake of 1978.

However, in some cases, although the improvement of the diaphragmatic stiffness of the gallery floors and of the roofs leads to a shortening of the structural fundamental period T and therefore to an increasing of the seismic effects, at the same time, it allows the participation of the transverse walls of the narthex and the apse to the transfer of the seismic action to the foundations through shear. In this respect, alternative intervention schemes should be examined before the final decision, namely, whether or not to stiffen the above diaphragms.

4.4.4 Domed circular buildings

As it was already mentioned earlier, Rotunda of Thessaloniki is the only building classified under this form in the Greek territory. This building, clearly, reminds us more closely of the Pantheon (Figure 4.44), the caldarium of Caracalla's Baths and the Tor de Schiavi in Rome. A rather recent example of this form is the orthodox church of St George (Mar Girgis church) in Cairo (Figure 4.45) constructed at the beginning of the 20th century above a Roman tower attributed to the Roman Emperor Traian built in the 1st century A.D. (Penelis, G. and Penelis, Gr., 2015, 2018).

The structural behaviour of the dome, the ring and the drum of these buildings has already been analysed thoroughly in a previous subsection dealing with domes as structural elements; thus, here only a few more details are going to be given for the structural behaviour and the vulnerability of this type of monument.

Figure 4.40 Acheiropoietos of Thessaloniki – internal view. (From Penelis S.A., Archives, 2019a.)

Figure 4.41 Acheiropoietos of Thessaloniki – dislocation of walls at one corner. (Courtesy: Ephorate of Antiquities of Thessaloniki.)

Figure 4.42 Acheiropoietos of, Thessaloniki – dislocations of external colonnades and transversal external walls.

Figure 4.43 Acheiropoietos of Thessaloniki – cracks at the crowns of the arches of external arcades.

Particularly the analysis of the Rotunda of Thessaloniki has given the following results (Penelis, G., et al., 1983, 1984; Penelis, G., 1992):

a. This monument did not exhibit any vulnerability to vertical loads. The thrust line doesn't pass anywhere out of the middle third of the critical sections and reaches at the bottom of the foundation with an eccentricity of about 10 cm for a foundation width of 8.6 m (Figure 2.42).
b. Concerning the earthquake loads, it also exhibited very low vulnerability. Its natural period, with the assumption of a ring and the drum without cracks was smaller than $T = 0.02$ s, meaning that there was no amplification of the peak ground acceleration

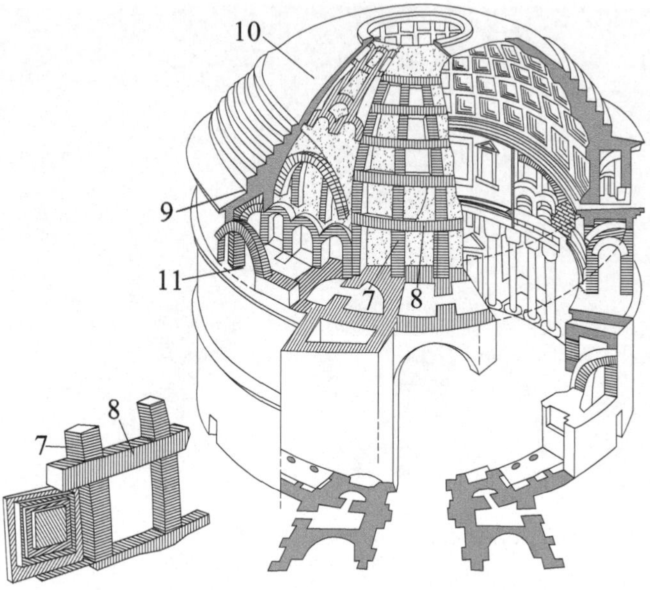

CONSTRUCTION OF THE DOME
7 MERIDIAN RIBS OF BRICKWORK.
8 HORIZONTAL RINGS OF BRICKWORK.
9 DISCHARGE ARCHES EMBEDDED IN THE CONCRETE..
10 MONOLITHIC ROMAN CONCRETE .
11 BUTTRESSING WALL OF THE DOME.

Figure 4.44 Axisymmetric cross section of Pantheon. (Adapted from Andrade, Fr. (1995). La construccion romana (II). Revista de Edificación. RE. No19 Teoriae historia de la arquitectura. p. 69. Courtesy of the author.)

on the monument masses. It is important to note that, even for a base shear of 12% of vertical loads, the tensile stresses in the ring of the dome were below the tensile strength of the building mortar.

c. The Christian intervention on the monument, which included the cutting of the ring of the dome and of the middle ring of the cylindrical wall drum, changed the structural behaviour of the monument making it vulnerable to seismic action. In other words, the geometrical relations of the dome, of its ring and of the cylindrical wall, safeguarded the safety of the monument to gravitational loads and seismic action, as it has happened to 'Pantheon' and 'Caracalla's Baths'. The above geometrical relations were not kept in the case of 'Mar Girgis church' where the thrust of the dome caused vertical cracks to the cylindrical part of the church soon after its construction.

4.4.5 Vaulted and domed basilicas

In the Greek area, but also generally in Southern Europe, the vaulted basilica is not a usual type of building. As known, the barrel vault causes significant outward directed thrusts, and the only way to transfer them to the foundation safely is by giving the vault either a continuous support of thick walls or a support of somewhat thinner walls buttressed at intervals

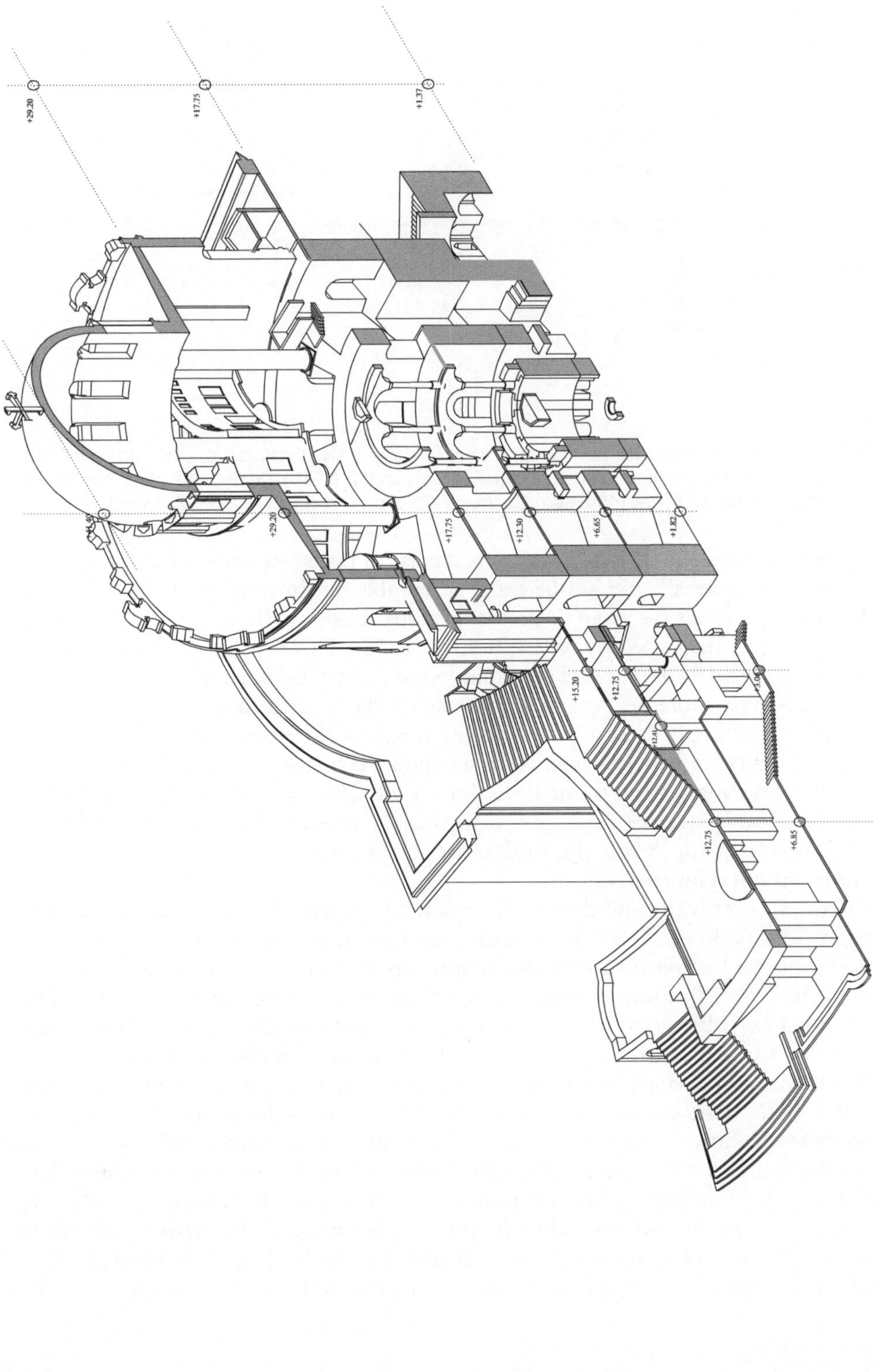

Figure 4.45 Axisymmetric cross section of Mar Girgis, Cairo. (Adapted from Penelis, G. and Penelis, Gr. (2018). Restoration of the Mar Girgis Church and Roman tower in Cairo. *Proceedings of the 4th Architectural and Urban Restorations BRAU4*, CI COP Italia. pp. 163–174.)

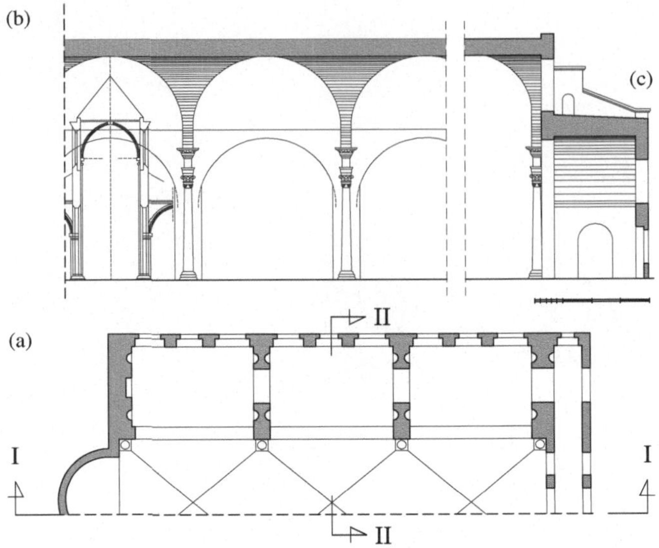

Figure 4.46 Basilica of Maximilian Rome (ruins), a vaulted three-aisled basilica: (a) plan; (b) longitudinal section I–I (c) cross section II–II. (Adapted from Hart, F. (1965). *Kunst and Technik der Wolbung.* Verlag George Callwey, München, Germany. Courtesy of Georg, D.W., Callwey Verlag.)

by projecting buttresses. Another alternative is the use of transverse ties at the springings of the vaults, solutions generally not aesthetically acceptable for the nave of the church (Figure 5.37). The same also holds for groin or domical vaulted Basilicas (Figure 4.46) (Mainstone, 1975; Hart, 1965; UNDP/UNIDO, 1984; Bouras, 1965). This type of support, illustrated above, may be regarded as downward continuation of the vault within which the thrust line becomes progressively more vertical as it approaches the foundation. An example of this type of structure existing in Greece is the Christian part of the Rotunda (Figure 4.47).

This type of structure is very sensitive to earthquake thrusts if the walls are not well buttressed. It is very interesting to mention that in the Christian part of Rotunda, where only two symmetric buttresses exist, the vertical walls appear to have a large drift of the order of 40 cm at their top, while the vault itself has extended cracking at its intrados, in the area of its crown (Figure 4.48).

To avoid thick outer walls and flying buttresses, which were not accepted aesthetically, even though they made up the basic structural and aesthetic element in the gothic form (Figure 4.49), the vaulted basilica with transept wings and a dome in the centre was developed during the early Byzantine period, known as the transept basilica (Krautheimer, 1965; Mango, 1975). In this type of structure, four-barrel vaulted arms radiate from a domed or groin-vaulted centre bay. However, the nave and the chancel are considerably longer than the transept wings. In this way, the transept wall of one arm acts as a buttress for the other. But, as the arms of the chancel and the nave are long enough, this type of buttressing was not sufficient to carry over all the thrusts of the barrel vault. So, although many transept basilicas with vaults and central dome were built, very few (e.g. Katopoliani, Paros) (Figure 4.50) continue to survive to date. It can be said that churches of this type were precursors of the cross-domed church *with a rather unsuccessful structural system.*

Eight centuries later (14th century A.D.), in Paleologian architecture, the form of vaulted transept basilica with domes was remade, but with some additional characteristics which

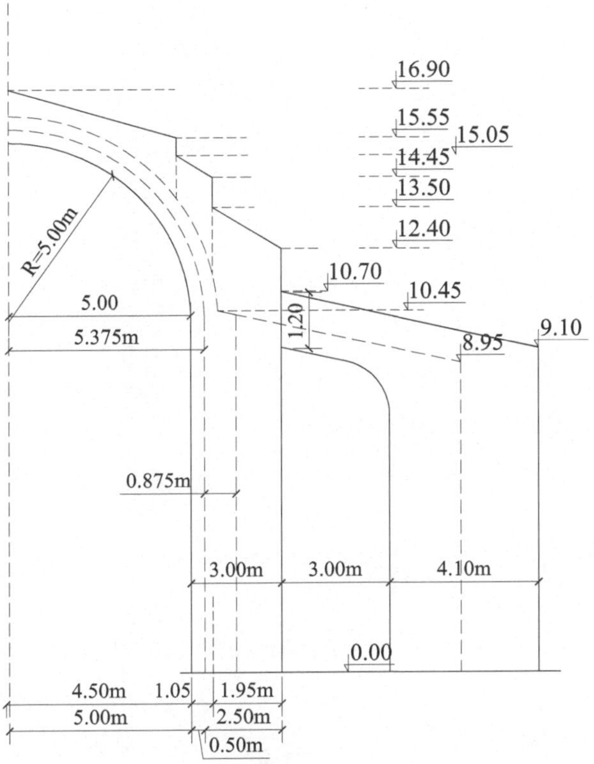

Figure 4.47 Rotunda of Thessaloniki: cross section of the Christian part, vaulted structure with two buttresses. (Adapted from Penelis, G. karaveziroglou, M., Stylianidis, K., and Leohdaridis, D. (1980a). Restoration and strengthening of Rotunda of Thessaloniki. Archives of Ephorate of Antiquities of Thessaloniki.)

Figure 4.48 Rotunda of Thessaloniki – cracks at the crown of the Christian vault.

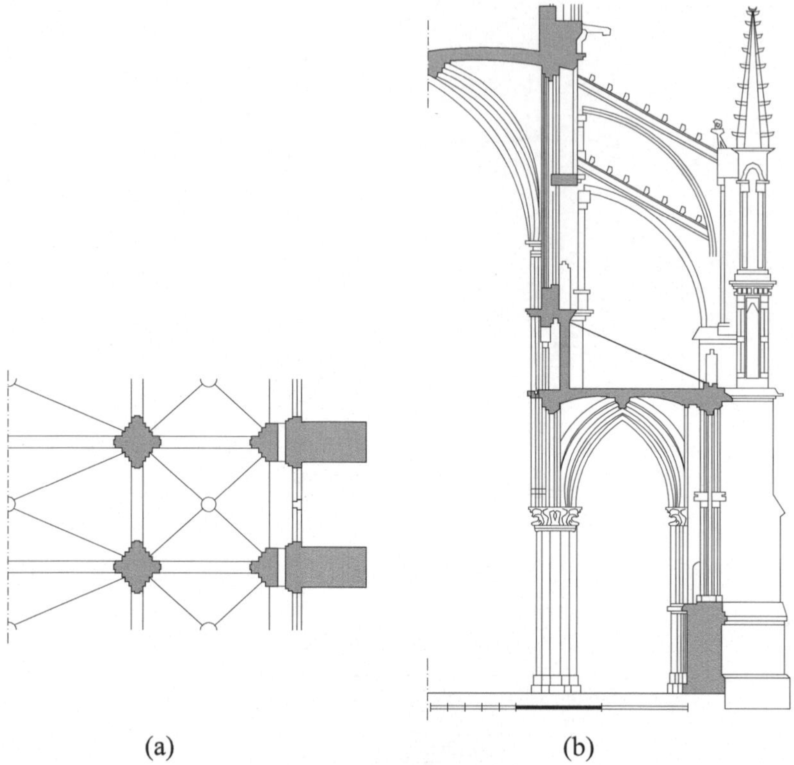

Figure 4.49 Reims Cathedral, groin-vaulted basilica with flying buttresses: (a) plan; (b) cross section. (Adapted from Hart, F. (1965). *Kunst and Technik der Wolbung*. Verlag George Callwey, München, Germany. Courtesy of Georg, D.W., Callwey Verlag.)

have improved the overall structural behaviour. The main Cross with the dome and the vaulted arms is accomplished with four additional vaulted halls which are the aisles of the basilica and whose vaults buttress the main vault (through their opposite thrusts). This type of structure is found in many churches such as Kato Panaghia, Porta Panaghia, Metrolopis of Mystras, Panaghia Odighitria and Pantanassa of Mystras (Figures 4.51 and 4.52).

4.4.6 Cross-domed central nucleum

This structural type covers the greatest part of Byzantine churches and the majority of Muslim mosques.

The basic structural system consists of (Krautheimer, 1965; Mango, 1975):

a. Four strong piers that mark the corners of the centre square. Solid or pierced in one or two directions and sheltering small groin vaulted bays, these piers carry the barrel vaults of four rather short cross arms of equal or nearly equal depth (Figures 4.53, 4.54 and 4.55). Sometimes, in case the piers are pierced in two directions, the small quarter pier that is formed thus in the corner of the square bay may be replaced by a column.
b. A dome on pendentives –large with a short drum, medium with a medium drum or small with a high drum – rises from apses or the barrel vaults.

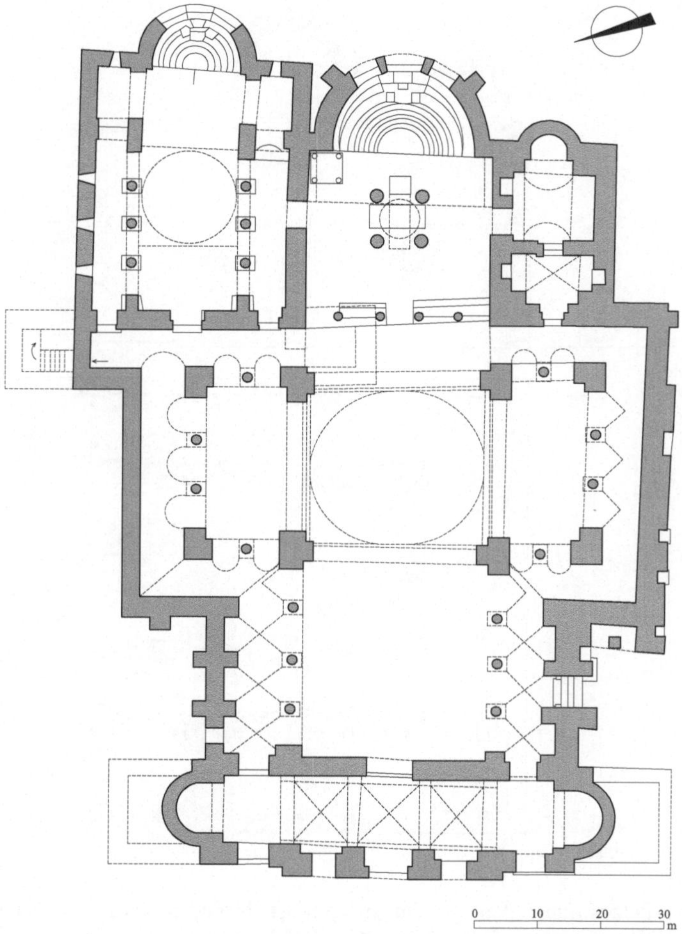

Figure 4.50 Ekatontapyliani, Paros Island cross-domed basilica: main part of plan view. (Adapted from Penelis, G. (1982). *The Masonry Roman and Byzantine Monuments in Greece. UNDP/UNIDO, PROJECT (RER/79/015) National Report of Greece.*)

The vertical loads of the dome flow through the ring of the drum as vertical and horizontal thrusts to the pendentives and through the barrel vaults smoothly to the solid or pierced piers (Penelis, G., et al., 1984; Mainstone, 1975) (Figures 4.2, 4.56 and 4.57). As the barrel vaults are not long, even in the case of pierced piers, the vaults are well buttressed by the piers themselves. It is self-understood that such a system is very robust and compact, which means that it does not amplify the ground accelerations. On the other hand, the existence in plan of a cross with shear walls at its arms or large piers is a very good layout for an ideal performance during an earthquake. From the seismic analysis of Hagios Andreas in Peristera (Penelis, G., et al., 1980b) and Vlatades Monastery (Penelis, G., et al., 1980c) in Thessaloniki, it has come out that these systems could carry safely a seismic base shear greater than 25% of vertical loads. It should be noted that the above value has been determined taking also into account the shear resistance of the system in combination with bending. This is *the reason why* so many churches and mosques, with a central-domed nucleum as their basic structural component, have survived for

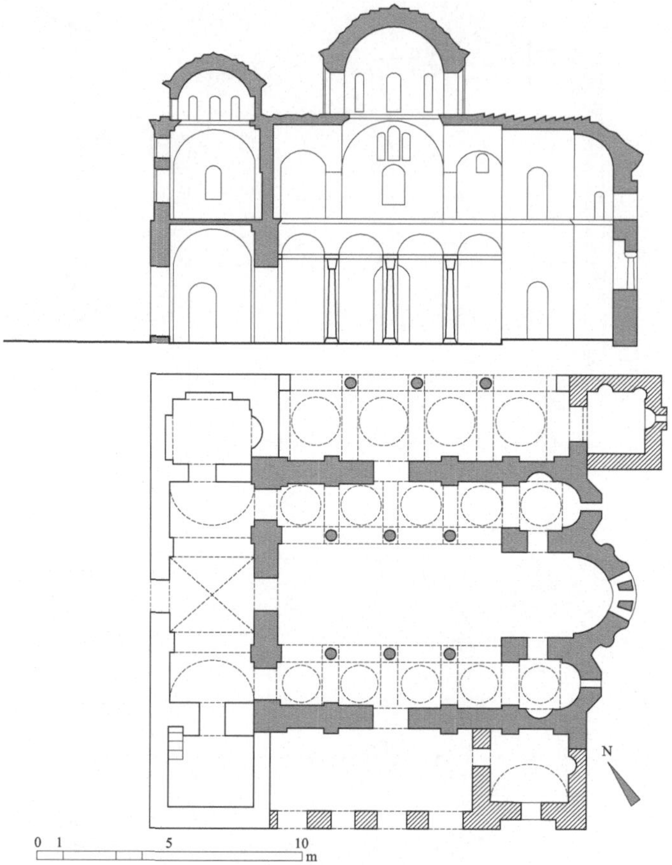

Figure 4.51 Mystras, Peloponnese, Panagia Odigitria, cross-domed basilica: main part of plan view and cross section. (Adapted from Mango, C., 1975.)

Figure 4.52 Mystras, Peloponnese, Panagia Odigitria: internal view. (From Penelis S.A., 2019b.)

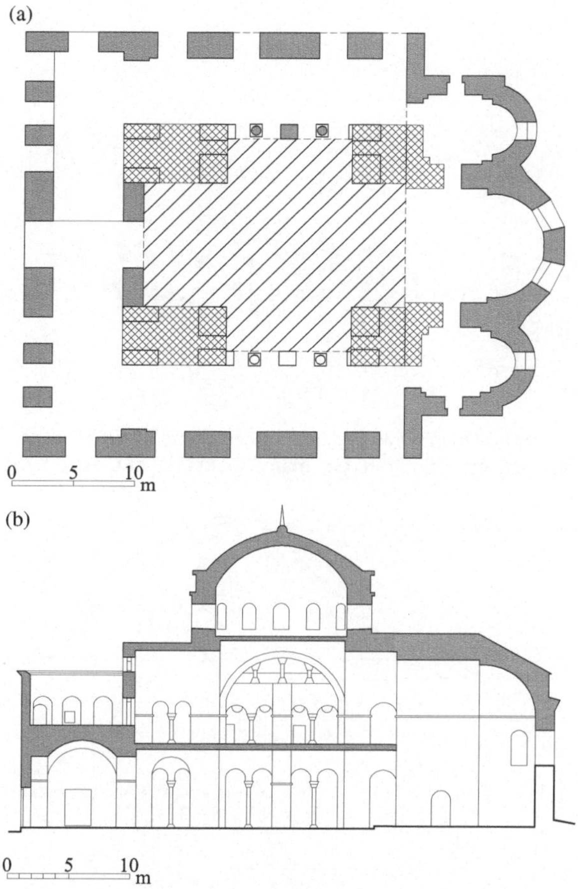

Figure 4.53 Hagia Sophia of Thessaloniki; cross-domed nucleum: (a) plan view; (b) cross section. (Adapted from Penelis, G. (1982). The Masonry Roman and Byzantine Monuments in Greece. *UNDP/UNIDO, PROJECT (RER/79/015) National Report of Greece.*)

centuries around Mediterranean basin, although they are subjected to very strong seismic action every now and then.

From the above qualitative analysis, together with the detailed structural analysis for vaults and domes, it is concluded that the following types of damage may be expected in a church or a mosque with a square-domed nucleum as basic structural system (UNDP/UNIDO, 1984; Penelis, G., et al., 1992; Guera, G., 1979):

a. Radial cracks in the drum-ring and the lower part of the dome (Figure 4.58).
b. Horizontal cracks at the basis of the drum in its interior, especially when the drum is high with windows. These cracks may become very dangerous for the structure, in case the timber ties at the top of the drum have been taken away (Figure 4.32).
c. Cracks at the crown of the vaults or arches (Figures 4.14 and 4.48). These are cracks of no importance. In most cases, these vaults were also tied with timber ties at their springings. The destruction of these ties has led to additional opening of the cracks at the crown of the vaults.

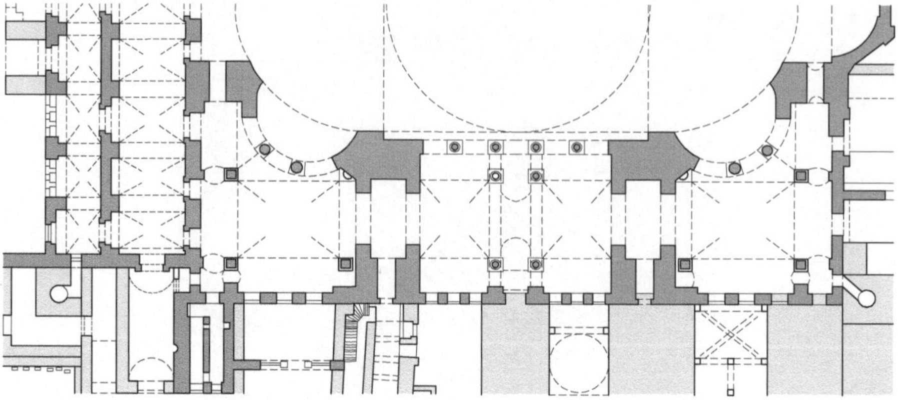

Figure 4.54 Hagia Sophia, Istanbul: plan view. (Adapted from Hart, F. (1965). *Kunst and Technik der Wolbung.* Verlag George Callwey, München, Germany. Courtesy of Georg, D.W., Callwey Verlag.)

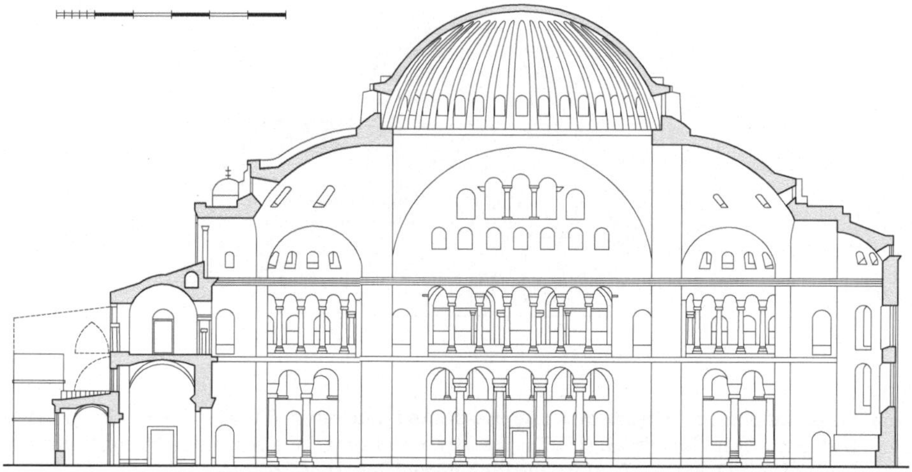

Figure 4.55 Hagia Sofia of Istanbul: cross section. (Adapted from Hart, F. (1965). Kunst and Technik der Wolbung. Verlag George Callwey, München, Germany. Courtesy of Georg, D.W., Callwey Verlag.)

d. Cracks at the spandrels of the pierced piers (Figure 4.59).
e. Diagonal cracks at the piers or at the shear walls of the cross at the central bay (Figures 4.15 and 4.60).

4.4.7 Secular buildings

4.4.7.1 The two main residential masonry structural systems

The prevailing structural systems of residential masonry buildings may be classified into two main categories (Penelis, G., 1998; Tomazevic, 1999):

a. The free-standing wall systems with intermediate floors and a roof
b. The wall systems with decks formed as diaphragms and a roof.

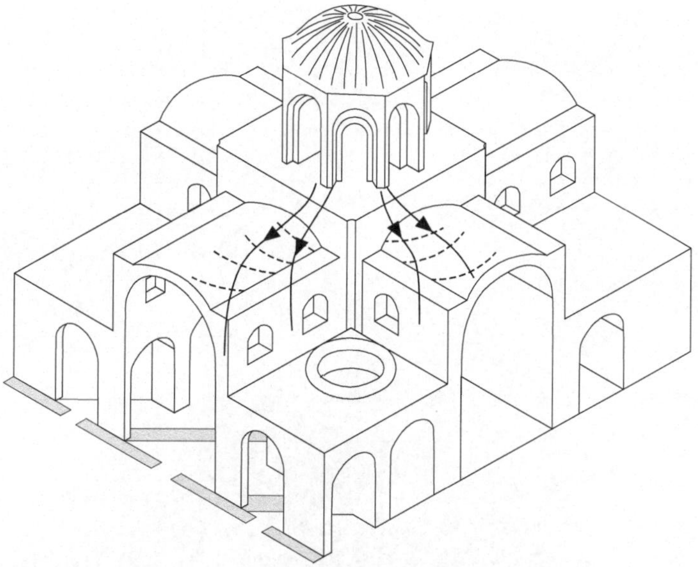

Figure 4.56 Thrust flow in the cross-domed nucleum in case of small dome with high drum in the form of piers. (Adapted from Penelis, G. (1998). Assessment of residual bearing capacity of masonry structures. 1st *International Course on 'the conservation of ancient masonry in Byzantine Architecture'* (CAMBA-98) ICCROM-AUTH Thessaloniki, Greece.)

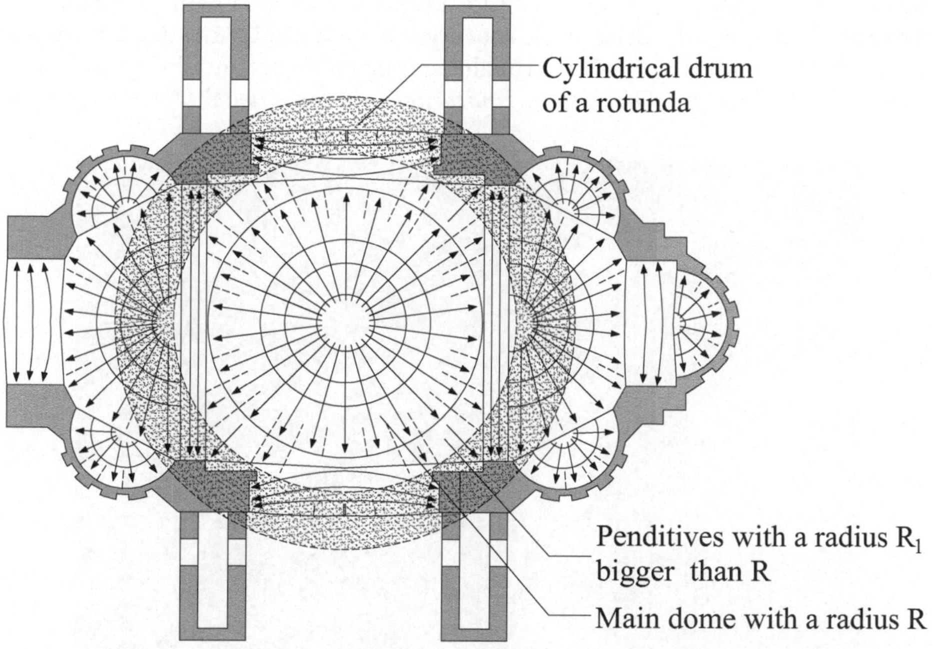

Figure 4.57 The development from the rotunda to the cross-domed central nucleum. (Adapted from Mainstone, R. (1975). Developments in structural form. Penguin Books Ltd, England.)

Figure 4.58 Hagia Irene, Istanbul: radial cracks at the ring pendentives and the lower part of the dome.

In the first category, the floor or the roof consists of an independent beam of truss elements capable of carrying gravity loads and transferring them to the free-standing walls but unable to transfer seismic horizontal loads in the deck plan, because of the fact that the joists or the trusses do not constitute a rigid structure in plan (Figure 4.61a).

In the second category, the deck or the roof constitutes a rigid plane diaphragm that can transfer horizontal loads to the walls through bending moments and shear forces acting in the plane of the diaphragm. *The diaphragmatic action* is very usual in masonry buildings

Figure 4.59 Hagia Sophia, Thessaloniki: cracks at the spandrels of the pierced piers.

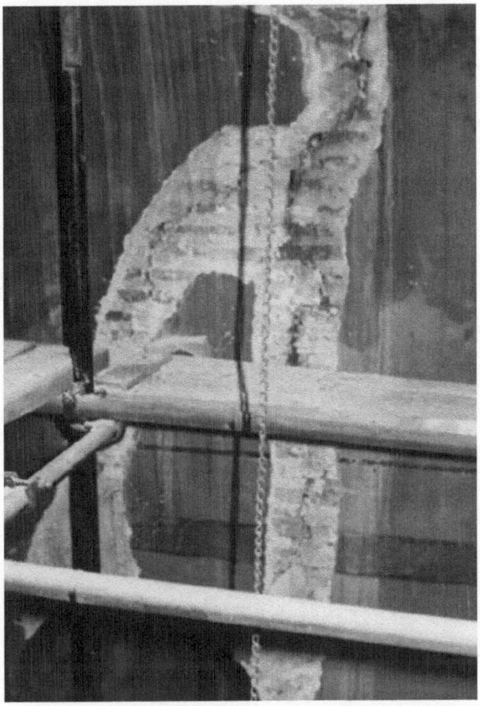

Figure 4.60 Hagios Panteleimon, Thessaloniki: shear cracks at the walls of the Cross supporting the dome.

of the last two centuries due to the fact that the decks are reinforced concrete slabs or solid systems of steel beams connected with concrete to a solid diaphragm (Figure 4.61b) (composite structures).

It should be noted that, even in the case of free-standing walls, the connection of the deck joists with the timber boards of the floor or the connection of the trusses of the roof with wind ties helps in the creation of a pseudo-diaphragmatic action.

4.4.7.2 The transfer of seismic action in the free-standing wall systems

In the case of buildings, without diaphragmatic action of the floors, each wall acts as a free cantilever element in the out-of-plane direction. At the same time, it is supported perpendicularly through its joints with the transverse walls (Figure 4.62). So, the usual modes of failure in buildings of this structural system are the following:

a. Out-of-plane overturning of external walls
b. Tensile cracks at the lintels (Figure 4.62b)
c. Bending cracks at the top and the bottom of the piers (Figure 4.62a)
d. Failures at the corners (Figure 4.63).

In other words, the main failure mode is the result of bending moments or tensile forces. In order that the masonry is in position to carry the horizontal tensile force in seismic regions, horizontal timber ties have been incorporated in the most of the walls, and iron connectors have been foreseen at the edges of these buildings at the level of their decks and roofs.

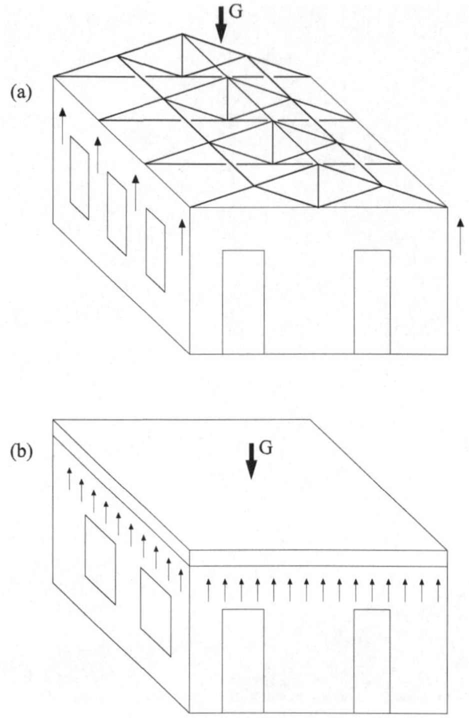

Figure 4.61 The two main structural wall systems of transferring gravity and seismic action: (a) free-standing wall systems; (b) diaphragmatic wall system. (Adapted from Penelis, G. (1998). Assessment of residual bearing capacity of masonry structures. 1st *International Course on 'the conservation of ancient masonry in Byzantine Architecture'* (CAMBA-98) ICCROM-AUTH Thessaloniki, Greece.)

4.4.7.3 The transfer of seismic action in the case of diaphragmatic action of the floors

In the case of buildings with diaphragmatic action of floors, due to the stiffness of the diaphragm, the seismic horizontal forces are transferred to wall elements of high stiffness, in other words, to walls parallel to the horizontal forces. The usual failure types in this case are the bending failure at the top and the bottom of the piers of these walls and the shear failure with X-type cracks at the same piers (Figure 4.64). This last type of failure is the most dangerous one, because the situation becomes unstable even for the gravity loads resulting in a local or general collapse as already explained (see Subsection 4.3.1). For this reason, in seismic areas, these walls are reinforced with timber trusses incorporated in the masonry.

4.4.8 Bell towers and minarets

Both forms are characterised by their large height in comparison to their base dimensions (Figures 4.65 and 4.66). They are usually of tubular form, tall and slender and with a stair ascending in the centre of the tube.

Usually, the bell towers have a quadratic cross-section constant all over their height with timber floors every 4–5 m. Usually, they have small windows at every storey for

(a)

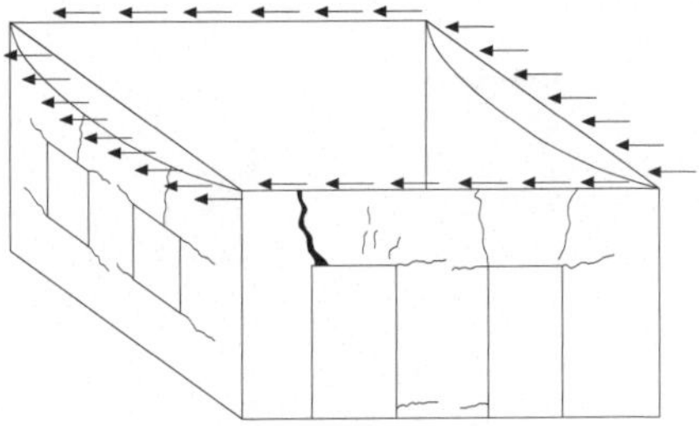

(b)

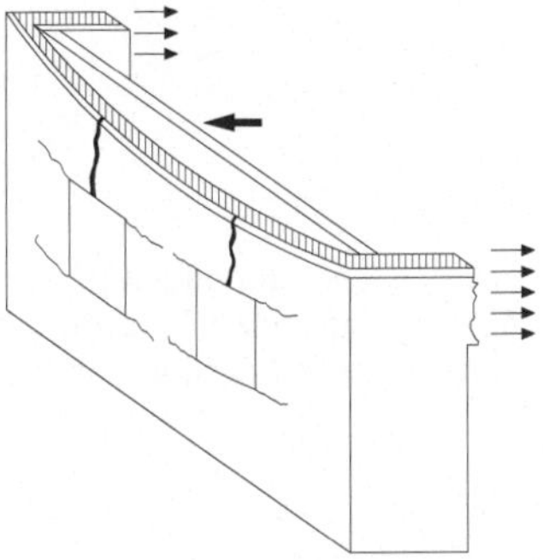

Figure 4.62 The transfer of seismic action at a free-standing wall system: (a) general layout; (b) detail of the left side wall. (Adapted from Penelis, G. (1998). Assessment of residual bearing capacity of masonry structures. 1st *International Course on 'the conservation of ancient masonry in Byzantine Architecture'* (CAMBA-98) ICCROM-AUTH Thessaloniki, Greece.)

lighting the staircase. At their top, a canopy is formed, with four to twelve columns and a roof where the bells are installed.

The minarets begin usually at their base with a square cross section and continue upwards as a conical tube. Minarets have one or more slightly projecting platforms (balconies).

Both bell towers and minarets have in addition concentrated masses, the minarets at the balconies due to their construction material, while bell towers have the additional problems of the bell, whose mass functions as a pendulum at the top of the tower (Selby and Wilson, 1997).

Both, due to their long natural period of vibration, are out of the predominant phase of a usual earthquake; that is, their eigenperiod is at the right side of the acceleration response spectrum of the seismic action. Therefore, their seismic excitation is reduced below the peak ground acceleration, and so they present, generally, rather a good behaviour during earthquakes (Diaferio and Foti, 2017). They present damages mainly at their upper part due to their higher frequencies combined with the low weight of the upper part. So, usual damages after a strong earthquake appear at the upper balcony of the minaret and the upper tubular part of it. The type of damage is destruction of the balcony and cracks at the base of tube on the balcony (rocking action) (Penelis, G., et al., 1993).

On the other hand, the bell towers present damages at the columns and the arches of the bell station.

Many times, the lower part of the bell tower or the minaret is incorporated in the church or in the mosque, respectively, making the structural behaviour of both more complicated (Figure 4.66).

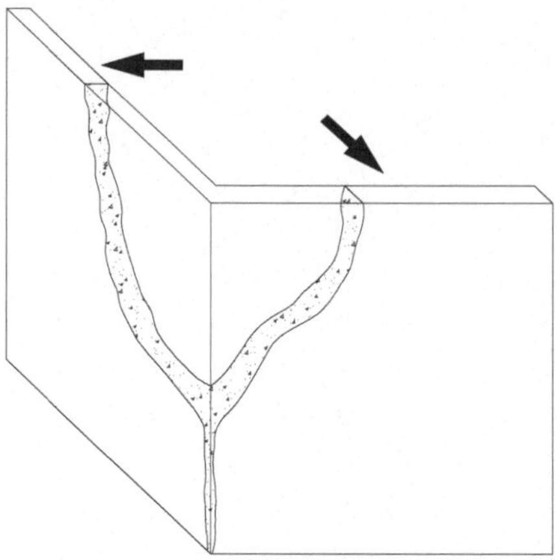

Figure 4.63 Local collapse at the corner of a free-standing wall system. (Adapted from Penelis, G. (1998). Assessment of residual bearing capacity of masonry structures. 1st *International Course on 'the conservation of ancient masonry in Byzantine Architecture'* (CAMBA-98) ICCROM-AUTH Thessaloniki, Greece.)

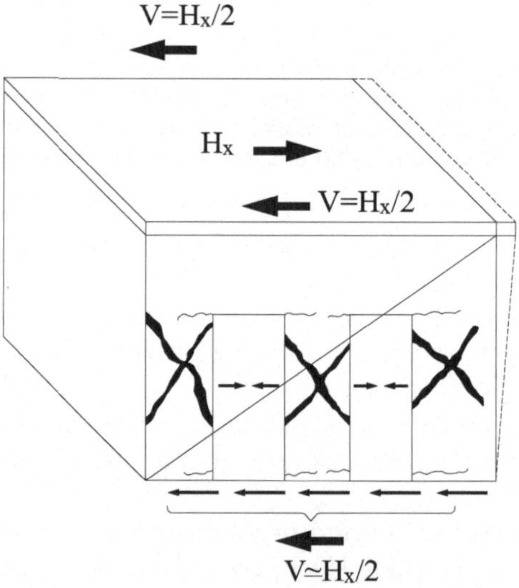

Figure 4.64 Diaphragmatic action of the deck for the transfer of the seismic action. (Adapted from Penelis, G. (1998). Assessment of residual bearing capacity of masonry structures. 1st *International Course on 'the conservation of ancient masonry in Byzantine Architecture'* (CAMBA-98) ICCROM-AUTH Thessaloniki, Greece.)

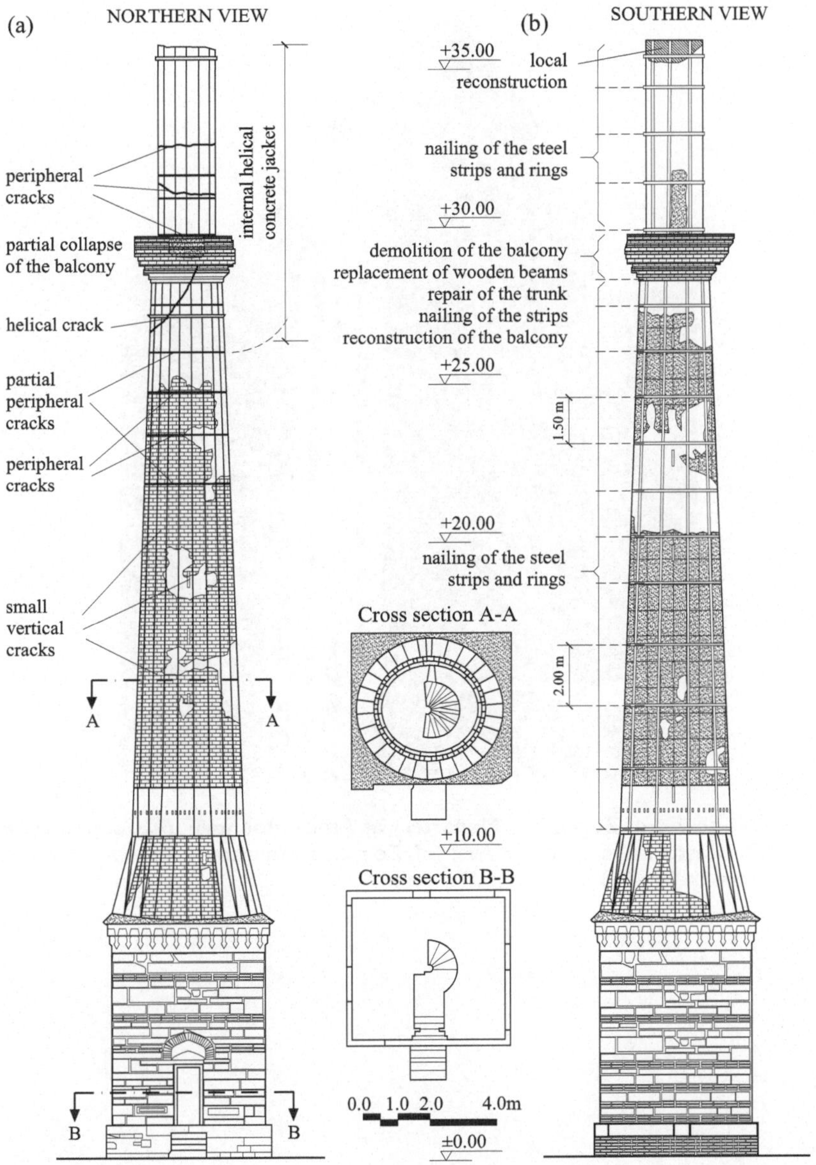

Figure 4.65 The minaret of Rotunda, Thessaloniki. (Adapted from Penelis, G., Papayianni, I., Stylianides, K., Ignatakis, C., and Athanassiadis, L. (1993). Strengthening of a 400-year-old Ottoman Minaret. *IABSE Proceedings of the IABSE Symposium 'Structural prevention of the architectural heritage'*, Rome, Italy. pp. 629–630.)

Figure 4.66 The bell tower of Zographos Monastery at Athos Mountain. (Adapted from Penelis, S.A. (2009). Restoration of bell tower of Zografos monastery, Mount Athos Penelis S.A., Archives. In Greek.)

Chapter 5

Techniques and materials in use for structural restoration

5.1 TECHNIQUES

5.1.1 Introduction

Before any involvement with the procedure of structural restoration of a monument, it is advisable to present techniques and materials in use for structural restoration, since it is a prerequisite for the structural engineer in charge to be informed well and in depth on these issues, so that he can propose an intervention scheme complying on one side with restoration principles and on the other with the structural safety of the building.

Bearing in mind what was already presented in Chapter 1, and what is included in Article 10 of Venice Charter and other international resolutions as well as the Document on Authenticity of ICOMOS (ICOMOS, 1994), it is concluded that the intervening procedure to a monument is an action of high sensitivity and complexity with many differences from a conventional one that is applicable to other existing structures and requires special techniques, and materials suitable in each case. Therefore, special reference must be made here to techniques and materials suitable in restoration.

5.1.2 Classification of techniques

The restoration techniques may be classified under two main categories (UNDP/UNIDO, 1984):

- Reversible ones
- Irreversible ones.

Generally speaking, reversible actions are preferable for the following reasons:

- In case they are proven inefficient or of low durability over time, they can be replaced without disturbing the original fabric.
- In case, in the future, better techniques or materials are developed, they can be replaced easily.
- The artistic or historical evidence is not falsified at all.

In the category of reversible actions, the following techniques may be classified (Penelis, G., et al., 1985; Tassios, 1985):

- External buttresses (Figure 5.1)
- Ties at the springings of the arches
- Rings at the base of the domes (Figure 5.2)
- Anastylosis of stone or marble monuments with dry joints
- External ties (Figure 5.3)
- Internal steel curbs for confinement (Figure 5.4).

The materials used in these techniques are usually subjected to very few restrictions. For example, the external rings or ties need special care only at the points of connection with the original fabric and face compatibility problems due only to an eventually different coefficient of thermal expansion. *So, in the reversible techniques, all modern materials may be used without serious limitations.*

Despite the above-mentioned advantages, *reversible methods* are not always realistic and applicable or, rather, it is not possible to solve an existing restoration problem only with reversible interventions. On the contrary, the interventions are mainly of *irreversible type*. By this term, we mean *interventions that cannot be easily undone without disturbing the original fabric*. We mention some of them (Penelis, G., et al., 1989; Lourenco, 2018):

- Groutings (Figure 5.5) (Siaxabani, 1985)
- Bonding-in of new bricks across cracks after grouting and cutting out some of those in each side (Figure 5.6b)
- Deep rejointings (Figure 5.6a)

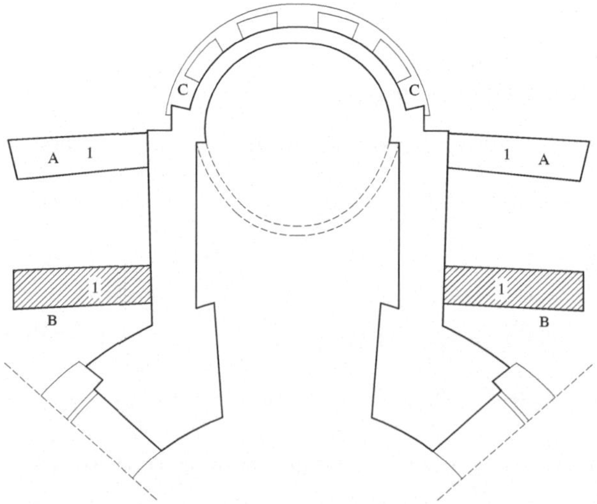

Figure 5.1 External buttresses at the Christian part of Rotunda, Thessaloniki (alternative proposal). (A) Existing ones; (B) proposal not approved. (Adapted from Penelis, G. Karaveziroglou, M. and Papayianni, I. (1989). Materials used in the restoration of historical buildings and monuments. *Seminario per Conservao Recuperaçäo e consolidacäo de monumentos e edificios histöricos*, Conselho Directivo National (ed) Lisbon.)

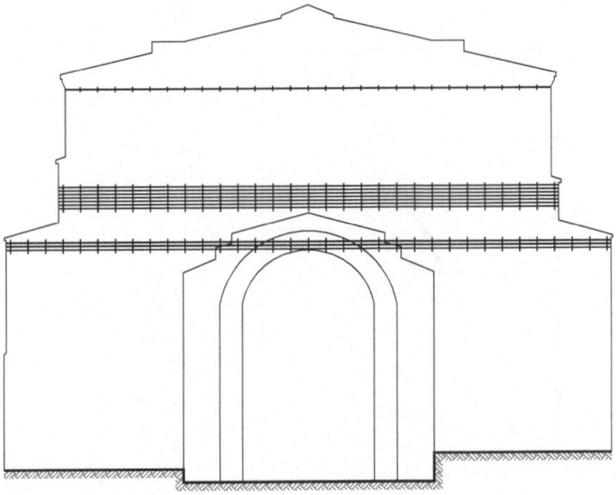

Figure 5.2 External elevation of posttensioned tie rings at the Roman part of Rotunda, Thessaloniki. (Adapted from Penelis, G. Karaveziroglou, M. and Papayianni, I. (1989). Materials used in the restoration of historical buildings and monuments. *Seminario per Conservao Recuperaçäo e consolidacäo de monumentos e edificios historicos*, Conselho Directivo National (ed) Lisbon.)

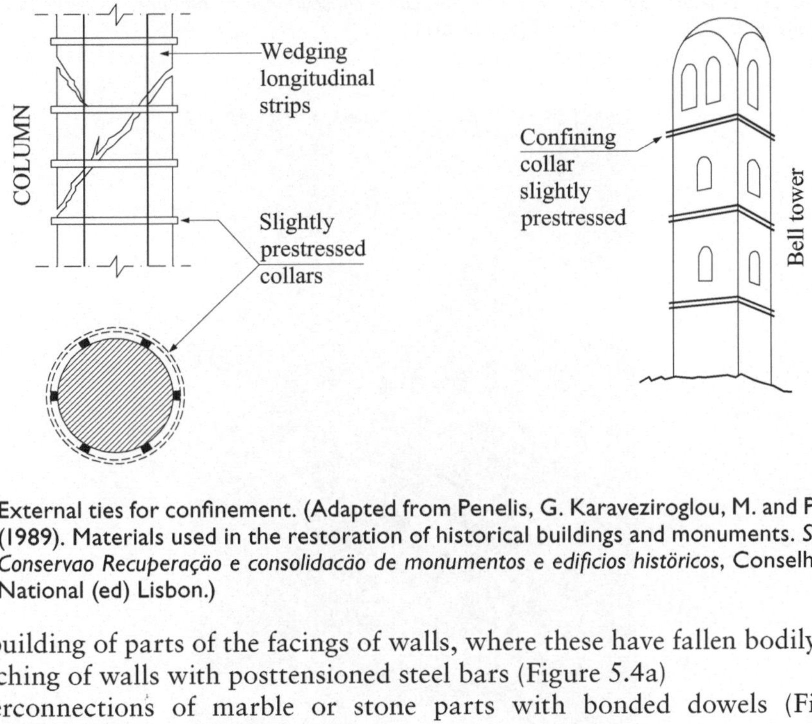

Figure 5.3 External ties for confinement. (Adapted from Penelis, G. Karaveziroglou, M. and Papayianni, I. (1989). Materials used in the restoration of historical buildings and monuments. *Seminario per Conservao Recuperaçäo e consolidacäo de monumentos e edificios historicos*, Conselho Directivo National (ed) Lisbon.)

- Rebuilding of parts of the facings of walls, where these have fallen bodily
- Stitching of walls with posttensioned steel bars (Figure 5.4a)
- Interconnections of marble or stone parts with bonded dowels (Figure 5.7) (Zambas, 1982)
- Skins of reinforced concrete on masonry walls, shotcreting etc.

Among the above-mentioned techniques, *groutings, bonding-in of new bricks* across cracks, *deep rejointings* and *rebuilding of masonry walls* are the most common and the most extensive interventions, since this is the only way for a masonry structure to regain

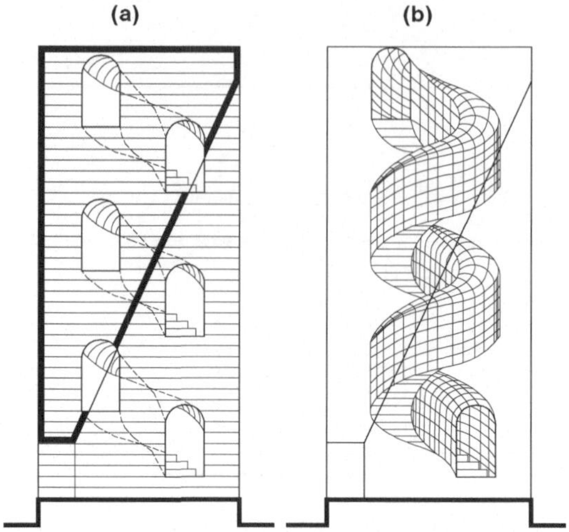

Figure 5.4 Confinement of piers P1 and P2 of the Roman part of Rotunda, Thessaloniki: (a) posttensioned stiches; (b) internal steel curb in the staircase. (Adapted from Penelis, G. Karaveziroglou, M. and Papayianni, I. (1989). Materials used in the restoration of historical buildings and monuments. *Seminario per Conservao Recuperaçäo e consolidacäo de monumentos e edificios histöricos*, Conselho Directivo National (ed) Lisbon.)

Figure 5.5 Roman part of Rotunda, Thessaloniki: injections in pier P8. (Adapted from Penelis, G. Karaveziroglou, M. and Papayianni, I. (1989). Materials used in the restoration of historical buildings and monuments. *Seminario per Conservao Recuperaçäo e consolidacäo de monumentos e edificios histöricos*, Conselho Directivo National (ed) Lisbon.)

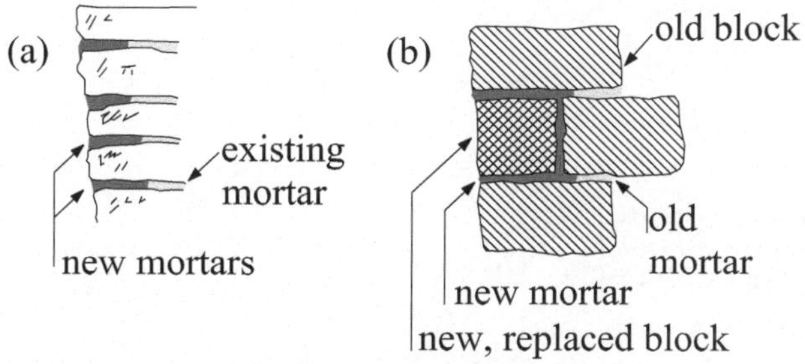

Figure 5.6 Local repairs. (a) Deep rejointing; (b) rebonding. (Adapted from Penelis, G. Karaveziroglou, M. and Papayianni, I. (1989). Materials used in the restoration of historical buildings and monuments. *Seminario per Conservao Recuperaçäo e consolidacäo de monumentos e edificios histöricos*, Conselho Directivo National (ed) Lisbon.)

its integrity without any serious violation of the principles and actions established by the Venice Charter.

For the whole range of irreversible interventions together with the principles applicable to the repair and strengthening of conventional structures, the following two principles should be taken into account:

- *Compatibility* of the materials for repair and strengthening with those of original fabric
- *Durability* of the new materials for a very long period of time.

The term *compatibility* refers to the chemical, physical and mechanical properties of the new materials. In addition, it refers to the aesthetic harmony with the whole monument. For example, strength, stiffness, bonding, coefficient of thermal expansion, permeability and the problem of efflorescence are points of reference for the term *compatibility*.

By the term *durability*, we refer to not only the necessity for the lifetime of the new materials to be at least equal to that of the original fabric, but also to the necessity for preservation of the compatibility between the new and original fabric for the same period of time.

These two additional principles, that is, the *compatibility* and the *durability* necessitate the deep knowledge of the properties of the original materials, on one side, and of the new ones, on the other. References to the properties of the original materials have been made in Chapter 2. In Section 5.2, reference will be made to materials in use for the restoration.

From what has been presented so far, it is evident that in repair or strengthening of a member or a construction as a whole, reversible and irreversible techniques are implemented simultaneously, *giving preference to the reversible ones*.

Finally, it should be noted that, apart from the classification of the type of intervention as reversible, or irreversible, interventions are also classified as *local* and *general* procedure types:

- *Local-type interventions* refer to repair or strengthening of local damage. Such an intervention affects the appearance, the usability and the strength of a member without affecting seriously the overall structural behaviour of the construction.

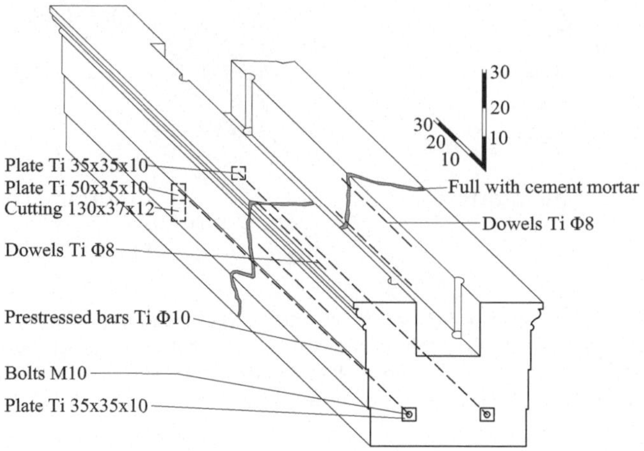

Figure 5.7 Interconnections of marble parts with bonded dowels and posttensioned bars at Erechtheion, Acropolis Athens. (Adapted from Zambas, K. (1982) *Articulated ancient Greek and Roman buildings*. UNDP/UNIDO, PROJECT (RER/79/015) National Report of Greece Courtesy of the author.)

- *General-type interventions* refer to strengthening of the overall structural system, so that it can withstand in a better way the various actions and particularly the accidental ones, like seismic actions, that have caused the damage under consideration.

In Subsections 5.1.3, 5.1.4, 5.1.5 and 5.1.6, an overview will be given of the techniques in use for structural restoration of characteristic elements, structural members and structural systems damaged mainly by earthquakes.

5.1.3 Local types of 'repair' and 'strengthening'

5.1.3.1 Masonry walls and piers

5.1.3.1.1 Introduction

Repair and strengthening of masonry walls are the most frequent types of intervention. Suitable techniques vary according to the type of masonry, the type of damage (cracks, compression splitting, buckling of the outer leaf of three-leaf walls etc.) and the decision of re-establishing of the structural interconnection between external leaves and the core, in addition to the visual rejointing and filling of cracks.

The main procedures for repair and strengthening of historical masonry walls are the following (Wenzel, F. and Maus, 1992):

- Deep pointing and rejointing of cracks (Figure 5.6a)
- Bonding-in of new bricks or blocks of stone across the crack to provide a structural interconnection in addition to rejointing (Figure 5.6b)
- Rebuilding of parts of the facings of the wall, where these have fallen bodily (Figures 5.8, 4.34 and 4.35)
- Injection of grouts in cracks (Figure 5.5)
- Internal or external reinforcement of the wall with stainless steel or FRP (Figures 5.4a and b).

It should be noted that the most usual form of historical masonry walls consists of three leaves, two external and a central core consisting of rubble and mortar (see Chapter 2, Figure 2.15). Special reference will be made in Subsection 5.1.4 to mass grouting of this type of wall, as this technique is of general-type intervention, aiming at the overall improvement of the structural behaviour of the masonry walls, which constitute the backbone of the structural system of masonry buildings.

In the following paragraphs, special reference will be made to *injecting* of cracks and *reinforcement*, as these two techniques are of special interest.

5.1.3.1.2 Injecting grouts in cracks

The injecting of grouts to fill cracks is the most usual and effective technique for masonry restoration. Grouting material should be compatible with the original mortar. The required properties and consistency will be examined in detail in Section 5.2. Typical mechanical grouting equipment consists of a pump, a mixing tank, hoses, gauges and nozzles (Figure 5.9).

Before grout injection, the following preliminary steps are required:

- All cracks and other voids should be cleaned out and washed well to promote adhesion of the grout.
- Wide cracks should be filled with coarse aggregate. For crack widths between 30 and 70 mm, an aggregate size of 3–7 mm is recommended; for wider cracks, a size of 7–15 mm might be used. Special care must be taken to ensure that no smaller particles are included, because their presence could prohibit the flow of the grout.
- All cracks should be sealed on surface to prevent the overflow of grout. For cracks wider than 20 or 30 mm, a mixture of cement, gypsum and sand (1:1:3 in weight) may be used. For narrower cracks, gypsum alone is preferred.
- Injecting ports of grout are provided with short pieces of plastic hose sealed at intervals along the cracks at distances of 0.30–0.50 m (Figure 5.10). Injection usually proceeds from the bottom upwards. The pressure at the nozzle should be about 0.5

Figure 5.8 Facings of defense walls that have fallen locally (Thessaloniki, Greece).

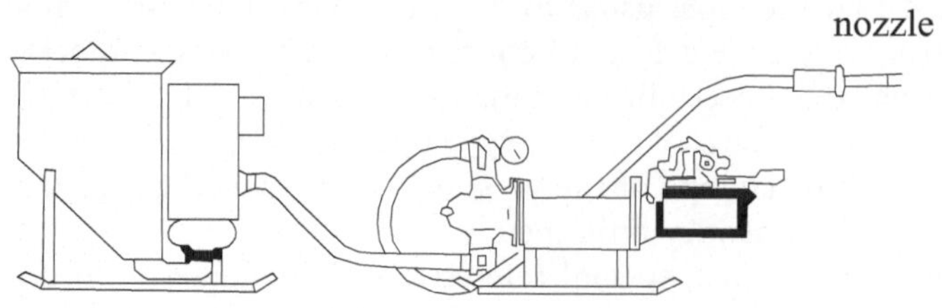

Figure 5.9 Grouting equipment. (Adapted from Penelis, G. and Penelis, Gr. (2019). *Concrete buildings in seismic regions.* CRC Press (A SPON Press Book), London.)

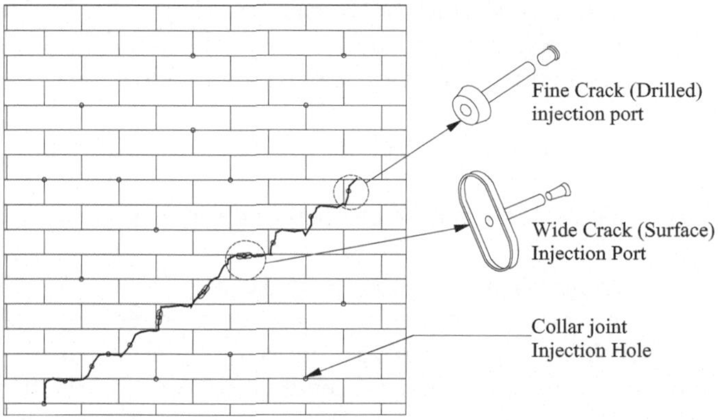

Figure 5.10 Injection ports raster. (Adapted from Shuller, M., Atkinson, R. and Borgsmiller, J. (1994). Injection grouting for repair and retrofit of unreinforced masonry. *10th 132 MaC, Galgary, Canada.* pp. 549–559.)

atmosphere and not more than 1.0 to avoid damage by excessive internal pressure. When injection is advancing vertically, it is important to avoid an excessive build-up of hydrostatic pressure in the grout already injected. Therefore, the advance should be less than 1.0 m per day (Figure 5.11). The verification of the successful execution of grouting is tested by drilling cores of 30 mm in diameter or by implementing ultrasonic measurements.

The effect of grouting on the lateral shear resistance of brick and stone masonry walls has been tested experimentally (Tomazevic, 1999; Shuler, et al., 1994; Ruth and Schuler, 1996; Saxiabani, C., 1986). All reporters above have found that by injecting the cracks, the original load-bearing capacity of the walls is recovered, but the rigidity, in most cases, is not. Typical results are found in Table 5.1.

5.1.3.1.3 Reinforcement

The insertion of stainless steel or fibre-reinforced polymer (FRP) bars along the bed joints allow an increase of the tensile strength and the control of the horizontal deformation of the masonry. Therefore, it constitutes a solution to stop the progression of damage (Oliveira, et al., 2012; Van Gemert and Toumbakari, 2003).

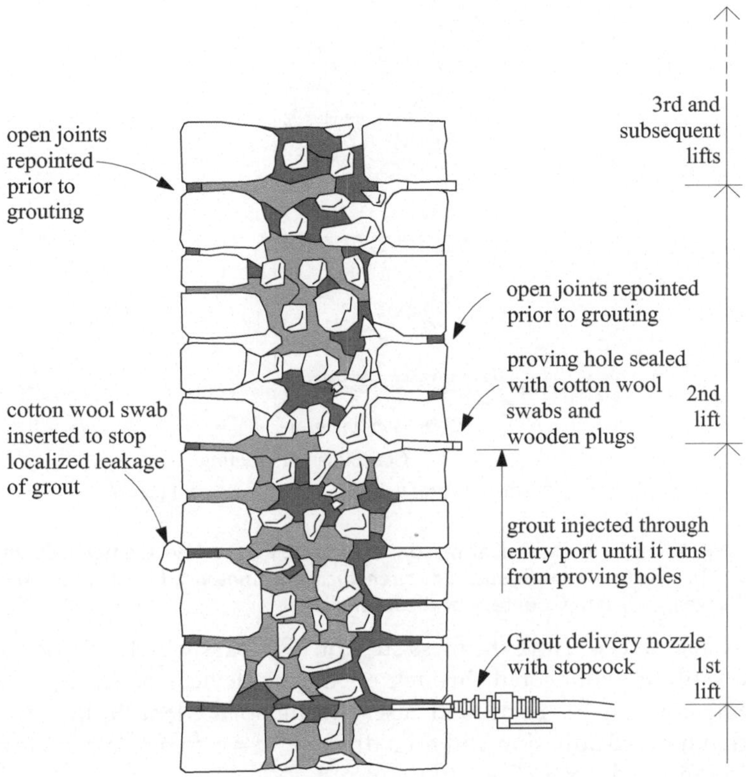

Figure 5.11 Grouting voids in a three-leaf masonry wall. (Adapted from Wood, C. and Burns, C. (2013). Grouting solid masonry walls. *English Heritage website, Conference of Exeter* UK. pp. 1–6.)

The transversal tying of the external leaves is aimed at improving the bond between them and the core (Figure 5.12). Thus, the risk for out-of-plane failure and collapse of the external leaves due to debonding from the core and buckling is reduced. According to experimental evidence, tying improves the compressive strength of the wall by about 55%. However, average young's modulus is of the same level or a little bit lower than that of unstrengthened walls (Oliveira et al., 2012).

Table 5.1 Effect of injecting the cracks on tensile strength and shear modulus

	Mortar	*Original*		*Grouted cracks*	
Masonry unit	f_m *(MPa)*	f_{tk} *(MPa)*	*G (MPa)*	f_{tk} *(MPa)*	*G (MPa)*
Brick (B 20)	0.5	0.07	-	0.11	-
Brick (B 20)	3.0	0.20	-	0.25	-
Ceramic block (B 20)	4.8	0.15	360	0.26	250
Ceramic block (B 20)	6.1	0.19	240	0.18	380
Light concrete block (B 7.5)	2.9	0.19	380	0.28	380
Fly-ash block (B 15)	1.3	0.14	370	0.14	230
Fly-ash block (B 15)	1.3	0.16	480	0.22	490

Source: Adapted from Tomazevic, M. (1999). Earthquake – resistant design of masonry buildings. Imperial College Press, London, UK. With permission of Imperial College Press.

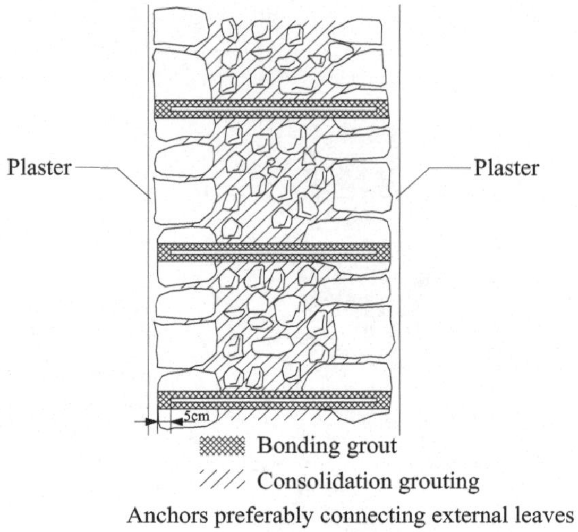

Figure 5.12 Grouted anchors in three-leaf masonry. (Adapted from Van Gemert, D. and Toumbakari, E. (2003). *Structural consolidation and strengthening of masonry. Historical overview and evolution.* KU Leuven, Belgium. Courtesy of the authors.)

Tying is implemented by using stainless steel or FRP bars which are inserted into holes drilled on the wall and connected through anchoring devices or simply locally injected with white cement grout. In the second case, the anchoring length, in case of reinforcement rods with improved adhesion and for a diameter $d = 8$ to $d = 16$ mm, ranges between $l_0 = 90 + 9d$ (in mm) and $l_0 = 120 + 12d$ (in mm).

In case of brick structures with no adequate wall resistance to bending or shear, a network of stainless steel bars is inserted into the wall thickness through injection holes and grouted with white cement grout. This type of intervention is known as *reticolo cementato* (cemented network) (Lizzi, 1982) (Figure 5.13). The net behaves as a steel skeleton with a function similar to that of the steel reinforced concrete creating a sort of *reinforced masonry*. In this case, the stiffness of the masonry is modified significantly. In this respect, *reticolo cementato* is an irreversible intervention with serious compatibility issues, and therefore, it should be implemented with serious concerns.

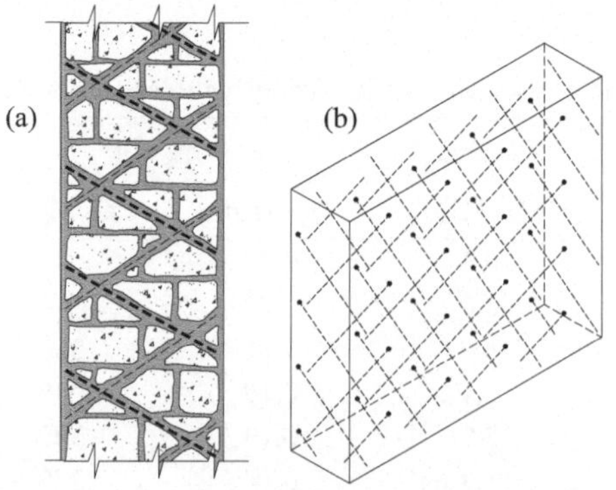

Figure 5.13 Reticolo cementato typical scheme: (a) vertical cross section; (b) isometric view of *reticolo cementato*. (Adapted from Lizzi, F. (1982). *The static restoration of monuments*. Sagep Editrice, Genova, Italy. With permission of Sagep Editrice.)

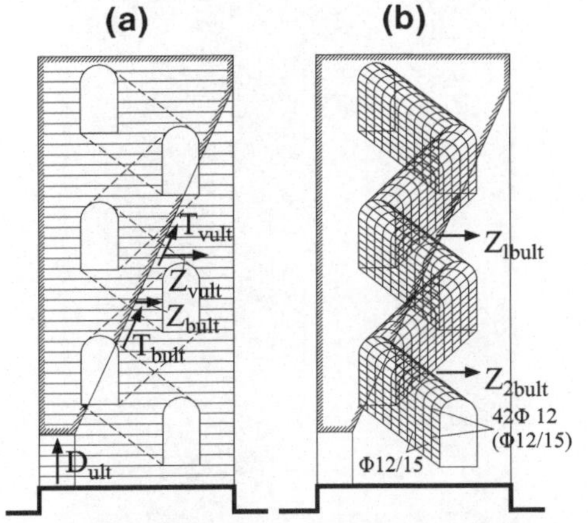

Figure 5.14 (a) Posttensioned stitches through the crack of pier; (b) confinement with R/C jacketing the staircases in the piers P2 and P3 (Rotunda Thessaloniki). (Adapted from Penelis, G. karaveziroglou, M., Stylianidis, K. and Leohdaridis, D. (1980a). *Restoration and strengthening of Rotunda of Thessaloniki*. Archives of Ephorate of Antiquities of Thessaloniki.)

Sometimes, in case of thick piers or shear walls, which have failed under shear–compression presenting one wide inclined crack or more thinner ones, the reinforcing bars should be posttensioned to a degree, in order to avoid additional widening of the cracks, during the reinforcement activation. Such a type of intervention requires anchorage plates, bolts and proper equipment for the posttensioning together with highly qualified personnel. The posttensioned tendons are usually injected with white cement grout (Figure 5.14a).

5.1.3.2 Strengthening of walls by applying skins of reinforced concrete (R/C)

The addition of skins of reinforced concrete to one or both sides of walls that have been badly cracked, or cannot easily be given the desired strength by grouting, has obvious attractions. But it should be adopted with strong reservations in historical monuments and other buildings of individual importance. Often the skin or skins can easily be hidden beneath subsequent plaster. But in this case, walls obtain very different stiffnesses and permeabilities (UNDP/UNIDO, 1984).

In the past years, there have been serious drawbacks even for use in conventional listed buildings, because the addition of these skins alleviates the texture of the masonries of the building. The problem of corrosion of reinforcement may be tackled with the use of stainless steel bars, of course, at a higher cost (about eight times higher than conventional structural steel rebars).

These drawbacks suggest that, in general, such skins should be applied only to the internal walls of buildings of the kind discussed here and chiefly just in buildings of fairly recent date and limited individual historical importance. When they are so applied, they should preferably be implemented in both faces of the walls, with interconnecting ties at regular intervals unless the walls are much thicker than the skins. However, even the implementation on one side of the wall has very favourable results on diagonal compression (Stylianidis and Ignatakis, 1995). In this case, the skin should be anchored in the wall by stainless steel anchors. In case of R/C skins on both sides of a wall, the result will be virtually a new wall of sandwich construction (Figures 5.15 and 5.33).

Figure 5.15 Gunite executed on masonry walls at Mangos mansion in Athens. (From Penelis S.A. 2014c.)

Careful stripping and cleaning of the wall surfaces is necessary, including the removal of all dust and fragments of brick or stone with air or water jets. Cracks must be filled according to the circumstances and the judgement of the engineer: Reinforcement is, then, fixed as shown in Figures 5.15 and 5.33 by means of metal dowels anchored in the wall or by means of bars through the thickness. Concrete is usually applied by the shotcrete or gunite process as seen in Figure 5.16. Typical thickness varies from 30 to 80 mm. While work is proceeding, some temporary support may be required for floor or roof members (UNDP/UNIDO, 1984).

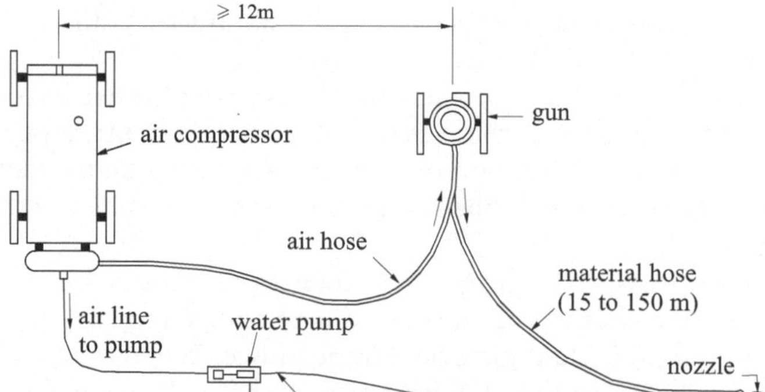

Figure 5.16 Typical arrangement of equipment for shotcreting. (Adapted from Penelis, G. and Penelis, Gr. (2019). *Concrete buildings in seismic regions*. CRC Press (A SPON Press Book), London.)

Recently, a tendency has been developed, to substitute the gunite skins by *reinforced cement plaster* with a thickness varying between 30 and 50 mm. The reinforcement is of nets of stainless steel or FRP bars of small diameter (2–5 mm) properly fixed on the wall as in the case of concrete gunite skin (Figure 5.17). This type of intervention is more easily accepted by the authorities; however, it is structurally less effective.

5.1.3.3 Strengthening of free-standing columns

Free-standing columns can be classified under three main categories:

a. Monolithic columns of stone or marble
b. Columns consisting of superimposed stone or marble drums
c. Columns built of masonry.

Experience shows that the first two categories behave very well under seismic loading unless the horizontal bearing area has, somewhere, been considerably reduced (UNDP/UNIDO, 1984). The best strengthening measure is the restoration of the original bearing area by fitting of a patch of similar stone followed by local circumferential confinement by means of an encircling stainless steel band eventually posttensioned. An early application of this technique is to be found in the use of bronze bands around the top and bottom of all monolithic marble columns in Hagia Sophia, Istanbul, and in Saint Demetrius church of Thessaloniki (Figure 5.18) (Shulz & Burnsley, 1917).

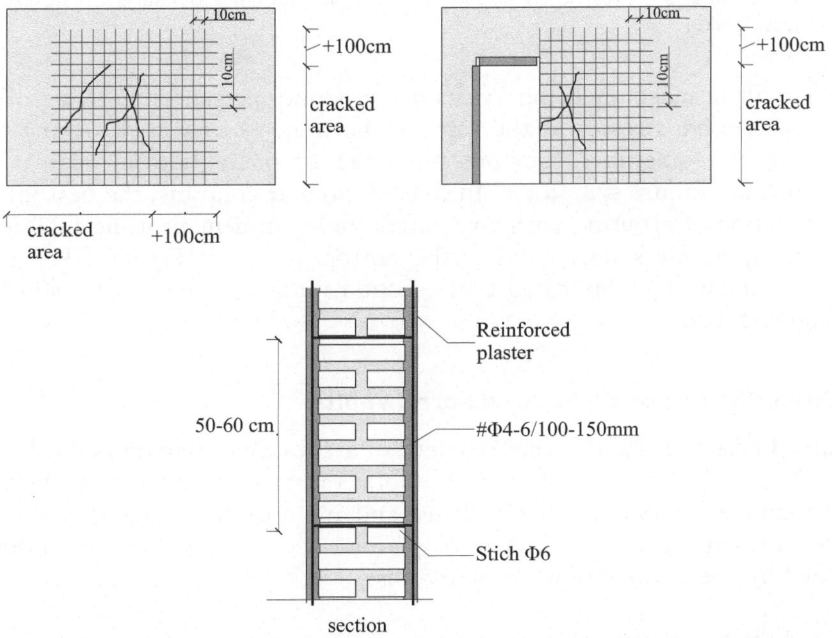

Figure 5.17 Reinforced cement plaster at Zardinides mansion in Thessaloniki. (Adapted from the Archives of Penelis S.A. (2014b). Archives of restoration project of Zardinides, mansion Thessaloniki, Greece.)

Figure 5.18 Hagios Demetrius church of Thessaloniki: restoration activities, 1930. (Adapted from Papasouliotis, G. (2018). Restoration activities at Hagios Dimitrios church of Thessaloniki, 1930. Thessaloniki.)

Columns built of masonry using bricks or cut stones present, due to seismic action, excessive compressive stresses at the top and bottom, as they rock during an earthquake. As a result, sometimes, they present at the top or the bottom vertical cracks or splitting which are failure symptoms. In case of circular columns, the best intervention is the combination of grouting with circumferential confinement using FRP warps and of an encircling stainless steel band at the bottom and top (Figure 5.19). In case of masonry columns with complicated cross section, *reticolo cementato* could be implemented (Figure 5.20).

5.1.3.4 Strengthening of arches and barrel vaults

It is presumed based on the inherent strength of arches that their supports do not move away (see Chapter 4). Damages of an arch or of a vault is a symptom that their supports under the combined action of the arch thrust and any other actions (e.g. seismic actions) have moved away from each other. This displacement of the support system outwards is accompanied by one of the following symptoms:

- Drift of the foundation of the supports
- Failure of the supports due to shear and compression (inclined cracks)
- Failure due to eccentric action of the compressive axial forces of the supports (bending moment–axial load, rocking)

Figure 5.19 Confinement of a masonry column at Mar Girgis church, Cairo. (Permission from Penelis, G. and Penelis, Gr. (2018). Restoration of the MarGirgis Church and Roman tower in Cairo. *Proceedings of the 4th Architectural and Urban Restorations BRAU4,* CI COP Italia. pp. 163–174.)

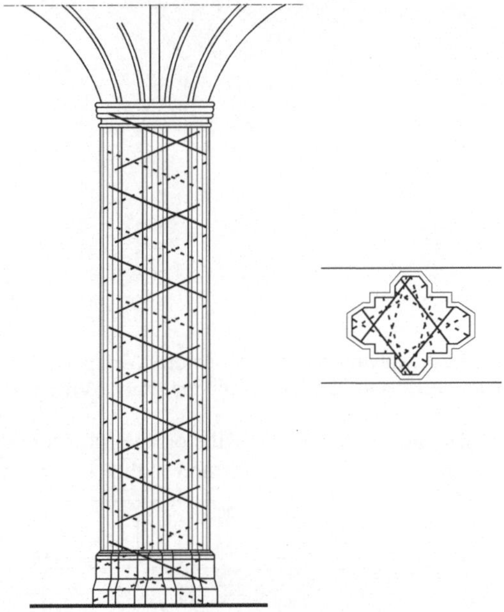

Figure 5.20 The San Lorenzo Maggiore church of Naples. Strengthening scheme of a column. (Adapted from Lizzi, F. (1982). The static restoration of monuments. Sagep Editrice, Genova, Italy. With permission of Sagep Editrice.)

- Besides cracks vertical to the arch axis at inactive regions, e.g. intrados of the crown, additional defects on an arch i.e. eventual local splitting at the compressed region or the crown (Figures 4.16 and 4.17), slipping of voussoirs and eventual collapses.

Therefore, any strengthening must be considered primarily in terms of the whole system, namely, arch plus supports. This is also true for barrel or groined and ribbed vaults which behave essentially as families of arches one by the other.

Local damages at the arch such as damaged or slipped voussoirs and tensile cracks usually at the crowns and at the springings must be repaired having in mind that they must be combined with general-type interventions such as installation of ties at the springings of the arch, stitching of the supports or consolidation of the foundations. These general-type interventions will be examined in Subsection 5.1.4.7.

Usual local interventions may be:

- Deep pointing and rejointing
- Rebuilding of parts after a proper shoring
- Groutings
- Reinforcement by applying *reticolo cementato* (Figure 5.21).

In case of shallow or thin vaults, it is possible for them to be strengthened with an R/C concrete skin combined with *reticolo cementato* (Figures 5.22 and 5.23) or an FRP warp reinforcement applied at the extrados of the vault.

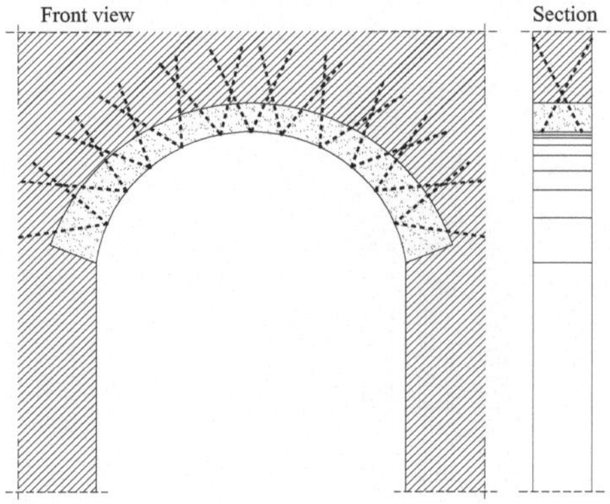

Figure 5.21 Arch strengthening by applying *reticolo cementato*. (Adapted from Lizzi, F. (1982). *The static restoration of monuments*. Sagep Editrice, Genova, Italy. With permission of Sagep Editrice.)

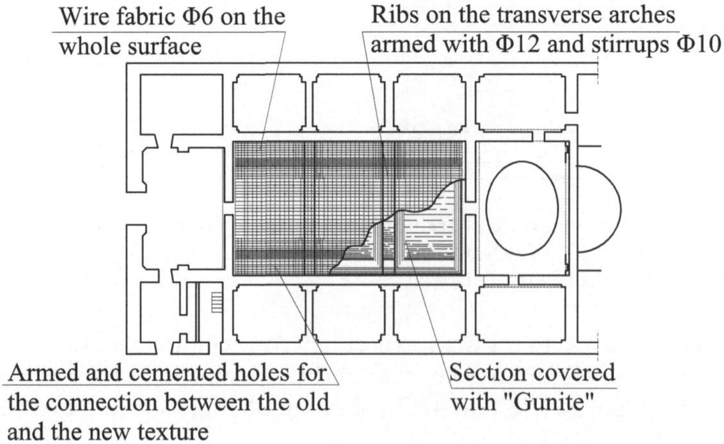

Figure 5.22 Strengthening of a vault with a concrete skin at the extrados with *reticolo cementato* (plan). (Adapted from Lizzi, F. (1982). *The static restoration of monuments*. Sagep Editrice, Genova, Italy. With permission of Sagep Editrice.)

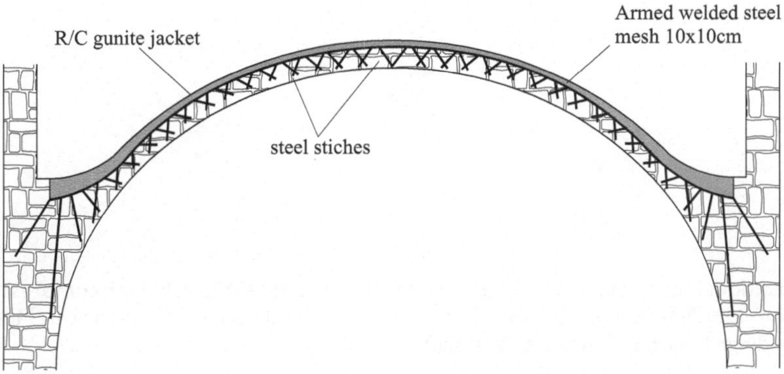

Figure 5.23 Strengthening of a vault with a concrete skin at the extrados combined with *reticolo cementato* (section). (Adapted from Lizzi, F. (1982). *The static restoration of monuments*. Sagep Editrice, Genova, Italy. With permission of Sagep Editrice.)

5.1.3.5 Strengthening of domes

The dome, being a structure of double curvature, differs from arches and arched vaults in that the geometry of its surface permits the transfer of all or part of the ring-type horizontal tensile stresses by circumferential reinforcement installed at the lower region of the dome.

The absence of effective reinforcement of this kind, together with the tensile inability of masonry, means, however, that domes of historical structures develop meridional cracks and consequently bigger thrusts outwards. Thus, it is possible to strengthen them significantly by adding circumferential reinforcement on the outside near the base (Figure 5.2).

The added reinforcement should be posttensioned to the full calculated circumferential tensile stresses for an idealised, equivalent, uncracked membrane state. However, attention is required in posttensioning because there will be a tendency for closing the existing cracks. Therefore, it would be better to precede grouting interventions at the cracks before the installation of the posttensioned rings.

The reinforcement should also be corrosion resistant, though this is less important in case it remains accessible, provided that regular inspection can be ensured.

5.1.3.6 Repair and strengthening of timber decks

Although replacement of a damaged old wooden member by a new one is easier than repairing the original, repair will be preferable if the damaged member is, in itself, of architectural artistic or historical importance (UNDP/UNIDO, 1984).

Repair of damaged wooden members should, where possible, be carried out by inserting new timber of the same kind with a minimum of cutting into of the original timber to permit making a good joint. One of the traditional scarf joints, as illustrated in Figure 5.24, will usually be appropriate, reinforced as necessary by metal plates or straps provided that these are protected against corrosion and are accessible for future inspection. Joints (a) and (c) are suitable for members subject to tension; and the scissor scarf (b), for members subject to compression.

A further possibility, where a member has failed at an end connection (as frequently happens at the feet of rafters, where they are housed in the ends of horizontal ties), is to remake the joint with metal straps, provided again that these are protected and will

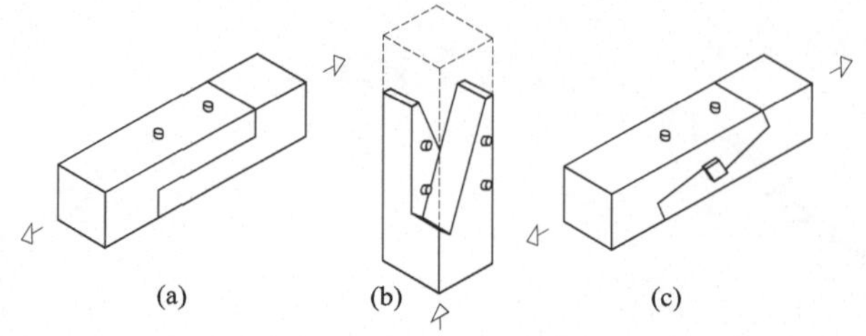

Figure 5.24 Traditional scarf joints: (a) and (c) suitable for tension; (b) suitable for compression. (Adapted from UNDP/UNIDO (1984). Repair and Strengthening of Historical Monuments and Buildings in Urban Nuclei, 6, Vienna.)

remain accessible. But it is always preferable in a historical structure to repair or replace the damaged member or members and retain the original form.

Unless there is to be a major change in the use of a building that will lead to significantly higher live loads, strengthening of timber floors will usually be necessary only to improve their capacity to resist in-plane bending and shear, aiming at distributing of horizontal shears between the vertical load-bearing elements of the structure. Only local strengthening for this purpose is considered here. Connections between floors and walls and associated horizontal tying are discussed in Subsections 5.1.4.5 and 5.1.4.6.

Diaphragmatic strengthening is best undertaken by adding an additional layer of boards at an angle to the existing board, after carrying out any necessary repairs to the existing floor. If the new boards are well nailed through to the existing boards at each intersection and to the joists beneath, the angle of intersection is not of great importance, but it is suggested that it should not be less than 45°.

Alternatively, the additional layer could consist of one or two sheets of plywood, installed one above the other in a chest arrangement (Figure 5.25). Closing, it should be noted that the diaphragm which is formed should be encased in its perimeter by a steel ring capable of carrying tensile forces caused by the in-plane response of the diaphragm as a deep beam (Figure 5.26).

5.1.4 General type of strengthening in structures above the ground

5.1.4.1 Introduction

General type of strengthening of a monumental building aims at the improvement of its structural response, as a whole, to gravity loads and particularly to seismic actions which constitute a long-term hazard.

This means that the synergy of the overall strength, stiffness and ductility of the building is in position to satisfy the following inequity at all critical regions of the structure.

$$[E_d] \leq \frac{[R_k]}{\gamma_m}, \tag{5.1}$$

where

E_d are the design effects of gravity loads combined with seismic action at all critical regions
R_k are the characteristic values of strength at all critical regions
γ_m are the partial safety factors of the materials (original and intervening ones).

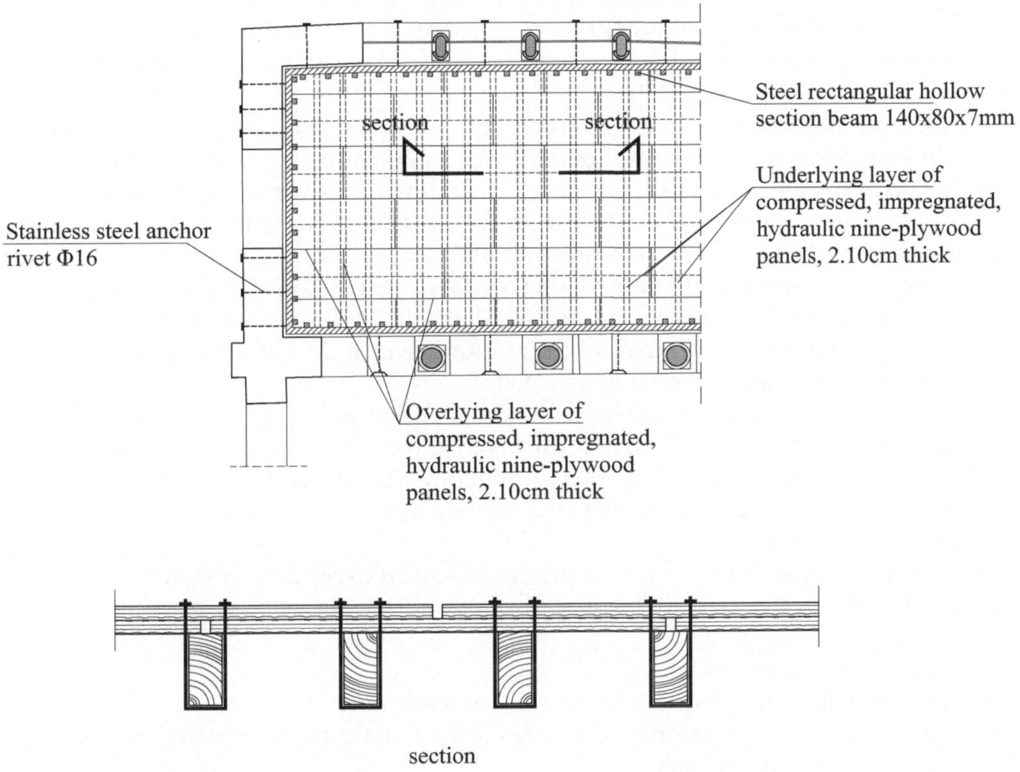

Figure 5.25 Diaphragmatic scheme composed of two layers of plywood: Acheiropoietos church Thessaloniki gallery floor. (Adapted from Penelis, G. and Stylianidis, K. (2001). Structural restoration of the Acheiropoietos basilica in Thessaloniki – Greece. *Proceedings of the 2nd International Congress on Studies in Ancient Structures.* (ed. Arun, G. and Seçkin, N), Istanbul. pp. 847–856.)

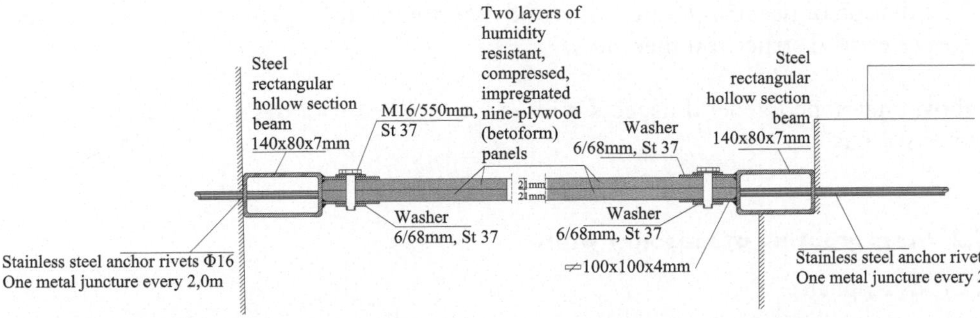

Figure 5.26 Steel ring at a diaphragmatic deck Acheiropoietos gallery floor. (Adapted from Penelis, G. and Stylianidis, K. (2001). Structural restoration of the Acheiropoietos basilica in Thessaloniki – Greece. *Proceedings of the 2nd International Congress on Studies in Ancient Structures.* (ed. Arun, G. and Seçkin, N), Istanbul. pp. 847–856.)

From the above, it is concluded that a general-type strengthening is decided when the extent and the type of damage shows that the structural resistance of the construction has been exceeded at many critical points and the system has approached *the life safety limit* (Penelis, G. and Penelis Gr., 2019).

In this case, all local repair or strengthening at various structural damaged elements should be considered as a part of the general-type intervention governed by equation (5.1) at each damaged element. In this inequity, the first part is governed by *the results of the analysis* of the structure for the element under consideration, while the second part is determined from the *redesign* of the rehabilitated element.

This approach is in effect not only for conventional contemporary buildings but also for monumental buildings. However as noted in Chapter 1 in the case of monumental buildings, solutions that result in extreme interventions, which do not comply with aesthetic and authenticity principles, should be avoided.

Therefore, in case of a monumental building, the general-type intervention aims at the improvement of its overall structural behaviour *by diminishing and smoothening* the design seismic action effects $[E_d]$ on one hand and *by improving of the structural capacity* $[R_k]$ of critical structural elements on the other, in a way that extreme structural interventions are avoided.

The following types of interventions may be classified in the first category (diminishing seismic action):

- Removal of excess masses
- Improvement of the distribution of mass and stiffness
- Avoidance of large torsional effects resulting from unfavourable relative distributions of mass and stiffness
- Improvement of ductility of the structure
- Diminishing seismic action with the help of seismic isolation and energy dissipation systems.

The following types of interventions may be classified in the second category (improvement of structural capacity):

- Strength improvement by extended mass grouting in masonry walls (not only repair of cracks)
- Installation of ties, rings, buttresses and diaphragms to improve the strength capacities of critical structural members.

The above-mentioned general aspects will be exemplified to a degree, in short, in some characteristic cases.

5.1.4.2 Mass grouting of masonry walls

5.1.4.2.1 Description

The use of grouting to consolidate loose or weak cores of rubble in three-leaf walls is the most usual and effective technique for masonry restoration. It contributes to rubble adhesion with the existing mortar and to the adhesion of the core to the external skins. In this respect, it contributes to the improvement of the core strength and stiffness without visual intervention to the walls and modifications of their texture. But, more important, it contributes to the reduction of the risk of the external leaves departing from the core to an out-of-plane direction and collapsing due to buckling.

Although the use of cement groutings for consolidation of engineering structures and soils was known for almost 150 years (Tomazevic, 1999), cement grouting and, particularly, pozzolanic grouts were first used on a large scale to earthquake-damaged masonry buildings in Slovenia and Italy in 1976 after the earthquakes of Friuli and, a little bit later in Greece, to the monumental buildings of Thessaloniki in 1978, after the earthquakes of Volvi (Penelis, G., 1982).

In case of mass grouting of walls for core consolidation, a raster of injecting ports at a distance of 0.50 × 0.50 m is drilled up to the middle of the masonry core. The equipment and execution procedure are the same as those in the case of crack injection (see Subsection 5.1.3.1.2) (Figure 5.10).

5.1.4.2.2 Structural effectiveness

Critical questions arise on the degree of strengthening that such a procedure ensures for masonry walls. During the last few years, extended experimental research has been carried out for the compressive behaviour of *three-leaf walls* (Vintzileou and Tassios, 1995; Toumbakari, 2002; Valluzi et al., 2004; Vintzileou, 2006, 2011; Hemeda, 2012; Papayianni and Pachta, 2012; Papayianni, 2014; Oliveira, et al., 2012; Vintzileou, 2018). Results of these works are summarised in Table 5.2 (Vintzileou, 2011). The main conclusions of the above research, referring to the compressive strength of a grouted three-leaf masonry wall, are more or less the following:

i. Grouted three-leaf masonry subjected to compression exhibited the same behaviour and the same failure mode in all tests.
ii. The positive effect of grouting to the compressive strength, as well as on masonry's overall behaviour, was proven by all experimental results. Three-leaf wall strength enhancement ranges between 30% and 100% (Table 5.2), while its stiffness $\left(E_{ws}/E_{wo}\right)$ enhancement ranges between 18% and 35%.
iii. The reasons of preference of ternary grouts (slacked lime–pozzolana–cement) were justified experimentally, as they offer significant enhancement of compressive strength with *substantial increase of the strain at failure.* On the contrary, in case of cement-based grouts, while their higher compressive strength *does not contribute to further increase of the compressive strength of masonry, they lead to a rather brittle behaviour.*

It should be noted that serious efforts have been made recently (Vintzileou, 2011) for the development of simple formulas for the calculation of the compressive strength of three-leaf masonry after grouting. A reliable formula for ternary grouts is given below:

$$\boxed{f_{wc,i} = f_{wc,o}\left(1 + \frac{V_i}{V_w} \cdot \frac{f_{i,s}}{f_{wc,o}}\right)}, \tag{5.2}$$

where

$$\boxed{f_{i,s} = 1.60 + 0.50 f_{gr,t}.} \tag{5.3}$$

Table 5.2(a) Three-leaf masonry in compression: Experimental results of mechanical properties before and after grouting

Wallette	*Dimensions (m)*	$t_e/t_i/t_e$	$f_{c\ units}$ *(N/mm²)*	$f_{m,c}$ *(N/mm²)*	$f_{i,c}$ *(N/mm²)*	$f_{gr}/f_{gr,t}$ *(N/mm²)*	V_{gr}/V_w *(%)*
Vintzileou et al. (1995)							
1[(1)]	0.6 × 0.4 × 1.2	130/140/130	100	1.70	0.15	30/2.5	43.8
2	0.6 × 0.4 × 1.2	130/140/130	100	1.70	0.15	-	-
3[(1)]	0.6 × 0.4 × 1.2	130/140/130	100	1.70	0.15	30/2.5	67.0
4	0.6 × 0.4 × 1.2	130/140/130	100	1.70	0.15	30/2.5	31.75
5	0.6 × 0.4 × 1.2	130/140/130	100	1.70	0.15	30/2.5	34.72
6	0.6 × 0.4 × 1.2	130/140/130	100	1.70	0.15	13/1.4	29.8
7	0.6 × 0.4 × 1.2	130/140/130	100	1.70	0.15	30/2.5	29.8
8	0.6 × 0.4 × 1.2	130/140/130	100	1.70	0.15	13/1.4	29.8
Toumbakari (2002)							
BC1	0.6 × 0.4 × 1.2	90/220/90	25	3.40	1.00	7.3/1.7	31.0
BC2	0.6 × 0.4 × 1.2	90/220/90	25	3.40	1.00	9.0/1.1	23.0
BC3	0.6 × 0.4 × 1.2	90/220/90	25	3.40	1.00	19.5/4.5	19.0
BC4	0.6 × 0.4 × 1.2	90/220/90	25	3.40	1.00	7.3/1.7	25.0
BC5[(1)]	0.6 × 0.4 × 1.2	90/220/90	25	3.40	1.00	7.3/1.7	-
SC1	0.6 × 0.4 × 1.2	130/140/130	55	3.40	1.00	9.0/1.1	50.0
SC2	0.6 × 0.4 × 1.2	130/140/130	55	3.40	1.00	19.5/4.5	46.0
SC3	0.6 × 0.4 × 1.2	130/140/130	55	3.40	1.00	7.3/1.7	-
SC4	0.6 × 0.4 × 1.2	130/140/130	55	3.40	1.00	7.3/1.7	30.0
Valluzzi (2004)							
511	0.8 × 0.5 × 1.4	180/140/180	164	1.57		5.1/0.8	54.2
611	0.8 × 0.5 × 1.4	180/140/180	164	1.57		5.1/0.8	54.2
1311	0.8 × 0.5 × 1.4	180/140/180	164	1.57		5.1/0.8	41.5
112	0.8 × 0.5 × 1.4	180/140/180	164	1.57		3.2/0.4	54.2
812	0.8 × 0.5 × 1.4	180/140/180	164	1.57		3.2/0.4	47.8
1612	0.8 × 0.5 × 1.4	180/140/180	164	1.57		3.2/0.4	47.8
1211T	0.8 × 0.5 × 1.4	180/140/180	164	1.57		5.1/0.8	54.2
1411R	0.8 × 0.5 × 1.4	180/140/180	164	1.57		5.1/0.8	60.6
1711RT	0.8 × 0.5 × 1.4	180/140/180	164	1.57		5.1/0.8	47.8
Vintzileou et al. (2006)							
1	1.0 × 0.45 × 1.2	190/120/140	25	4.35	0.15	4.5/2.5	32.8
2	1.0 × 0.45 × 1.2	190/120/140	25	4.35	0.15	8.16/2.2	40
3	1.0 × 0.45 × 1.2	190/120/140	25	4.35	0.15	4.5/2.5	36.4

Note: (1) Specimens with transverse masonry leaves.

(Continued)

Table 5.2(b) (Continued) Three-leaf masonry in compression: Experimental results of mechanical properties before and after grouting

Wallette	$f_{wc,0}$ (*N/mm²*)	$f_{wc,i}$ (*N/mm²*)	E_{w0} (*N/mm²*)	E_{w8} (*N/mm²*)	$\varepsilon_{wu,0}$ (%)	$\varepsilon_{wu,s}$ ($^{0}/_{00}$)
Vintzileou et al. (1995)						
1[1]	2.10	3.10	7000	6250	0.00104	0.0006
2	1.30	-	2706	-	-	-
3[1]	2.40	4.30	5000	5971	0.00141	0.0011
4	1.60	-	4442	-	0.0008	-
5	1.70	4.20	5670	7778	0.0028	0.0012
6	1.35	4.05	5625	8438	0.00058	0.0010
7	-	3.70	-	15413	-	0.0009
8	-	3.00	-	3333	-	0.0009
Toumbakari (2002)						
BC1	-	5.04	-	2238.2	-	0.00242
BC2	2.41	3.15	729.6	1564.9	0.00312	0.00254
BC3	2.09	2.91	1018.3	1404.8	0.00185	0.00155
BC4	2.18	3.00	1097.6	1040.4	0.00234	0.00294
BC5[1]	2.28	3.86	1144.9	1170.2	0.00231	0.00314
SC1	2.02	3.25	720.4	1622.2	0.00142	0.00355
SC2	2.09	3.36	1138.7	1558.6	0.00165	0.00233
SC3	2.65	3.51	1374.8	1187.6	0.00173	0.00245
SC4	2.71	3.29	1443.3	1014.5	0.00211	0.00349
Valluzzi (2004)						
511	1.45	2.49	2390	2273	0.00363	0.00726
611	1.95	2.49	2029	3093	0.00457	0.00571
1311	-	2.54	-	3992	-	0.00991
112	1.97	2.57	1450	3449	0.0062	0.00625
812	1.91	1.82	1559	2367	0.00622	0.0072
1612	-	2.48	-	1223	-	0.0107
1211T	-	2.59	-	1336	-	0.00818
1411R	-	2.14	-	1617	-	0.00821
1711RT	-	3.06	-	1772	-	0.00824
Vintzileou et al. (2006)						
1	1.82	3.00	1000	1200	-	
2	1.74	3.75	1440	1550	0.0016	0.0025
3	2.26	3.73	1500	1300	0.0025	0.0039

(Continued)

Table 5.2(c) (Continued) Three-leaf masonry in compression: Experimental results of mechanical properties before and after grouting

Wallette	W_{vert} *(mm)*	W_{trans} *(mm)*	$W_{trans}, f_{wc,0}$ *(mm)*	W_{trans}, f_{wci} *(mm)*	$E_{wo}/f_{wc,0}$	$E_{ws}/f_{wc,s}$	$f_{wc,i}/f_{wc,0}$	E_{ws}/E_{wo}	$\varepsilon_{wu,s}/\varepsilon_{wu,o}$
Vintzileou et al. (1995)									
1	1.00	0.231	0.00	0.00	3333	1048	1.48	0.89	0.58
2		0.27	-	-	2081				
3	3.00	-			2083	1389	1.80	1.19	0.78
4		2.533	-	-	2776				
5	1.50	7.59	0.00	1.035	3335	1852	2.47	1.37	0.43
6		1.81	0.00	1.325	4167	2083	3.00	1.50	1.72
7		-	-	1.3		4166			
8		-	-	0.125		1111			
Toumbakari (2002)									
BC1				1.17		444			
BC2		7.07	0.15	5.04	302.7	497	1.31	2.14	0.81
BC3		2.69	0.86	2.59	487.2	483	1.39	1.38	0.84
BC4		2.32	0.38	4.03	503.5	347	1.38	0.95	1.26
BC5		2.00	0.03	1.80	502.1	303	1.69	1.02	1.36
SC1		2.89	0.03	2.19	356.6	499	1.61	2.25	2.50
SC2		6.9	0.01	0.60	544.8	464	1.61	1.37	1.41
SC3		3.63	0.51	2.33	518.8	338	1.32	0.86	1.42
SC4		2.84	1.01	2.98	532.6	308	1.21	0.70	1.65
Valluzzi (2004)									
511		2.25	0.002	4.59	1648	913	1.72	0.95	2.00
611		9.2	0.0015	8.55	1040	1242	1.28	1.52	1.25
1311			0.06	13.85		1572			
112		3.97	0.16	3.67	736	1342	1.30	2.37	1.00
812		5.9	0.04	4.95	816	1300	0.95	1.52	1.16
1612			0.11	9.45		493			
1211T			0.02	7.95		516			
1411R			0.01	9.8		756			
1711RT			0.0005	10.95		579			
Vintzileou et al. (2006)									
1	0.78	8.2	0.0	3.2	550	400	1.65	1.20	
2	0.26	4.8	0.0	1.3	828	413	2.16	1.08	1.56
3	1.55	4.0	0.0	2.6	655	349	1.65	0.87	1.36

Source: Adapted from Vintzileou, E. (2011). Three-leaf masonry before and after grouting: A review of literature. *International Journal of Architectural Heritage*, Vol. 5, Taylor and Francis Group, LLC. pp. 513–538. Courtesy of the Author.

Notation: t_e, thickness of external leaf (leaves); t_i, thickness of filling material; $f_{c,\,units}$, compressive strength of masonry units (stones or bricks); $f_{m,c}$, compressive strength of mortar; $f_{i,c}$, compressive strength of filling material before grouting; $f_{gr,c}/f_{gr,t}$, compressive strength of grout/flexural strength of grout (measured at the age of testing the respective wallettes); V_{gr}/V_w, volume of grout reported to the volume of masonry; $f_{wc,0}$, compressive strength of masonry before grouting; $f_{wc,i}$, compressive strength of masonry after grouting; E_{w0}, modulus of elasticity of masonry before grouting, measured at ~30% of the respective compressive strength; E_{ws}, modulus of elasticity of masonry after grouting, measured at ~30% of the respective compressive strength; $\varepsilon_{wu,0}$, strain at $f_{wc,0}$; $\varepsilon_{wu,s}$, strain at $f_{wc,i}$; W_{vert}, opening of vertical cracks at mid-height of the face of ungrouted wallette, at $f_{wc,0}$; W_{trans}, opening of transverse cracks at mid-height of ungrouted wallette at $f_{wc,0}$; W_{trans}, $f_{wc,0}$, opening of transverse cracks at mid-height of grouted wallette at $f_{wc,0}$; W_{trans}, f_{wci}, opening of transverse cracks at mid-height of grouted wallette at f_{wci}.

The meanings of the various symbols in the above equations are given below:

$f_{wc,i}$ is the compressive strength of masonry after grouting (MPa)
$f_{wc,o}$ is the compressive strength of masonry before grouting (MPa)
V_i is the volume of infill material (core volume) (m^3)
V_w is the volume of masonry (m^3)
$f_{i,s}$ is the compressive strength of the grouted infill material (MPa)
$f_{gr,t}$ is the tensile strength of the grout (MPa).

The calibration of the results of the above formula using some experimental results given in Table 5.2 shows that the analytical results lie on the safe side (Figure 5.27) (Vintzileou, 2011).

It should be noted that the output of the above formulas should be used in design with much care, as there are many parameters that might influence the prediction of the compressive strength that have not been investigated in detail yet.

The effect of grouting on the *shear strength* of three-leaf walls was investigated either by testing walls under simultaneous vertical and step-by-step increased horizontal loading (Tomazevic and Apih, 1993) (Figure 5.28 and Table 5.3) or by testing wallettes under diagonal compression (Table 5.4) (Vintzileou, 2011). The existing results are limited and therefore the outcoming conclusions of limited reliability. These results may be summarised as follows:

i. Grouting leads to a substantial shear strength enhancement varying between 110% and 250% associated with a rather slight decrease of the horizontal displacement at maximum resistance (between 10% and 30%).
ii. The strength increase is definitely not proportional to the compressive strength of the grout (Figure 5.29). This may be attributed to the fact that the grout does not

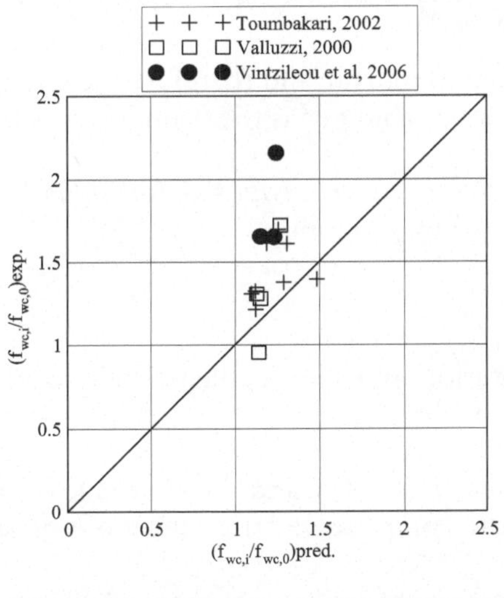

Figure 5.27 Comparison between predicted and experimental results based on equations (5.2) and (5.3). (Adapted from Vintzileou, E. (2011). Three-leaf masonry before and after grouting: A review of literature. International Journal of Architectural Heritage, Vol. 5, Taylor and Francis Group, LLC. pp. 513–538. Courtesy of the author.)

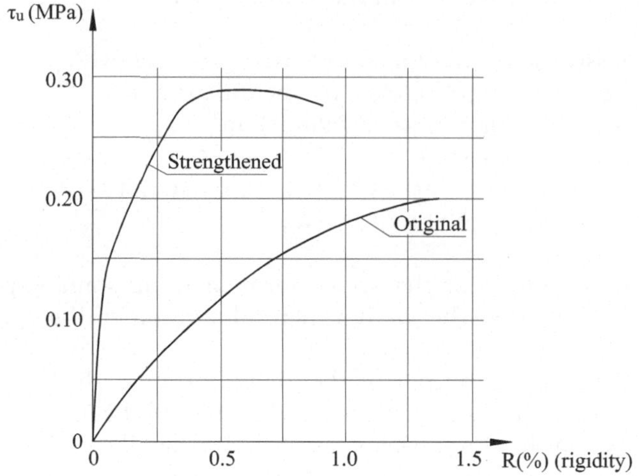

Figure 5.28 Effect of cement grouting. (Adapted from Tomazevic, M. (1999). Earthquake – resistant design of masonry buildings. Imperial College Press, London, UK. With permission of Imperial College Press.)

Table 5.3 Tensile strength and shear modulus of existing and cement-grouted stone masonry

	Original		*Cement-grouted*	
Masonry type	f_{tk} *(MPa)*	*G (MPa)*	f_{tk} *(MPa)*	*G (MPa)*
Two-leaf, uncoursed stone (rural)	0.08	90	0.18	160
Two-leaf, uncoursed stone (urban)	0.12	150	0.23	300
Compact, uncoursed stone and brick (urban)	0.21	-	0.38	-

Source: Adapted from Tomazevic, M. and Apih, V. (1993). The strengthening of stone-masonry walls by injecting the masonry friendly grouts. *European Earthquake Engineering*, Vol. 7, No. 2. Patron, Bologna. pp. 10–20. With permission of Imperial College Press.

penetrate into the original mortar and therefore the strength of the original mortar does not change. What has changed is the connection between bricks and stones via grout adhesion.

iii. For the time being, *there are no proposed formulas* for the estimation of shear enhancement of the wall by grouting.

5.1.4.2.3 Conclusions

From the preceded presentation and the detailed review of the relevant references, the following conclusions may be drawn:

i. Grouting improves the compressive and shear strength of a three-leaf masonry wall.
ii. For the estimate of the compressive strength, more or less reliable formulas have been developed.
iii. Walls with low percentage of voids (less than 4%) seem unsuitable to be injected (Binda, et al., 2006b; Oliveira, et al., 2012). As indicated by experience on site, the quantity of the dry part of the grout needed to fill voids in stone-masonry walls does not exceed 50–150 kg/m^3 of the wall (Tomazevic, 1999).

Table 5.4 Test results of wallettes subjected to diagonal compression

		Mechanical properties of ungrouted specimens			*Mechanical properties of grout*		*Mechanical properties of grouted specimens*			
Specimen	*Dimension (m)*	f_t *(MPa)*	ε_v *(%)*	w_{hor} *(mm)*	f_c *(MPa)*	$f_{t,fl}$ *(MPa)*	$f_{t,f}$ *(MPa)*	$\varepsilon_{v,f}$ *(%)*	W_{hor}, f_t *(mm)*	λ_f
Toumbakari [3]										
SDC1	0.8 × 0.8 × 0.4	0.47	0.9	0.85	15.9	3.4	0.50	1.36	0.47	1.1
SDC2	0.8 × 0.8 × 0.4	0.34	1.2	0.71	19.5	4.5	0.68	1.17	0.05	2.0
SDC3	0.8 × 0.8 × 0.4	0.28	1.5	1.20	7.8	2.5	0.59	2.65	0.02	2.1
BDC1	0.8 × 0.8 × 0.4	0.44	0.8	0.50	7.8	2.5	0.60	1.56	0.17	1.4
BDC2	0.8 × 0.8 × 0.4	0.34	0.3	1.07	15.9	3.4	0.73	1.07	0.02	2.2
BDC3	0.8 × 0.8 × 0.4	0.35	0.6	0.50	19.5	4.5	0.75	1.02	0.02	2.1
Vintzileou et al. [2]										
9		(0.15)			30		0.64			
10		(0.15)			30		0.64			
Vintzileou et al. [4]										
4	1.0 × 1.0 × 0.45	0.10	2.1	5.2	4.5	2.52	0.22	1.00	0	2.2
5	1.0 × 1.0 × 0.45	0.10	1.1	0.2	8.2	2.29	0.34	0.80	0	3.4
6	1.0 × 1.0 × 0.45	0.10	1.0	1.5	4.5	2.52	0.23	1.40	0	2.3

Source: Adapted from Vintzileou, E. (2011). Three-leaf masonry before and after grouting: A review of literature. International Journal of Architectural Heritage, Vol. 5, Taylor and Francis Group, LLC. pp. 513–538. Courtesy of the Author.

f_t and ε_v: tensile strength and respective vertical strain of ungrouted specimen, f_c and $f_{t,fl}$: compressive and flexural strength of grout, $f_{t,f}$ and $\varepsilon_{v,f}$: tensile strength and respective vertical strain of grouted specimen, w_{hor}: sum of vertical crack openings on the face of the ungrouted specimen (at f_t), w_{hor},f_t: sum of vertical crack openings on the face of the grouted specimen (at f_t) $\lambda_f = f_t/f_t$.

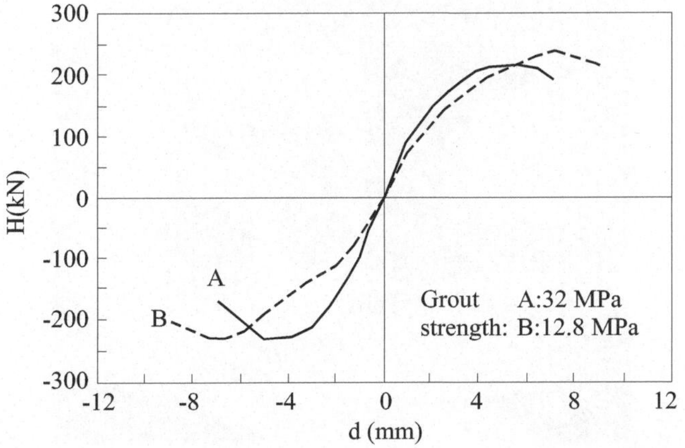

Figure 5.29 Comparison of hysteresis envelopes for grouts of various strength. (Adapted from Tomazevic, M. (1999). *Earthquake – resistant design of masonry buildings*. Imperial College Press, London, UK. With permission of Imperial College Press.)

iv. Ternary grouts are more preferable than the cement ones, as they present a much better compatibility with the original masonry mortars. Additionally, their influence on the masonry strength seems not to be significantly lower than that of a cement grout.
v. The existing experimental results and the formulas for the strength estimate of grouted masonries are limited. Therefore, the degree of the strength enhancement is decided by the structural engineer in charge of the structural restoration, making use of his engineering judgement in parallel.
vi. In case of ternary grouts, the stiffness of the grouted structural element taken into account in the analysis should not be enhanced.
vii. Grouting may be highly effective, but it is certainly not a low-cost option. The process is likely to cost 1,300–1,700€ per square metre of intervention, including material, equipment, labour and all other costs (Wood and Burns, 2013).

5.1.4.3 Removal of masses

Excess mass is most frequently found as a filling over vaults to give a plane floor surface or as an earth covering to the roofs of certain traditional types of dwellings primarily to give weather protection. Because this filling is at levels well above the ground, it leads to considerable increase in seismic inertial loads, without adding anything to the strength of the structure.

Heavy fills over vaults should be removed, substituted by lighter floors (Figure 5.30). Heavy roof coverings of earth should be substituted by lighter types of roofing providing also a better form of insulation.

Figure 5.30 Replacement of the heavy infill by lightweight concrete over the side aisle and narthex vaults of Hagia Sophia, Thessaloniki. (Adapted from UNDP/UNIDO (1984). Repair and Strengthening of Historical Monuments and Buildings in Urban Nuclei, 6, Vienna.)

5.1.4.4 Improvement of the distribution of mass and stiffness

Most historical monuments and other buildings in urban nuclei had fairly a rational form with uniform mass and stiffness distribution when first built. Changes and additions that followed to meet changing and growing needs have, however, resulted in less satisfactory structural forms. In other cases, new buildings were constructed as annexes to existing ones, even in churches, like new chapels, bell towers etc. All these have changed the logic of the original structural form, in relation to mass and stiffness distribution.

In all these cases, the rearrangement of the walls aiming at the returning to the original form would be advisable, if such interventions would be in an agreement with the concept of restoration of the building. Another approach would be the introduction of separating joints between the old building and eventual annexes (e.g. between the church and the bell tower) (Figure 5.31), so that the two structures respond independently in case of an earthquake. Of course, the joint should be wide enough to eliminate the risk of pounding during the earthquake (Athanassiadou, et al., 1995).

5.1.4.5 Improvement of structural response of basilicas with nonthrusting timber or steel roofs

The most useful overall strengthening in structures of these types may be achieved via the following three steps:

- The improvement of interconnection of the longitudinal walls and the shorter transverse ones at the end of the hall, using proper stitching with tensile bars (e.g. stainless steel, titanium, or FRP bars) (Figure 5.32).
- Improvement of the flexural stability of the longitudinal walls. This may be achieved by improvement of the transverse interconnections between the walls and the decks or roofs, through the horizontal tie members of the roof trusses and through the joists of the gallery decks in case such galleries exist (Figure 5.26).
- Improvement of the diaphragmatic action of the roofs and the gallery decks so that they may act in their plane as deep beams transferring a considerable part at the horizontal inertial seismic forces of the longitudinal walls, which are vulnerable in out-of-plane direction, to the transverse walls which have considerable shear strength in-plane (Figure 5.25).

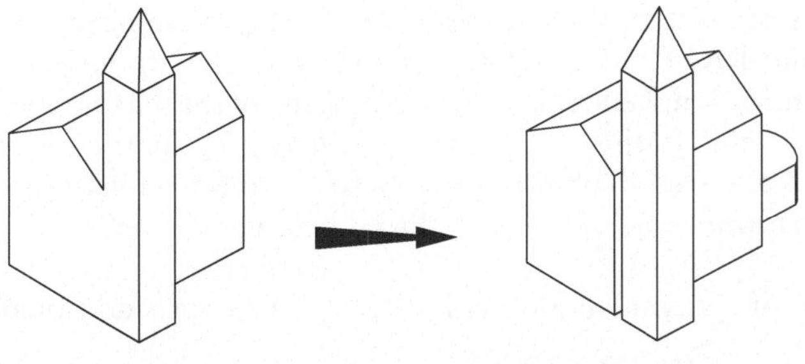

Figure 5.31 Improvement of mass and stiffness distribution by the introduction of separating joints. (Adapted from Croci, G. (2007). The conservation and structural restoration of architectural heritage. Advances in architecture series. WIT Press, Southampton. Courtesy of WIT Press.)

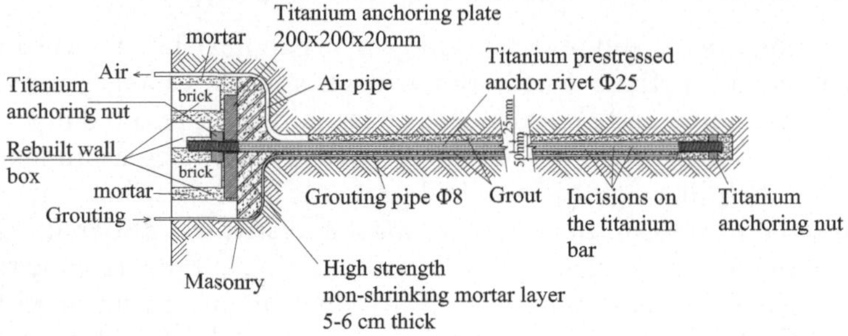

Figure 5.32 Acheiropoietos church of Thessaloniki: corner connections with titanium anchors. (Adapted from Penelis, G. and Stylianidis, K. (2001). Structural restoration of the Acheiropoietos basilica in Thessaloniki – Greece. *Proceedings of the 2nd International Congress on Studies in Ancient Structures.* (ed. Arun, G. and Seçkin, N), Istanbul. pp. 847–856; Raptis, K. and Zombou, A. (2010). The consolidation and restoration project of Acheiropoietos basilica in Thessaloniki. *Proceedings of the 8th International Symposium on the Conservation of Monuments in the Mediterranean Basin Monubasin 8,* Patras, Greece, pp. 411–428.)

5.1.4.6 Improvement of the structural response of multistorey residential buildings

In this type of buildings with walls arranged transversally to each other and usually in a raster, a good response to seismic actions is essential to have good cooperation between them so that they all participate in the transfer of horizontal seismic actions with safety. This means that the following issues should be carefully ensured:

- Acceptable strength of the masonry
- Good interconnections where walls meet
- Floor systems strong and stiff enough to transfer horizontal shears to the walls according to their stiffness and capacity
- Adequate tying of external walls to the adjacent transverse ones to prevent them from falling outwards
- Adequate tying of the floor joists to external walls to prevent them from falling outwards.

The strength of the walls may be improved with mass grouting in the core of the wall. However, this intervention does not improve its ductility. In case the implementation of gunite R/C skins is permitted, both strength and ductility are improved.

In case the gunite R/C skins are not permitted and the strength improvement of the masonry using grouting is not enough, inner reinforcing of the masonry may be implemented with the 'reticolo system'. In this case, apart from strength, ductility is also improved.

Many techniques are available for improving the connection between walls which meet each other at a junction; some of them are listed below:

- Local repair of existing bond with the insertion of additional stitching blocks (Figure 5.34)
- Tying of walls together with reinforcement in the form of stitching (Figure 5.32)
- Tying of walls together with gunite skins with continuous reinforcement across the junction (Figure 5.33).

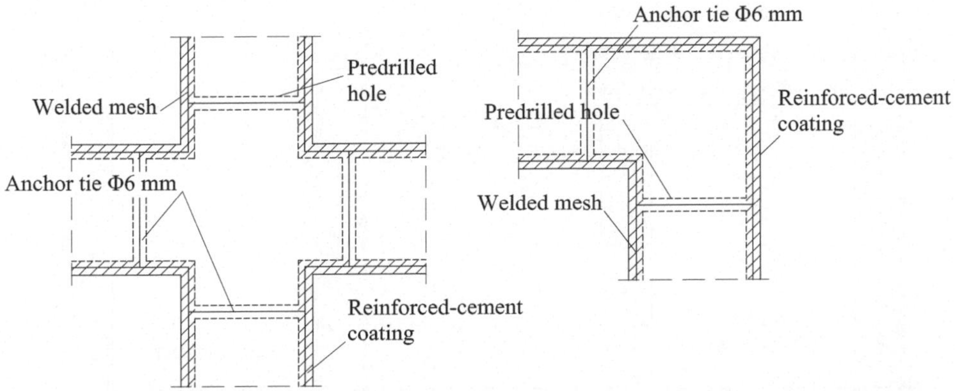

Figure 5.33 Anchoring of R/C coating at the corners and wall intersections. (Adapted from Tomazevic, M. (1999). *Earthquake – resistant design of masonry buildings.* Imperial College Press, London, UK.)

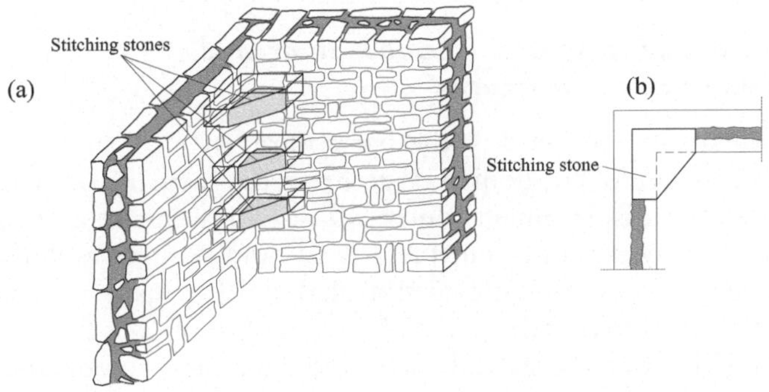

Figure 5.34 Strengthening of a corner with stone stitching. (a) Axisymmetric view; (b) horizontal section. (Adapted from Tomazevic, M. (1999). *Earthquake – resistant design of masonry buildings.* Imperial College Press, London, UK.)

Choice between them must be made in the light of the historical importance of the structure.

Sometimes systems should be improved by transforming their decks to diaphragmatic discs as analysed in Subsection 5.1.3.6.

The most widely applicable and valuable technique for the structural improvement of multistorey buildings is that of the insertion of metal ties for linking external walls of opposite sides of the building across its width (Figure 5.35). If these ties are very long, in case of seismic excitation, they are imposed to large elongations allowing the development of cracks and gaps *between facades* and the transverse walls. In order to keep these elongations at an acceptable level, either the ties are posttensioned to a degree or their cross sections are overdimensioned, so that the expected stresses and, therefore, elongations can be kept at an acceptable low level. Otherwise they must be bonded in the masonry so that the tensile stresses are absorbed, thanks to bonding at a short distance from the facades.

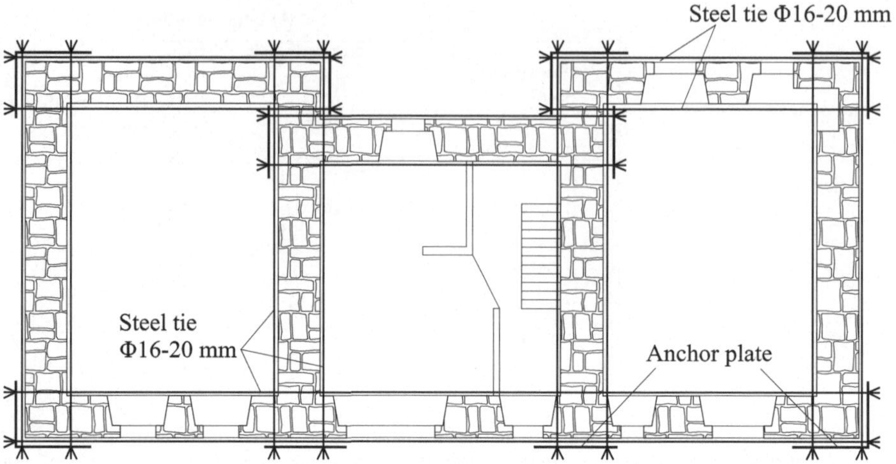

Figure 5.35 Position of steel ties in plan of a stonemasonry house. (Adapted from Tomazevic, M. (1999). *Earthquake – resistant design of masonry buildings*. Imperial College Press, London, UK.)

5.1.4.7 Improvement of structural response in arched, vaulted and domed structures

As explained in detail in Section 4.3, the basic parameter for any adequate structural response of an arched, vaulted or domed structure is the securing of their supports (piers) from developing relative displacements outwards at their joints to arch springings.

The whole structural system of Byzantine, Islamic and Western architecture is based on this concept, namely, on developing of counterbalanced thrusts from successive arches, semiarches and buttresses so that the paths of thrusts of the arches finally end up smoothly in the foundations. This was successfully achieved, and many important and magnificent monuments of this type have survived to date, like Hagia Sophia and Blue Mosque in Istanbul, St. Peter's Basilica in Rome etc.

However, in many cases, this balance is violated, particularly by seismic actions. In this case the vertical component of the seismic action increases the thrusts and therefore increases the relative displacements of the tops of the supports (piers). Additionally, the horizontal seismic actions, due to the absence of a stiff diaphragm at the top of the supports, allow them to move independently to a degree, causing damages or even local or overall collapse of arched structural systems (Figure 5.36). These deficiencies may be tackled with four main interventions:

i. Mass grouting of masonry walls
ii. Implementation of tensile rings of proper materials (steel, FRP rods etc.) at the dome base and its drum (Figure 5.2)
iii. Installation of ties at the springings of the arches (Figure 5.37)
iv. Installation of buttresses (Figure 5.1).

The first two intervention types have been examined already in Subsection 5.1.4.2.

For free-spanning ties, there are some reservations, because they are visible, in the inner space of the building. However, there has long been a tradition of using them. The use of high-strength steels or FRPs decreases the cross section of the tensile ties, and, therefore,

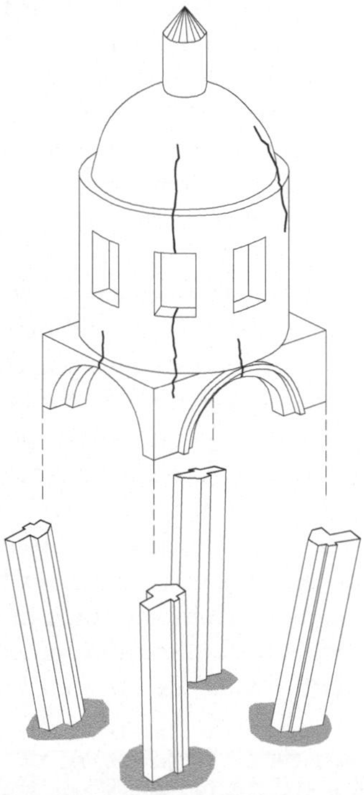

Figure 5.36 Cracks in the drum of a dome caused by outward rotation of the supporting piers. (Adapted from Croci, G. (2007). The conservation and structural restoration of architectural heritage. Advances in architecture series. WIT Press, Southampton. Courtesy of WIT Press.)

Figure 5.37 Hagia Anastasia monastery, Vasilika Thessaloniki. Replacement of the wooden rotten ties by pretensioned steel ones to stop the outward movement of the external piers (see Figures 4.21). (Drosou, K. (2014) Archives of the Ephorate of Antiquities of Halkidiki, Greece. Courtesy of the Ephorate of Antiquities of Halkidiki, Greece.)

the visual disturbance becomes very low (Figure 5.38). It should be noted that, in case of large spans, ties should develop considerable elongations for an effective activation, and, in this context, additional displacements of the piers would develop. Therefore, this fact, in connection with the use of high-strength steel, makes *the posttensioning of the ties inevitable*, so that these additional outward displacements can be avoided.

Ties should be capable of taking the full horizontal thrusts of the arches or vaults, possibly with some allowance for the additional thrusts for vertical seismic loadings, when the spans are big.

Sometimes, in case of arches that carry domical vaults that rise above the arches to support galleries in churches, or mosques, or upper floors in important public buildings, ties may be installed in the framework of a diaphragmatic truss above the crown of the domical vaults within the floor thickness (Figure 5.39). Of course, this is possible only if the supports continue upwards to this level and are in position to transfer the tie forces down in the form of shear and bending to the level of the springings where the thrusts of the arch or vaults act.

The addition of buttresses is another alternative, often used in the past to resist the outward thrusts or one-sided soil settlements. Buttresses being much more massive than ties are more obstructive, even if they are in the form of semiarch as in the case of the Christian part of Rotunda in Thessaloniki. So, nowadays, structural improvement solutions of this type are examined with strong reservations by the archaeological authorities

Figure 5.38 Internal view of Rotunda of Thessaloniki: strengthening ties of the vaulted altar. (From Penelis S.A., Archives, 2019d). Archives of photographic material of Rotunda of Thessaloniki.)

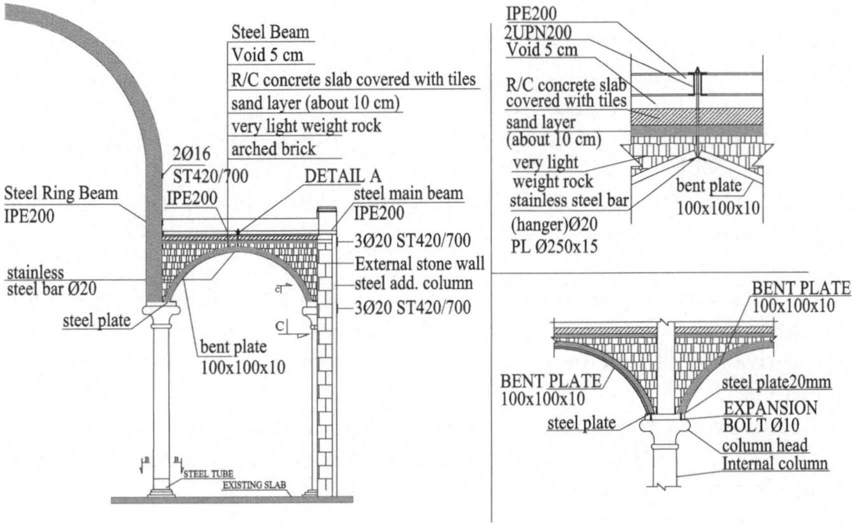

Figure 5.39 Diaphragm above groin vaults of Mar Girgis church of Cairo. (Adapted from Penelis, S.A. (2009). Restoration of bell tower of Zografos monastery, Mount Athos Penelis S.A., Archives. In Greek.)

in charge of a restoration. Even if such a solution is approved, much attention should be given to the way used for activating them, namely, how they could react to large thrusts for small displacements. The best way would be the installation of jacks between buttresses and the building. In this case, a data acquisition system should be installed for a period of 2–3 years, for monitoring the relaxation of the jacks, due to shrinkage and creep of the buttress masonry and the soil consolidation below buttress foundation.

5.1.5 Strengthening of foundations

5.1.5.1 Introduction

Foundation problems have as a sequence the appearance of settlements' drifts and cracks on the building. The source of such damage can be easily recognised by the direction of the cracks or the settlements. The reason for this damage may be attributed to:

- Soil failure
- Consolidation of clayish substrata due to lowering of water table
- Land sliding due to nearby excavation for the foundation of a new work deeper than the historical building
- Soil liquifaction during a strong earthquake
- Underground excavations for the construction of an infrastructure project (subways, sewage ducts etc.).

For monumental buildings with long history, most of the above phenomena had occurred in the past and, therefore, their consequences on the building have been stabilised in the past. However, some of the above reasons are contemporary and, therefore, they call for interventions in the soil, the foundation or both. Specialist advice is recommended, however, before any of the techniques described below is adopted.

5.1.5.2 Consolidation of soil by means of injections

This method is attractive, especially in the light of the most recent developments in the technique of chemical grouts, including the injections of water glass. This aims at achieving the solidification of loose and saturated sands, while injections of emulsified asphalt or the injections of cement groutings aim at the consolidation of large grained sands (UNDP/UNIDO, 1984; Lizzi, 1982).

Nevertheless, methods of consolidation of soil by means of injections have some limits to be taken into consideration:

- The soil must have a proper permeability, namely a coefficient $k \geq 10^4$ cm/s, and in this respect, clay and silty soils cannot be injected.
- The effectiveness of this method calls for uniform spreading of groutings below the foundations.
- Grouting improves the behaviour of the soil under compression *but not under tension*. So, in case of towers with foundation regions with inactive areas due to overturning tendency, the contribution of the injections is limited, only in the regions under compression.

5.1.5.3 Underpinning with additional foundation

Inadequate existing footings may be strengthened by casting new reinforced concrete straps alongside them, after providing special measures for the shear transfer as shown in Figure 5.40a. An efficient method for safeguarding the shear transfer is the use of post-tensioned stiches as shown in Figure 5.40b. Alternatively, the existing footing may be underpinned with wider reinforced concrete straps, following the normal underpinning procedure. However, it should be noted that the normal underpinning, as it is carried out by sections, gives rise to a general relaxation of the structure (Figure 5.41).

5.1.5.4 Underpinning with micropiles

Where it seems desirable to carry the loads of the superstructure to a lower and firmer soil stratum, piles must be driven. Usually in this case, micropiles are preferred with a diameter of 10–30 cm. This type of pile is preferred because the mechanical equipment

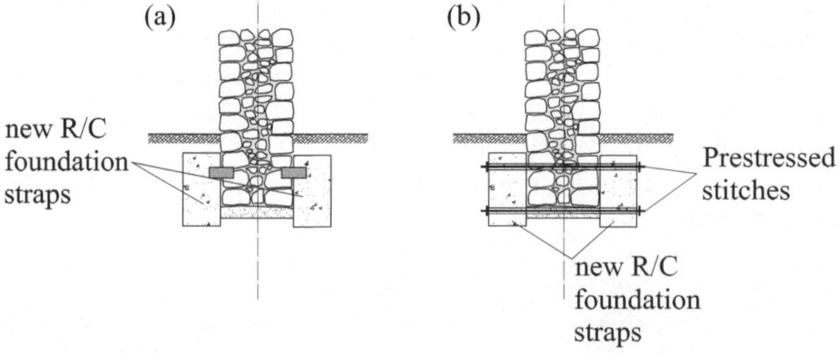

Figure 5.40 New reinforced foundation straps alongside existing foundation: (a) special measures for shear transfer; (b) prestressed stiches for shear transfer.

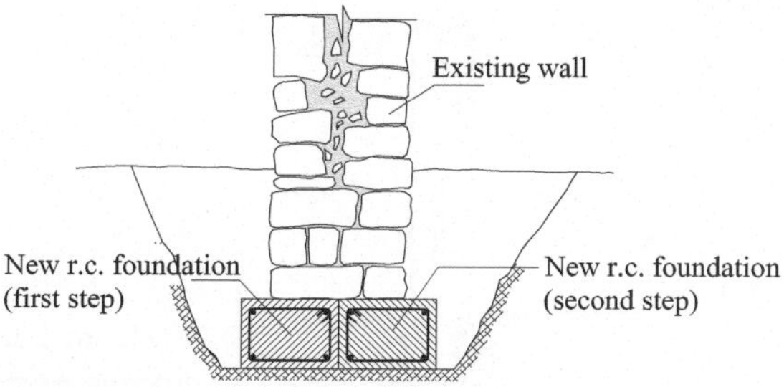

Figure 5.41 Underpinning of an existing foundation by a wider R/C strap, construction procedure of successive steps.

for drilling can function even in the basement of an existing building because the required free height for works may be even at the limit of 2.50 m.

Usually, micropiles are arranged alongside of the existing foundations. This again calls for casting of new reinforced concrete straps alongside the existing footings with adequate shear connections as depicted in Figure 5.42.

For soft soils	≅ 30 m
For soils with average consistency	≅ 20 m
For very stiff soils	10–15 m
For gravels or sands	6–10 m
For clay	10–15 m

An alternative approach is the implantation of micropiles in the form of *root piles* (Figure 5.43), (Lizzi, 1982). In this case, the drilling is carried out by rotation methods through the existing structure to penetrate the soil below. The usual diameter is 10–15 cm with a single bar at the centre as reinforcement. In case of bigger diameters, bar frames or tubes properly protected against corrosion may be used for reinforcement. The pile is automatically bonded with the upper structure, as the concrete is casted in the hole.

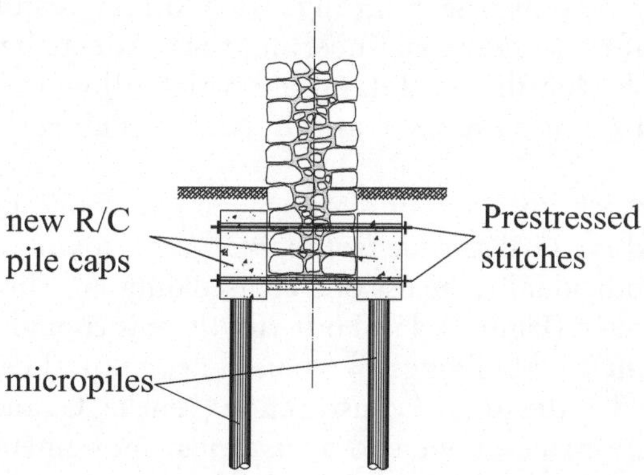

Figure 5.42 Underpinning of an existing foundation with micropiles.

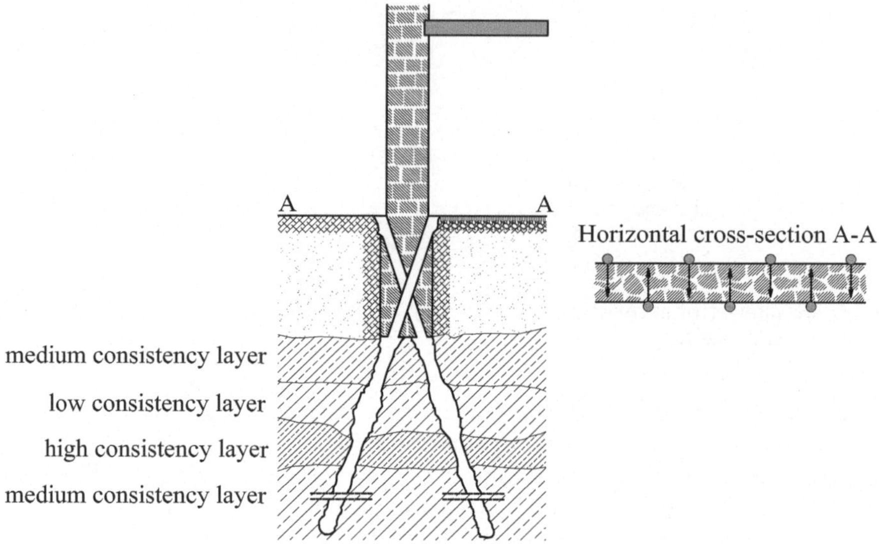

Figure 5.43 Typical scheme of root piles. (Adapted from Lizzi, F. (1982). The static restoration of monuments. Sagep Editrice, Genova, Italy. With permission of Sagep Editrice.)

Therefore, there is no need of additional structures for underpinning (R/C straps). The concrete of the piles is made of 600–800 kg of cement per cubic metre of sand. It is therefore a high-strength concrete.

The basic mechanical characteristic of the micropiles is that they function as *friction piles,* and, in this respect, they respond quickly for very small settlements in the range of some millimetres.

From existing experience, it is presumed that these piles develop their maximum bearing capacity equalised with the concrete failure over the following lengths.

Detailed information for the capacity design may be found in various handbooks (Lizzi, 1982).

5.1.6 Seismic isolation and energy dissipation systems

Earthquake effects are felt on the buildings as a direct result of ground shaking. During the evolution of earthquake engineering, these effects have been controlled by the concepts of strength, ductility and damping which allow the dissipation of energy through the accumulation of damage in appropriately engineered structural members and regions.

In an alternative approach, the concept of disengaging the building from ground shaking has been introduced and is called *seismic isolation* or *base isolation* (Figure 5.44). In addition to this concept, or individually, the use of increased dumping through specially designed devices may be implemented (Figure 5.45). Hitherto, there is enough experience from applications to very important new buildings like museums, cultural centres etc. in which the isolators are incorporated at the time of construction (Penelis, G. and Penelis Gr., 2019).

If these methods were to be applied to a historical monument of high significance, major underpinning operations should be carried out, aiming at the separation of the building from the ground for the interlocation of the isolators. The costs, the construction

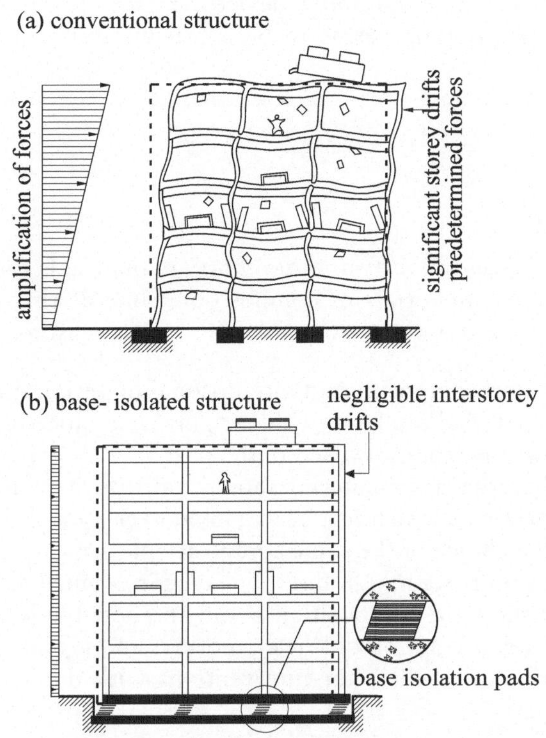

Figure 5.44 Comparative response to seismic action: (a) a conventional earthquake-resistant building; (b) a building with seismic isolation. (Adapted from Penelis, G. and Penelis, Gr. (2019). Concrete buildings in seismic regions. CRC Press (A SPON Press Book), London.)

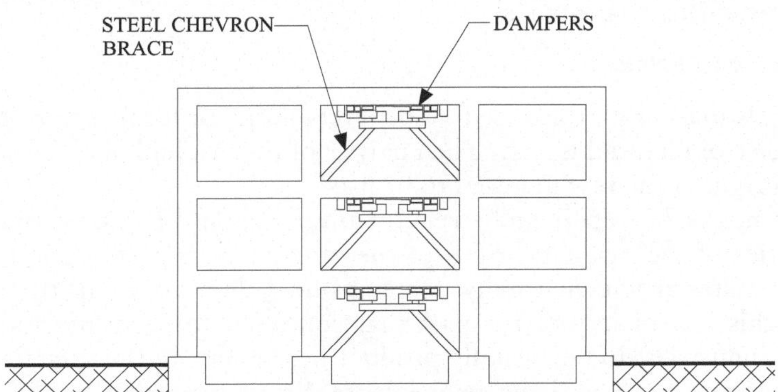

Figure 5.45 Passive energy dissipators (dumpers) in a building. (Adapted from Penelis, G. and Penelis, Gr. (2019). *Concrete buildings in seismic regions.* CRC Press (A SPON Press Book), London.)

difficulties and risks of such an intervention together with functioning difficulties of the building in relation to its surroundings, as it must be in position to move at its base at about 20–30 cm in both directions, make questionable the acceptance of such an intervention in an existing building of high cultural importance (UNDP/UNIDO, 1984).

However, the use of increased damping devices (Figure 5.45) without simultaneous implementation of base isolation seems to be a feasible intervention method even for historical buildings.

5.2 MATERIALS

5.2.1 Introduction

The best way to ensure *compatibility* and *durability* of materials for repair and strengthening is to use traditional materials of composition same as that of the original ones. However, this procedure is not always possible for various reasons:

- Difficulties arise in recomposition of a traditional material, due to either the inability to perform a detailed analysis of the original (e.g. additives) or the inability to find the sources of the materials used in the past.
- The necessity that the new material should gain, in a relatively short time, the strength of the original, which has been gained over a much longer period of time, does not allow the choice of the same composition.
- The necessity of grouts with high fluidity results in a much lower strength than the original even for the same composition of old and new binder.
- The necessity of a combination of bricks, mortars and grouts with reinforcement or posttensioned stitches requires for interventions with modern materials (stainless steels, FRPs etc.).

So, the choice of the proper materials for repair and strengthening is a complicated procedure requiring many factors to be taken into account.

5.2.2 Nonmetallic materials

5.2.2.1 Stones and bricks

Concerning cut stones or marble, used in local rebuilding or local repair of defective elements, the choice of material is easy. The analysis of the original one allows the determination of the origin of the raw material to be used.

Concerning bricks for repair and strengthening, it should be noted that the analysis of the properties of the original bricks of the monument is a very good guide for the choice of raw material which usually comes from a clay quarry in the surroundings. Many times, this is a quarry of the past that continues to be in use today. However, modern brick industries do not usually produce bricks having the specified form of the old handmade ones, and the firing temperature differs from that used for the ancient bricks. So, a special order should be given to small industries using the traditional way of production. Of course, these bricks cannot fulfil the homogeneity and the standard quality of conventional bricks for modern buildings. So, mechanical and chemical properties must be regularly tested. This procedure has been followed for many years in Greece with success. Results of sonic tests on $30 \times 40 \times 5$ cm bricks used for repair of Rotunda are shown in Table 5.5. These can be accepted as sufficient ones when they are compared with the results of sonic testing of old bricks of the monument under repair (Karaveziroglou, 1989). These 'traditional' bricks have a relatively high cost because of the production procedure.

Table 5.5 Sonic test results on bricks of 50 mm thickness

Bricks		*Distance (mm)*	*Time (μs)*	*Velocity (km/s)*	*Statistic evaluation of velocity* *Average val. (km/s)*	*St. deviation (km/s)*	*Variation (%)*
Repair	1	395	209	1.900	1.930	0.0525	2.7
		395	210	1.881			
		316	158	2.000			
		316	163	1.939			
	2	396	217	1.825	1.823	0.0207	1.1
		396	220	1.800			
		320	176	1.818			
		320	173	1.850			
	3	395	224	1.763	1.956	0.1699	8.7
		395	212	1.863			
		320	153	2.092			
		320	152	2.105			
Christian part of Rotunda	1	382	206	1.854	1.879	0.0741	3.7
		382	209	1.828			
		382	191	2.000			
		286	145	1.972			
		286	167	1.713			
		286	150	1.907			
	2	362	182	1.989	1.999	0.1047	5.6
		363	187	1.941			
		364	170	2.141			
		307	153	2.007			
		307	158	1.943			
		308	156	1.974			

Source: Adapted from Karaveziroglou, M., Papayianni, I., and Penelis, G. (1989). Criteria for selecting materials for repair brick masonry. *ICCROM, Proceedings of International Technology Conference on Structural Conservation of Stone Masonry*, Athens, Greece. pp. 335–342.

5.2.2.2 Mortars and grouts

5.2.2.2.1 Requirements and types

Mortars and grouts in use for repair and strengthening of monuments should accomplish the following requirements (Peroni, et al. 1981):

- They should have mechanical properties in the same range or a little higher than those of the original mortar.
- Pore size distribution should be comparable with that of the original one.
- Mortar or grout should contain the smallest possible amount of soluble salts (particularly sodium or potassium salts).
- Mortar should be easily workable, while the grout should have adequate fluidity to be pumped.
- Mortar or grout should set with sufficient rapidity and reliability, both in a dry and in a wet environment.
- They should present limited shrinkage.
- Grouts should have proper grading of constituents of their solid part, and special care should be taken for the maximum grain size.

- Grouts should have a low degree of bleeding and segregation.
- Grouts and mortars should have a coefficient of thermal expansion similar to that of the original material.

Based on the above, various types of mortars and grouts eventually proper for restoration will be examined next.

a. *Portland cement mortars and grouts*

From the point of view of the general principles discussed in Chapter 1, Portland cement has the following disadvantages for restoration activities:

- Portland cement is too strong (high strength and high modulus of elasticity). In case of differential displacements, stresses will be transferred to the older texture which will fail.
- Portland cement mortar and grout have a large thermal expansion coefficient. Thermal cycles result, therefore, in stress and damage of the 'weak', old masonry.
- Portland cement forms soluble salts while setting, causing extended efflorescence.
- Portland cement mortars have a low porosity causing damage because of accumulation of moisture behind cement layers that do not allow evaporation.

As the extended use of cement in the conservation of ancient structures in the last 100 years has resulted in quite a number of defects listed above, modern trend in recent times has been to restrain its use.

b. *Lime mortars and grouts*

Since most ancient masonry in Europe is made of lime-based mortars, there has been a tendency to use traditional lime mortars or grouts for masonry repair (Zacharopolou, 1998). This in turn has met with other difficulties.

- The strengths of these mortars are always below than those of the original ones. Water addition in grouts to gain fluidity results in unacceptable low strength.
- Lime mortars are slow-setting and require a relatively dry environment. In ancient thick masonry, which is frequently moist, lime may set very poorly, or not at all. The situation is worse for the grouts that are injected into the body of the masonry for reasons presented in detail in Chapter 2.
- Setting of lime depends upon carbon dioxide from the air. In thick structures, carbonation of the core is extremely slow and unreliable.

If we oversee their low strength, the lime mortars and grouts present characteristics that may be regarded as desirable. In particular, they show a relatively high deformability and a good proportion of large pores.

c. *Lime–cement mortars and grouts*

In an attempt to combine the hydraulic properties of cement with the low strength, high deformability and porosity of lime, mixed lime–cement mortars have been brought into use and have gained considerable acceptance for conventional buildings.

As has already been noted by Holmstrom in 1977 (Holmstrom, 1977), even this combination is not exempt from defects. Firstly, lime and cement do not react chemically with one another. So, they need controversial setting conditions (moist conditions for cement, dry conditions for lime). So, the mechanical properties obtained in any real situation will not be exactly predictable. Secondly, if the cement proportion is rather high, the amount of soluble salts formed will also be high. Therefore, their use for restoration is questionable.

d. *Pozzolanic mortars and grouts (ternary mortars and grouts)*

Lime mortars and grouts are usually improved by adding pozzolanic material in the mixture (see Chapter 2). These materials react chemically with hydrated lime to hydraulic constituents without needing carbon dioxide. In this respect, they can set under moisture conditions in very thick walls and piers. Additionally, due to the development of hydraulic constituents, they present higher strength than the lime mortars and grouts. Such mortars include pozzolana, fly ash, truss, silica fume and underfired pulverised bricks. Very often, a percentage of Portland cement (at about 30% of mass of binder materials) must be added to increase the early strength, particularly in grouts with high ratio of water/binder (at about 1.0), where fluidity is a prerequisite.

Pozzolanic mortars (ternary mortars) and grouts show a remarkably good behaviour as restoration materials, and they have gained general acceptance.

Particularly for grouts, high expertise is needed in the selection of materials and in elaborating their composition. Main properties that they must fulfil are the following (Vintzileou E, 2006):

i. *Permeability*

For a grout to be injectable into a cracked or porous material, the maximum grain size of the mix as well as the distribution of various grain sizes should be compatible with the size of cracks and voids.

ii. *Fluidity*

The fluid grout should be able to move easily within the voids and crack of masonry without high pump pressure.

iii. *Stability*

The third main property that must be fulfilled is stability, i.e. avoidance of excessive bleeding (<5%).

All the above parameters are accomplished by a proper design of the grout in the lab by a material expert.

e. *Epoxy resin mortars and grouts*

Recent development of organic resins with a wide range of strengths, elasticities etc. has presented a further choice, particularly since these resins do not shrink or dry, penetrate well and have excellent bond properties. It is possible to choose either a pure resin or a resin mixed with a fine aggregate to give, apparently, all that is required for any particular application. But the high cost is not the only drawback. There is, as yet, no long-term proof of their suitability, so it would be risky to use them for such irreversible interventions as grouting masonry, rejointing or local rebuilding. Therefore, while their use for injections in cracks of R/C structures is acceptable, their use in restoration is questionable.

5.2.2.2.2 Composition of mortars and grouts

Chemical, mineralogical and mechanical tests of the old mortar are the guides to the composition of the mortar or grout for restoration of a monument. They are usually of pozzolanic type with the main ingredients being hydrated lime, pozzolana and aggregate. Very often underfired brick powder with pozzolanic action is used, particularly in mortars to give the same chromatic gradation with the old ones. The recomposition of old mortars, based on their laboratory analysis (binder/aggregate: 1:2 to 1:3) with new materials, as mentioned above, does not succeed in reaching the 'high' strength of some of them – as *opus caementicium* of Romans – although the same gradation of grains is also used.

It is obvious that many conditions, such as the way of slacking the lime, the additives used etc., were different from the ones in use nowadays. Often, *a percentage of Portland cement must be added in order to increase their early strength.* This cannot be avoided particularly in grouts, where a high water/binder ratio is necessary for the material fluidity, a fact that results in lower strength. Based on experimental evidence (Penelis, G., 1982, Toumbakari, 2002), it is recommended that Portland cement should be at about 30% of mass of the binder system (cement–lime–pozzolana).

From the above, it follows that, for the choice of a proper mortar or grout for the restoration of a monument, a wide range of mixes with various constituents and proportions must be designed and tested. Workability for the mortars and fluidity for the grouts are basic requirements. In the last 20 years for the fluidity and segregation improvement, *fly ash* or *silica fume* and *superplastifiers* have been introduced in the grout design (Papayianni, 2014; Miltiadou, et al., 2005).

It would be interesting here to give some information from the authors' experience in Greece.

Tables 5.6 and 5.7 present the mixing proportions and the test results of a series of mortars for repair of Rotunda and some other Byzantine monuments in Thessaloniki (Penelis, G., 1982).

Tables 5.8 and 5.9 present the mixing proportions, the properties of grout mixtures and the mechanical properties of a series of grouts designed in the Laboratory of Reinforced Concrete and Masonry Structures of Aristotle University of Thessaloniki (AUTH) for repair and strengthening of Byzantine monuments in Thessaloniki. Table 5.10 presents the choice of grouts for the monuments of Rotunda, Hagios Andreas, Peristera and Vlatades Monastery.

In the case study of Acheiropoietos church of Thessaloniki (7th Century A.D.), the following composition for the stabilisation grouts was used (Papayianni, I., 2014).

Hydrated lime	40%
White cement	30%
Ground pozzolana	15%
Brick dust	15%
Water/binder ratio	$\cong 1.0$

5.2.2.3 Concrete

The use of modern concrete (including shotcrete) is better for application in the construction of new elements that are capable of working in harmony with the existing structure or in positions where their influence to the form of the monument is insignificant (e.g. foundations). As the concrete technology is well known, no further reference to this subject will be made here.

5.2.2.4 Timber

Timber has many excellent characteristics, and there is a long tradition of its use with reliable structural performance. Its drawbacks are the fire risk and the susceptibility of some species to fungal and insect attack. If timber is damaged or decayed or, if a whole timber floor or roof has proved to be structurally inadequate, the possibility of substituting it with steel or reinforced concrete will obviously arise. However, for historical monuments,

Table 5.6 Mixing proportions of mortars for Byzantine monuments of Thessaloniki, Greece (Penelis, G., 1982)

Pozzolana origin			*Mixing proportion by weight*									
						Sand max size		*Crushed max*	*Bricks size*			
		Mortar No	*Lime*	*Pozzolana*	*Cement*	*2 mm*	*6 mm*	*2 mm*	*6 mm*	*Water*	*Density kg/dm³*	*Water L+R+C*
	Santorin	K1	1	1	-	6	-	-	-	2.24	1.86	1.120
		K2	1	1	-	3	-	3	-	2.70	1.88	1.350
		K3	1	1	-	3	-	-	3	2.42	1.87	1.210
		K4	1	1	-	-	6	-	-	1.92	1.94	0.960
		K5	1	1	-	-	6	-	-	1.86	1.93	0.930
		K6	1	1	-	3	-	-	3	2.20	1.84	1.100
		K7	1	1	-	-	5	-	1	1.97	1.90	0.985
		K8	1	1	-	-	3	-	3	2.24	1.86	1.120
		K9	1	0.8	0.2	6	-	-	-	2.16	1.90	1.080
		K10	1	0.8	0.2	3	-	-	3	2.14	1.90	1.070
		K11	1	0.6	0.4	3	-	-	3	2.45	1.89	1.225
		K12	1	0.4	0.6	3	-	-	3	2.33	1.90	1.165
		K13	1	0.2	0.8	3	-	-	3	2.25	1.82	1.125
		K14	1	0.8	0.2	-	6	-	-	1.83	1.95	0.915
		K15	1	0.6	0.4	-	6	-	-	1.78	1.98	0.890
		K16	1	0.4	0.6	-	6	-	-	1.62	2.07	0.810
		K17	1	0.2	0.6	-	6	-	-	1.53	2.08	0.765
	Skydra	K18	1	1	-	6	-	-	-	2.17	1.94	1.085
		K19	1	1	-	-	6	-	-	1.67	2.02	0.835
		K20	1	0.8	0.2	6	-	-	-	2.17	1.84	1.085
		K21	1	0.8	0.2	3	-	-	3	2.08	1.93	1.040
		K22	1	0.2	0.8	3	-	-	3	2.25	1.92	1.125

Table 5.7 Test results (28 days) of mortars of Table 5.6 (Penelis, G., 1982)

		Strength		*Modulus of elasticity*	
Mortar No	*Density (kg/dm³)*	*Flexural (N/mm²)*	*Compressive (N/mm²)*	*Dynamic (N/mm²)*	*Static (N/mm²)*
K1	1.59	0.166	0.410	1118	553
K2	1.47	0.234	0.859	1244	634
K3	1.51	0.261	1.318	1807	940
K4	1.68	0.421	0.697	1505	760
K5	1.59	0.258	0.928	2390	1231
K6	1.54	0.214	0.770	1900	965
K7	1.63	0.194	0.784	1820	917
K8	1.51	0.269	1.249	2446	1272
K9	1.55	0.311	1.438	2431	1288
K10	1.48	0.519	2.605	3038	1686
K11	1.62	0.433	3.581	4846	2791
K12	1.62	9.859	4.589	5747	3460
K13	1.65	1.810	5.573	6912	4341
K14	1.65	0.264	1.270	2774	1442
K15	1.72	0.487	1.678	4184	2238
K16	1.76	0.638	2.035	4703	2540
K17	1.83	0.839	3.771	7300	4271
K18	1.60	0.193	0.336	1541	709
K19	1.74	0.142	0.652	1640	823
K20	1.50	0.303	0.520	2103	1051
K21	1.56	0.816	3.066	3461	1973
K22	1.61	1.409	5.018	6853	4215

this change in their structural character is undesirable. Repair or replacement with timber is preferable. But attention must be paid to the choice of species, seasoning and moisture content. Treatment with a suitable preservative may be desirable. It should be noted here that interventions with timber belong to the category of the reversible ones.

5.2.3 Metallic materials

5.2.3.1 Introductory remarks

Where high tensile strength or high compressive forces are required, steel must usually be used in spite of its liability to corrosion; alternatives would be too expensive. Many different steels are available, varying in cost, corrosion resistance (stainless steel) and strength. Many protective treatments of varying value are also available for steels that are corrosion resistant and would otherwise be vulnerable.

There are instances, however, where risk of future corrosion cannot be tolerated. The serious damage caused in the beginning of the 20th century by the rusting of iron cramps and other reinforcements embedded in the marbles on the Acropolis of Athens led, in the recent restoration activities, to the choice of *titanium*.

Table 5.8 Mixing proportions of grouts for Byzantine monuments of Thessaloniki, Greece

	Constituents									*W*	
No	*Lime (L)*	*Santorin earth (P)*	*Skydra earth (P)*	*Portland cement (C)*	*Sand max size < 1mm*	*Brick powder max size < 25mm*	*Water*	*Expanding Agent*		*L+R+C*	*Time of sinking DIN 4227 Tell 5 sec*
E1	1	0.2	-	0.8	-	-	1.515	-		0.75	34
E2	1	0.6	-	0.4	-	-	1.417	-		0.70	31
E3	1	0.2	-	0.8	2	-	1.750	-		0.87	33
E4	1	0.6	-	0.4	2	-	1.563	-		0.78	34
E5	1	0.6	-	0.4	1	1	1.875	-		0.93	35
E6	1	0.8	-	0.2	1	1	1.869	-		0.93	39
E7	1	-	0.8	0.2	1	1	1.875	-		0.93	34
E8	1	-	0.6	0.4	1	1	1.875	-		0.93	37
E9	1	-	0.2	0.8	2	-	1.637	-		0.81	33
E10	1	-	0.6	0.4	2	-	1.530	-		0.76	35
E11	1	0.6	-	0.4	-	2	2.187	-		1.09	32
E12	1	2	-	0.4	-	-	1.720	0.034		0.50	34
E13	1	2	-	0.4	-	-	1.788	-		0.52	35
E15	1	0.8	-	0.2	1	1	1.670	0.02		0.83	35
E16	1	-	0.8	0.2	1	1	1.610	0.02		0.80	33
E17	1	0.6	-	0.4	1	1	1.660	0.02		0.83	37
E18	1	-	0.6	0.4	1	1	1.690	0.02		0.84	35

Source: Adapted from Penelis, G., Karaveziroglou, M., and Papayianni, I. (1989). Materials used in the restoration of historical buildings and monuments. *Seminario per Conservao Recuperaçäo e consolidacäo de monumentos e edificios histöricos*, Conselho Directivo National (ed) Lisbon.

5.2.3.2 Conventional steels

5.2.3.2.1 Steel corrosion

- It is important to recall here the various types of corrosion:
  - General
  - Pitting
  - Galvanic
  - Intergranular
  - Crevice
  - Stress corrosion etc.
- It is of practical use to classify atmospheric corrosion into three typical *environments*:
  - Marine (mainly chlorides acting)
  - Industrial (sulphurous gases acting together with chlorides)
  - Rural (minor actions).

Monuments may be situated in any of the above three environments. Besides, humidity insulation of steel parts incorporated in a monument is a much more difficult task, compared to modern structures.

Table 5.9 Test results for strength of grouts of Table 5.8 of various ages

No	Age days	*Tensile strength (N/mm²)*	*Compressive strength (N/mm²)*	Age days	*Tensile strength (N/mm²)*	*Compressive strength (N/mm²)*	Age days	*Tensile strength (N/mm²)*	*Compressive strength (N/mm²)*	Age days	*Tensile strength (N/mm²)*	*Compressive strength (N/mm²)*
E1	8	0.51	1.05	14	0.25	1.45	22	0.58	2.25	27	0.57	2.45
E2	8	0.16	0.34	14	0.20	0.42	22	0.37	0.80	27	0.28	0.96
E3	10	0.16	1.01	13	0.23	1.16	21	0.37	1.68	28	0.67	2.25
E4	10	0.39	0.50	13	0.25	0.58	21	0.33	0.85	28	0.44	1.10
E5	9	0.20	0.43	17	0.22	0.71	21	0.43	0.94	28	0.52	1.30
E6	8	0.10	0.21	16	0.11	0.34	21	0.16	0.64	29	0.22	0.72
E7	8	0.12	0.25	16	0.10	0.35	21	0.20	0.42	28	0.26	0.61
E8	7	0.20	0.44	13	0.37	0.75	21	0.45	0.98	28	0.66	1.55
E9	7	0.40	0.92	15	0.49	1.60	21	0.98	2.28	28	1.14	3.22
E10	7	0.12	0.38	13	0.39	0.67	21	0.37	0.81	28	0.34	1.18
E11	9	0.34	0.42	14	0.45	0.97	21	0.64	1.91	29	0.50	2.65
E12	9	0.07	0.28	14	0.20	0.53	21	0.40	0.90	29	0.42	1.30
E13	8	0.09	0.32	14	0.23	0.52	21	0.44	1.00	28	0.53	1.44
E15	8	0.04	0.20	14	0.25	0.44	21	0.48	1.06	28	0.44	1.39
E16	8	0.05	0.24	14	0.29	0.49	21	0.37	0.88	27	0.39	1.20
E17	8	0.23	0.62	14	0.59	1.44	21	0.66	2.07	27	0.83	2.42
E18	8	0.21	0.57	14	0.53	1.18	21	0.55	1.54	27	0.92	2.10

Source: Adapted from Penelis, G., Karaveziroglou, M., and Papayianni, I. (1989). Materials used in the restoration of historical buildings and monuments. *Seminario per Conservao Recuperação e consolidacão de monumentos e edifícios históricos*, Conselho Directivo National (ed) Lisbon.

Table 5.10 Grout and mortar composition for some Byzantine monuments of Thessaloniki

Monument		Position	Grouts									Monument's mortar	
				Materials (kg)						Mechanical properties		Mechanical Properties	
			Type	Pozzolana	Lime	Cement	Sand	Water	Additive	Compressive Strength	Ed	Compressive Strength	Ed
Rotunda Thessaloniki	Roman Part	Cracks >10 mm	$KT_{3\text{-}1}$	10	5	6	21	20	0.23	42	60.760	28.76	18.905
		<10 mm	ET_6	10	5	6	14	-	0.21	56	66.574		
	Christian Part	>10 mm	$KT_{4\text{-}1}$	10	5	8	23	19.5	0.23	69	102.421	38.14	28.412
		<10 mm	ET_6	10	5	6	14	-	0.21	56	66.774		
Hagios Andreas Thessaloniki		Cracks >10 mm	$KT_{2\text{-}1}$	10	5	4	19	18	0.19	26	43.576	13.90	9.212
		Cracks <10 mm	ET_2	10	5	2	12.5	-	0.17	30	30.848		
Vlatades monastery Thessaloniki		Cracks >10 mm	$KT_{2\text{-}1}$	10	5	4	19	18	0.19	26	43.576	18.23	10.583
		Cracks <10 mm	ET_2	10	5	2	12.5	-	0.17	30	30.848		

Source: Adapted from Penelis, G., Karaveziroglou, M., and Papayianni, I. (1989). Materials used in the restoration of historical buildings and monuments. *Seminario per Conservao Recuperação e consolidacão de monumentos e edificios històricos*, Conselho Directivo National (ed) Lisbon.

Remark: units of mechanical properties: Kg/cm^2 (0.1 MPa)

- What really matters in the restoration having to face steel corrosion is the set of its consequences (Tassios, 1985):
  - The most important consequence is *cracking* and spalling of building elements due to the swelling of corroded steel.
  - Another important structural consequence is the loss of strength of external ties and rings or bond capacity of incorporated steel bars.
  - Stains of rust.
- The advice of a corrosion expert should be mobilised in case of the use of metallic materials for restoration activities.

5.2.3.2.2 Coated steel

a. *Zinc coating*

Zinc is one of means for passivation of steel. Coating is applied only on ready-made cut and bent elements to be galvanised by hot-dip galvanisation.

Coating offers a 'sacrificial' behaviour. It corrodes in the same way as it does a solid metal until the base metal is exposed. Although zinc is highly resistant to marine exposures, it is sensitive in sulphurous atmosphere. So, a loss of 5 μ/year of coating is expected in an industrial environment. Taking into account that coatings are in a range of 100–120 μ, it can be concluded that the protection lasts for almost 20–25 years in the atmosphere of an industrial city (Korvest galvanisers, 2018).

According to the Zinc Development Association, zinc-coated steel, as reinforcement embedded in concrete, should be considered as having very high resistance to corrosion (Rehm, et al., 1974). However, the results of bonding after 50 years of lifetime seem to be controversial. The cost of zinc-coated steels is 2–2.5 times higher than that of the structural steel.

Taking all the above into account, galvanised steel should be considered as a *questionable* choice for items embedded in the masonry for restoration. However, it could be used without any drawback for external reversible applications.

b. *Lead coating*

It is a very old method for steel protection in atmosphere. We are remind of the lead-coated cramps in Acropolis Athens and in Samothrace, Greece, that have survived to date (Harris, 1992).

Lead is highly resistant to sulphur and chlorine atmospheres. So, lead-coated bars or ties can be used in restoration exposed to the atmosphere without any restriction (reversible interventions). The only problem that arises has to do with the connections, as lead is a soft metal. Lead is very sensitive in alkaline environment of $Ca(OH)_2$. So, lead-coated steel is unsuitable for reinforcement in concrete, since cement has alkaline reaction. On the other hand, lead weakens the bonding between concrete and steel. So, lead is suitable only for external use in restoration.

The cost of lead-coated steel is 2–2.5 times higher than that of the structural steels. Finally, it should be noted that, while zinc-coating procedure is industrialised for many contemporary uses (e.g. electric energy transfer towers), lead-coating procedure for steel bars seems to be questionable.

c. *Resin-coated rebars*

Resin powder is applied on steel-reinforced bars under electrostatic conditions, after a steel-grid abrasive blasting. A resin film deposition of 150–300 μ is ensured, fusion bonded to steel under 200°C (Tassios, 1985). The film remains flexible enough to withstand on-site formatting processes without weakening.

This technique seems to be rather promising for restoration of monuments; however, our experience to date is not yet sufficient for any recommendation to be made.

5.2.3.3 Stainless steel

Special ferrous alloys with Cr (percentage more than 10%–15%) or Ni or even Mo are very corrosion-resistant, especially against atmospheric corrosion (marine, industrial, urban and rural). The most suitable for building construction are the austenitic steels with chromium, nickel and molybdenum, in the following composition (Penelis, G., 1982):

$C \leq 0.7\%$, $Cr = 16.5\%–18.5\%$, $Ni = 10.5\%–13.5\%$, $Mo = 2.0\%–2.5\%$, $Si \leq 1.0\%$, $Mn \leq 2\%$, $N = 0.15\%–0.25\%$.

The high corrosion resistance of these steels is due to the produced coating of chromium oxide which is a highly protective coating for the rest of metal. Their corrosion resistance in the atmosphere environment is 100 to 1,000 times higher than that of the structural steel. This means that they are practically inoxidable. They have the following satisfactory mechanical properties.

0.2% Yield strength	205 N/mm^2
Tensile strength	500–700 N/mm^2
Unit elongation at failure	45%
Hardness HB	130–210

Recently, high strength stainless steels have also been produced, in smooth or ribbed bars suitable for stitchings, anchorages, prestressings etc. These steels have the following mechanical properties.

0.2% Yield strength	630–880 N/mm^2
Tensile strength	90–990 N/mm^2
Unit elongation at failure	25%

Although stainless steels of the Cr–Ni–Mo type show a very high resistance to general and intercrystalline corrosion, they are sensitive to stress corrosion in a high chloride environment. So, prestressed rings or ties of this type of steel should be used with caution in the restoration of monuments near the sea. When these steels are embedded in concrete or cement grout, intercrystalline or stress corrosion does not appear at all, particularly if sulphate-resistant white cement is used.

On-site tests involving structural and prestressed stainless austenitic Cr–Nr–Mo steels in marine environment under water and at the height of the sea wave, after 5 years of observations, have not presented any indication of any type of corrosion (Parkin and Phiun, 1977).

From the above discussion, it is concluded that austenitic stainless steels of the type Cr–Ni–Mo are suitable for restoration. The only drawback is their cost (about eight times higher than that of structural steel).

5.2.3.4 Titanium

As it has already been mentioned, there are instances where the value of the monument does not allow any risk of future corrosion. In this case, we might be obliged to use the very expensive solution of titanium (at a cost about thirty-five times higher than that of structural steel) (Zambas, 1982; Tassios, 1985; Skoulikidis, 1977). Titanium has the following mechanical properties.

0.2% Yield strength	200–400 N/mm^2
Unit elongation at failure	15%–25%
Modulus of elasticity	1.1×10^5 N/mm^2

Titanium is a relatively light material ($d = 4.5$g/cm^3), and it is readily cold formed. Its most important feature is its exceptionally good corrosion resistance versus all types of corrosion and under high temperatures as well. It resists the actions of strong acids, such as sulphuric, nitric and hydrochloric acids.

5.2.3.5 Case studies from Greece

It would be interesting to present here some case studies from Greece.

- For the monuments Rotunda; Hagios Andreas, Peristera and Vlatades monastery in Thessaloniki, the following decisions were made at first (Penelis, et al., 1980a):
  - External rings and ties: lead-coated structural steel
  - Posttensioned stitches or reinforcement: austenitic stainless steel of the type Cr–Ni–Mo.

Later, due to unexpected difficulties in producing lead-coated bars to the length of 12.0 m in Greece, it was decided to use stainless austenitic steel of the type Cr–Ni–Mo for the posttensioned rings and ties too.

- For the monument of Hagios Panteleimon in Thessaloniki, stainless austenitic steel was also used for all metallic elements of intervention (Ignatakis, et al., 1989).
- For the restoration of Erechtheion at Acropolis in Athens, titanium was used for all metallic connections (clamps, dowels etc.) (Tassios, 1985; Zambas, 1982; Skoulikidis, 1977).
- For the restoration of Acheiropoietos church of Thessaloniki, titanium anchors were used for the corner stiches (Penelis, G. and Stylianidis, 2001; Raptis, K., Zombou, A. 2010).

5.2.4 Fibre-Reinforced Plastics (FRPs)

5.2.4.1 Introduction

The strengthening of masonry members with externally bonded FRP systems, laminates and sheets, and internally drilled ducts with FRP rebars properly bonded, is a relatively new method of intervention.

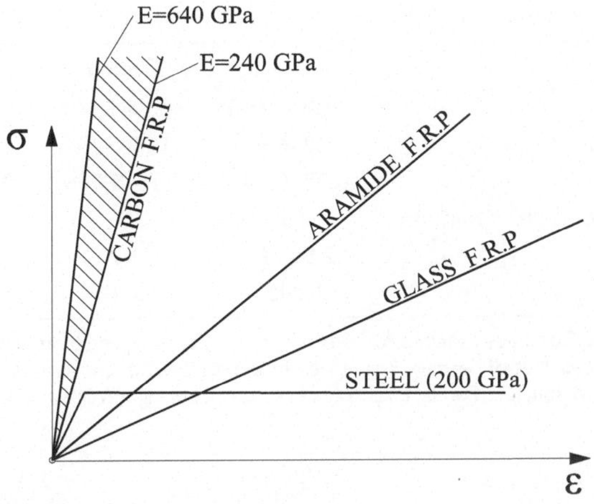

Figure 5.46 Constitutive law of FRP. (Adapted from Penelis, G. and Penelis, Gr. (2019). *Concrete buildings in seismic regions*. CRC Press (A SPON Press Book), London.)

In the last years, there has also been a tendency to use these systems in strengthening of monumental buildings since these materials exhibit the following advantages:

- They do not corrode.
- They have very high strength accompanied by a linear elastic behaviour up to failure (Figure 5.46).
- They are light in weight.
- They are available in large dimensions.

On the other hand, they exhibit a series of disadvantages:

- They have a brittle type of failure.
- They are susceptible to fire.
- They lose a large percentage of their initial strength under permanent loading ranging from 15% to 60% (Penelis, G. and Penelis, Gr., 2019 Table 5.11).

FRPs are formed by embedding continuous fibres into a resin matrix that binds the fibres together.

Table 5.11 Mechanical properties of FRPs

Material	*Elastic modulus (GPa)*	*Tensile strength (MPa)*	*Failure deformation (%)*	*Loss of strength under permanent load (%)*
GFRP	50	1700–2100	3	60
AFRP	65–120	1700–2100	2–3	50
CFRP	165–600	1400–3000	0.5–1.7	15
Steel	200	220–400	0.2[a]	

Source: Adapted from Penelis, G. and Penelis, Gr. (2019). Concrete buildings in seismic regions. CRC Press (A SPON Press Book), London.

[a] Yield deformation.

Table 5.12 Characteristics of fibers of FRPs

Type of fibers	*Glass*	*Carbon*	*Aramid*	*Steel*
Tensile strength (MPa)	1,300–3,400	2,000–5,600	2,500–3,620	1,500–3,500
Modulus of elasticity (GPa)	22–62	150–325	48–76	185–120
Ultimate strain (mm/mm)	0.03–0.05	0.01–0.015	0.02–0.036	0.04
Coefficient of thermal dilation (10^{-6} m/m K)	5.5	0.0	−0.5	6.5
Density (g/cm^3)	2.5–2.6	1.7	1.4	7.9
Melting point (°C)	1,100	310	420	1,300

Source: Based on ACI Manual Concrete Practice ACI 544 1R-96 (2002); Adapted from Papayianni, I. (2014). Retrofitting and strengthening masonries of heritage structures: materials used. in Encyclopaedia of Earthquake Engineering, Beer, M., Kougioumtzoglou, I.A., Patelli, E., and Au, S.K. (eds.). Springer-Verlag, Berlin. pp. 1–18. With permission of Springer Verlag.

The common fibres are

- Carbon fibres (CFRP)
- Glass fibres (GFRP)
- Aramid fibres (AFRP).

Some physical and mechanical characteristics of FRPs are given in (Table 5.12) (Papayianni, 2014).

The most usual resins are the epoxy ones.

The most common FRP in practice for seismic repair and strengthening is the CFRP. They are basically used for

- Confinement of piers and columns
- Shear strengthening of walls
- Prestressing of walls at their ends to stabilise them against rocking.

The externally bonded laminates or sheets provide an unacceptable look, and this aspect presents a difficulty for their acceptance in the restoration of monumental buildings, unless they are covered with plaster or are used for nonvisible surfaces, like the extrados of arches, vaults or domical vaults.

5.3 CONCLUSION

From the above analysis, the following conclusions can be drawn:

i. The choice of materials for repair and strengthening of monuments and historical buildings is strongly influenced by the general principles of restoration embodied in the Venice Charter.
ii. Keynote for the choice of materials is the classification of the restoration techniques under two main categories:
   a. Reversible ones
   b. Irreversible ones.
iii. Materials used in reversible interventions put usually very few restrictions.

iv. On the contrary, materials used for irreversible interventions put the following two additional restrictions:
   a. Compatibility of the new materials with those of original fabric
   b. Durability of the new materials for a very long period of time.

v. The above-mentioned restrictions necessitate a deep knowledge of the original materials of the monuments under restoration. This requires extended *in-situ* and laboratory research.

vi. The knowledge of the original materials must be the guide for the choice of the materials for repair and strengthening, particularly for irreversible interventions.

vii. It is generally accepted that the best way to satisfy compatibility and durability is the choice of 'traditional' materials for restoration as there is no long-term experience on the behaviour of the modern ones.

viii. However, in many instances when tensile strength or posttensioning forces are required, the use of steel or other metallic alloys or even FRPs is unavoidable. In this case, a specialist should be consulted, so that corrosion and its consequences may be avoided.

ix. From all the above, it is concluded that the process of restoration is a highly specialised operation, and, as such, it calls for collaboration of specialists of many scientific disciplines such as archaeologists, architects, structural engineers, materials engineers etc. In order to accomplish their task, all of them must be strongly supported by well-equipped laboratories.

Chapter 6

In-situ investigations and laboratory tests

6.1 INTRODUCTION

The capacity assessment of an existing structure is based on the convolution of two main sets of knowledge: the knowledge of the characteristics of the existing structural system on one hand and that of the structural behaviour of the system under expected loading on the other, based on structural analysis and design.

For the formation of the second set of knowledge, however, it is necessary to obtain in advance the first one. Particularly for historical buildings, which have survived for centuries, this information of its structural system is of major importance and not easily acquired, as it refers to the long history of the building, and therefore, it is based on *in-situ investigation, historical documentation* and *laboratory tests.*

In more detail, the investigation of the structural system of a historical building includes the following steps:

i. Investigation of existing files, drawings and written evidence of the various phases of the monument
ii. Visual inspection supported by the above information
iii. Geometrical survey
iv. Survey of existing damage (pathology)
v. *In-situ* destructive tests (DTs) and nondestructive tests (NDTs)
vi. Laboratory tests
vii. Monitoring of the time history of critical defects (crack width, drifts etc.)
viii. Soil investigation.

The extent of the above investigation and tests depends mainly on the importance of the monument and the seriousness of the damage.

It should be noted that visual inspection is crucial for planning of site investigation and laboratory tests and, therefore, must be carried out with special care.

6.2 INVESTIGATION OF EXISTING FILES

It should be noted from the beginning that this investigation is carried out mainly by the archaeologist or the architect who is responsible for the building's restoration. The structural engineer in charge studies all this material and tries to extract useful information for the structural system from a careful interpretation of the existing material. This material usually includes (Croci, 2007; UNDP/UNIDO, 1984):

- Drawings from the corpus of the monument or the listed building under consideration
- Written or pictorial evidence for the various phases of the monument's history
- Information about eventual damaging events like fire, earthquakes, war damage, soil settlements, floods etc.
- Old travel diaries and guides
- Archaeological reports or archaeological studies of the past
- Official accounts referring to restoration, modification activities etc.

All this information must be considered as useful documentation for the structural restoration provided that it is cross-referenced with site evidence or at least it does not contradict this.

6.3 *IN-SITU* INVESTIGATION

6.3.1 Visual inspection

After an accident that has caused damage to a monumental building, like a strong earthquake or a fire, or even after the decision for an eventual structural restoration of a monumental building that has presented structural defects due to long-term causes like soil settlements, decay of timber elements etc., the first step of the restoration procedure is a detailed visual inspection on site.

The first inspection is usually carried out having as a guide existing drawings of the monument or even sketches (UNDP/UNIDO, 1984). On these drawings, the basic cracks and the drifts of the walls and piers, i.e. all important information about deformations and dislocations of the monument, are plotted. A good engineering knowledge, together with reports of gathered information about the behaviour of monument's structural elements and structural systems prepared in this stage, might possibly be the best way to trace the points where damage is expected and to make a correct interpretation of this damage from the structural point of view. At the same time, an extended photographic documentation is carried out combined with existing drawings or prepared sketches for the case. The structural engineer, based on the previously mentioned background and on his engineering judgement, tries to evaluate the significance of the damage and *to assess the necessity or not of temporary shoring*. Very useful material for his work are undoubtedly existing documents of the condition of the monument in the past, i.e. cracking patterns, photographs, information about drifts etc. Usually, such information is not available in many cases, as it presupposes continuous structural survey of the monument, a process which has been established only for the most important ones.

The above visual inspection is a very important step for the whole procedure, because it constitutes the basis, for the planning of the extended and in-depth *in-situ* investigation and laboratory tests and for the localisation of the critical regions with significant damage or structural importance, where the site investigation and laboratory tests should be focused.

6.3.2 Geometrical and constructional survey

For the geometrical survey of the superstructure of a monument, the standard methods of surveying are usually used, as well as the architectural surveying procedures, the method of photogrammetric surveying, and the Laser Scanning processing (Shan, J. and Toth, C., 2018). For the constructional surveying of the superstructure, i.e. the determination of domes and vaults thickness, the determination of roofing material etc., usually sections are made. Sometimes, if only thickness is required, radio-ray measurements may be carried out to avoid interventions on the monument's body (UNDP/UNIDO, 1984).

For the geometrical and constructional surveying of foundations, sections are made through the ground at the immediate neighbourhood of the foundation and down at the level where it is resting. Ground penetrating radar studies may characterized as a useful tool for a rather indicative investigation of the foundations without excavations.

At this stage of investigation, the photographic survey is complemented and combined with survey drawings with reference on them of the register number of the photo and the point and direction of shots (Figure 6.1). At the same time, the building materials and the masonry texture are described and designated at the area they refer to.

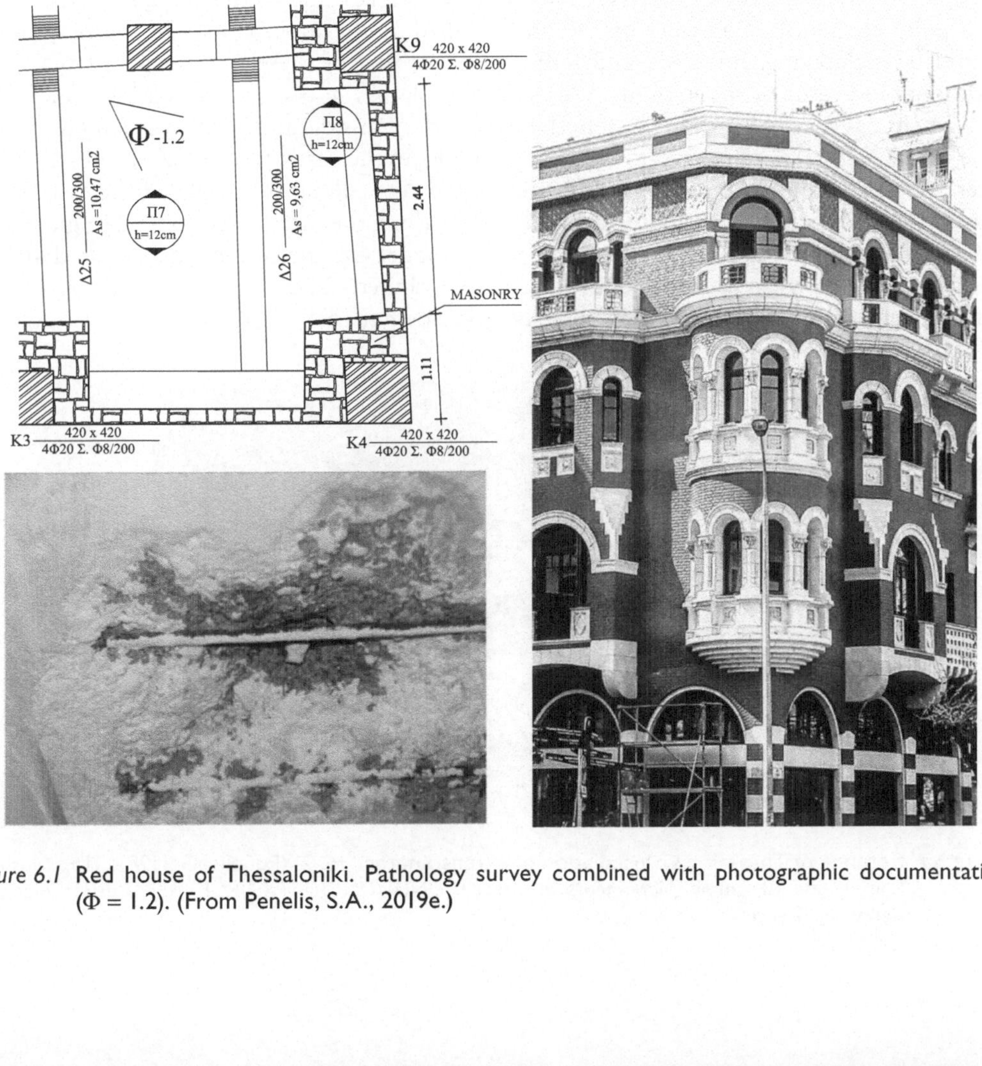

Figure 6.1 Red house of Thessaloniki. Pathology survey combined with photographic documentation (Φ = 1.2). (From Penelis, S.A., 2019e.)

6.3.3 Survey of damages and deformations

The survey of damage should include, first, a detailed surveying of the cracks, their width, the relative displacement of their edges both in the direction of the crack and vertically to it, as well as their depth. It should also include a photographic surveying of the main cracks. The survey of the crack pattern is made usually by means of geometric or photogrammetric methods. The depth is determined using special rods that are forced into the cracks. If the cracks do not pass through the walls, ultrasonic tests may be used for the determination of their depth. The survey of the cracking pattern is multipurpose:

- It serves as a document for the future condition of the monument in relation to its condition prior to the repair.
- It provides better conditions for studying the methodology of repair of each crack.
- It provides the possibility to have a rough estimate of the grouting work that will be required.
- The most important, from the structural point of view, it provides a very useful visible indication of the cause of cracking, the trend to collapse and the suitable repair procedure.

Together with the survey of damage, the continuous recording of the changes of the crack widths by means of velometers or LVDTs installed at the main cracks of the monument should be included (Figures 6.2 and 6.3). The recording of these movements serves as an additional indicator of the monument's trend to collapse. It also serves as a basic indicator for the effectiveness of the final strengthening under consideration if this is similar to the type of an eventual temporary shoring. For example, in the case of Rotunda of Thessaloniki, after the establishment of temporary posttensioned rings along the height of the cylindrical part of the building, the development of the cracks along the meridians stopped (Figure 6.3).

Finally, in the procedure of surveying the damage, the measurements of the drifts of the piers and walls from the vertical should be included. These measurements should be carried out at several points along their height, so that their deflection curve is derived.

Figure 6.2 Rotunda of Thessaloniki: installation of extensometer No. 2. (Penelis, G. (1982). *The Masonry Roman and Byzantine Monuments in Greece. UNDP/UNIDO, PROJECT (RER/79/015) National Report of Greece.*)

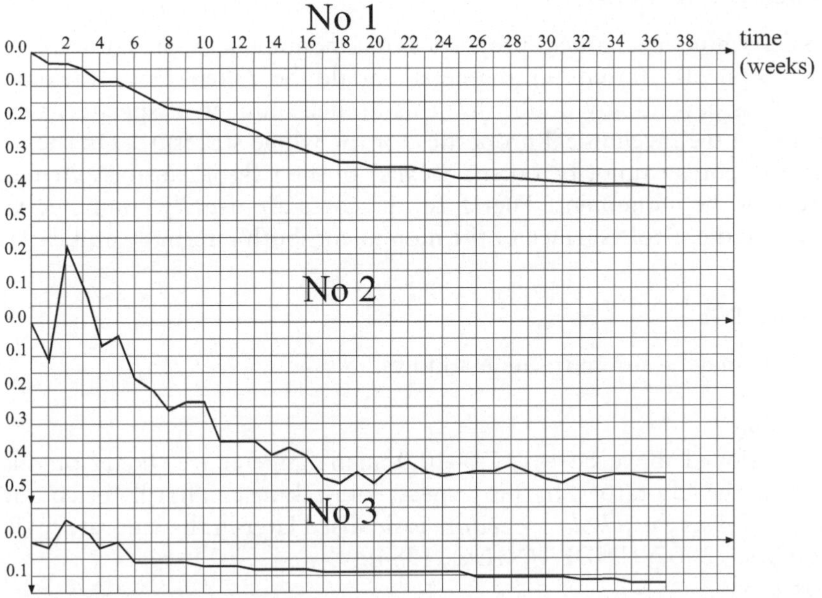

Figure 6.3 Rotunda of Thessaloniki: time-history recording of widening of cracks in mm per week. The stabilisation of the cracks occurred at the time of installation of temporary external post-tensioned steel rebars. (Adapted from Penelis, G. (1982). *The Masonry Roman and Byzantine Monuments in Greece. UNDP/UNIDO, PROJECT (RER/79/015) National Report of Greece.*)

The shape of this curve is used as an additional indication of the way in which the structural system is functioning in relation to its foundation, namely whether the deformations are due to a rotation of the foundation or to a deformation of the superstructure caused by horizontal thrusts (Figure 6.4).

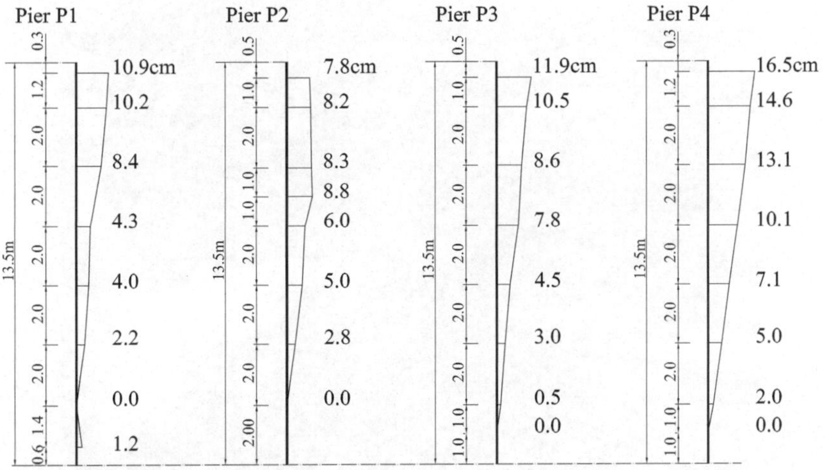

Figure 6.4 Rotunda of Thessaloniki: horizontal outward displacements of piers. (Adapted from Penelis, G. (1982). *The Masonry Roman and Byzantine Monuments in Greece. UNDP/UNIDO, PROJECT (RER/79/015) National Report of Greece.*)

6.3.4 *In-situ* DTs/NDTs

In order to determine mechanical properties of a monument in the laboratory, a limited amount of cores 7.0–10.0 cm in diameter should be taken from various locations (Figure 6.5). It is beyond any doubt that this limited number of specimens cannot give credible values of the mechanical properties for the whole structure. On the other hand, a big number of cores cannot be taken, because of the risk of causing extended damage to the texture of the monument. Therefore, the cores are usually combined with NDTs which extend on the whole surface of the monument, both internally and externally.

6.3.4.1 *In-situ DTs*

The *in-situ* DTs may be summarised as follows:

i. *Core-taking*
   As mentioned above, cores of 7.0–10.0 cm in diameter are taken from the masonry for tests in the laboratory. In this respect, core-taking can hardly be characterised as an *in-situ* DT as the *whole procedure* is finalised later in the laboratory (Penelis, G., 1982; Croci, 2007; UNDP/UNIDO, 1984).
ii. *Endoscopy*
   This method requires the drilling of small diameter holes (about 25 mm) to even tens of metres deep for the investigation of the interior of a structure. This small specimen can also serve as a material for testing in the laboratory. A mobile instrument in the form of a bar with a head composed of two lenses can estimate distances with accuracy, thanks to the stereoscopic effect of the lenses used for the investigation (Vintzileou, et al., 2005). The main advantage of endoscopy is that it allows access to nonvisible points inside the masonry body in a natural way providing visual and diagnostic information (Figures 6.6 and 6.7).

Figure 6.5 Core-taking. (From Penelis, G. (1982). *The Masonry Roman and Byzantine Monuments in Greece. UNDP/UNIDO, PROJECT (RER/79/015) National Report of Greece.*)

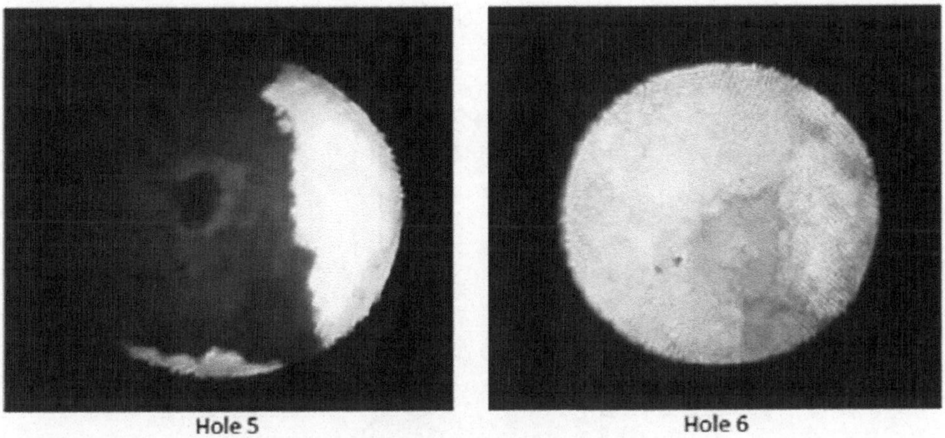

Figure 6.6 Galata Greek School renovation and reuse (Istanbul). Project: endoscopy in a depth of 90 cm. (Courtesy of Omer, U. (2019). Ulker Consulting Engineers. Archives of Galata Greek School renovation and reuse in Istanbul.)

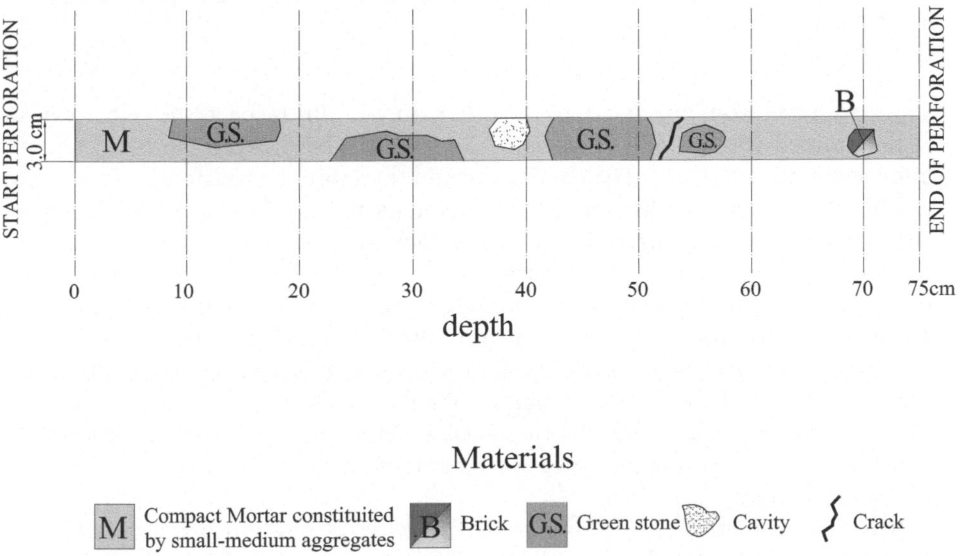

Figure 6.7 Typical stratigraphic section resulting from endoscopy.

6.3.4.2 *In-situ* NDTs

The *in-situ* NDTs are summarised as follows:

i. *Ultrasonic tests*

This test is based on the concept of introducing an ultrasonic longitudinal wave on the face of a masonry wall, using an impulse generator, and receiving the sonic wave with a piezoelectric receiver placed at a defined distance on the same wall. Both devices are connected to an oscilloscope which measures the time required for the

Figure 6.8 Rotunda of Thessaloniki: ultrasonic measurements. (From Penelis, G. (1982). *The Masonry Roman and Byzantine Monuments in Greece. UNDP/UNIDO, PROJECT (RER/79/015) National Report of Greece.*)

sound to travel from one device to the other, and, in this respect, the velocity of sound is determined (Figure 6.8) (Penelis, G., 1982; UNDP/UNIDO, 1984). These tests allow estimating the stiffness, the density and therefore the resistance of the masonry.

The correlation of ultrasonic measurements to the corresponding compressive strengths determined through DT at the locations of core-taking did not provide satisfactory results (correlation ratio $r < 0.500$) (Penelis, G., 1982). So, according to the authors' opinion based on extended investigations, it is rather incredible to correlate the compressive strength to the ultrasonic measurements. Anyhow, these measurements *give very good correlation results, concerning the dynamic modulus of elasticity*; therefore these measurements are used in the statistical evaluation of the dynamic and, indirectly, the static modulus of elasticity. Additionally, this method may be used for the detection of cracks and their depth.

ii. *Hammer testing*

It is a sclerometric type of test which provides a relation of the rebound of the hammer of a relevant device with the compressive strength of the mortar of the masonry (Figure 6.9) (Penelis, G., 1982). An alternative procedure is based on the mortar resistance to the penetration of a nail. In this device, the number of blows is correlated with the degree of penetration of the nail.

In the case of Rotunda of Thessaloniki, a very satisfactory correlation was obtained between the hammer test results and the corresponding measurements of compressive strength in the laboratory (correlation ratio $r = 0.867$). This correlation ratio was used for the statistical evaluation of mortar strength of the monument which was based on hammer test results (Table 6.1).

iii. *Flat-jack tests*

The jack consists of a plane metallic reservoir with two openings for the fluid input and output. Jacks are inflated with oil through a pumping system instrumented with

Figure 6.9 Hammer testing. (From Penelis, G. (1982). *The Masonry Roman and Byzantine Monuments in Greece. UNDP/UNIDO, PROJECT (RER/79/015) National Report of Greece.*)

Table 6.1 Rotunda of Thessaloniki: Hammer test results

Construction phase	*Location of Tests*	*Number of Tests*	*Mean value μ* (kg/cm²)	*Standard deviation σ* (kg/cm²)	*Lower 5% fractile μ–1,645σ* (kg/cm²)	*Upper 5% fractile μ+1,645σ* (kg/cm²)	*Minimum value* (kg/cm²)	*Maximum value* (kg/cm²)
Roman	Π_1	54	28.33	4.61	20.75	35.91	17.80	39.40
	Π_2	54	29.96	6.91	18.59	41.33	19.19	44.94
	Π_3	54	26.04	5.54	16.92	35.16	15.16	42.93
	Π_4	54	32.36	5.89	22.67	42.05	15.16	42.93
	Π_5	54	29.67	6.64	18.75	40.59	15.16	49.08
	Π_6	54	28.23	6.35	17.78	38.68	17.13	39.99
	Π_7	54	28.13	6.44	17.54	38.72	13.90	39.04
	Π_8	54	27.34	7.57	14.89	39.79	11.51	48.03
	Top	144	32.44	6.34	22.01	42.87	15.16	49.08
	Mid-height	144	28.27	5.04	19.98	36.56	17.80	39.04
	Bottom	144	25.56	6.13	15.48	35.64	11.51	42.93
	Inside	216	27.36	5.25	18.72	36.00	15.16	42.93
	Outside	216	30.16	7.29	18.16	42.16	11.51	49.08
	Total	432	28.76	6.50	18.07	39.45	11.51	49.08
Christian	Top	54	36.96	11.60	17.87	56.05	15.81	60.16
	Mid-height	54	37.39	11.15	19.05	55.73	19.19	59.00
	Bottom	60	39.88	14.32	16.33	53.43	13.90	69.77
	Inside	72	33.91	11.25	15.40	52.42	13.90	59.00
	Outside	96	41.32	12.51	20.74	61.90	15.81	69.77
	Total	168	38.14	12.50	17.57	58.71	13.90	69.77

Source: Adapted from Penelis, G., Karaveziroglou, M., and Papayianni, I. (1989). Materials used in the restoration of historical buildings and monuments. *Seminario per Conservao Recuperação e consolidacão de monumentos e edificios históricos*, Conselho Directivo National (ed) Lisbon.

the necessary gauges for pressure measurements (Gregorezyk and Lourenço, 2000; Cucci, et al., 1995). These devices are used for the evaluation of the present state of compressive tress in a given masonry zone, the compressive and shear resistance, the stress–strain relationship and related modulus of elasticity.

One or two flat jacks or one cylindrical hydraulic jack may be used for the execution of the following tests:

- Relaxation test (existing compressive stress)
  Equipment: one flat jack and three deformometers (Figures 6.10 and 6.11)
  Procedure: Measure the distance between the deformometer points.
  Make one cut.
  Insert the flat jack (Figure 6.10).
  Measure the distance between the deformometer points (measure the displacement at the cut).
  Apply pressure into the flat jack until the displacement at the cut is restored (Figure 6.11).
  Elaborate the relevant P–δ diagram.

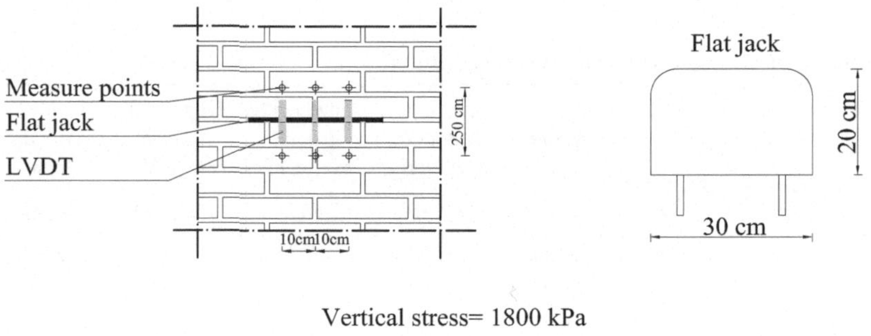

Figure 6.10 Typical relaxation test (estimate of the existing compressive stress).

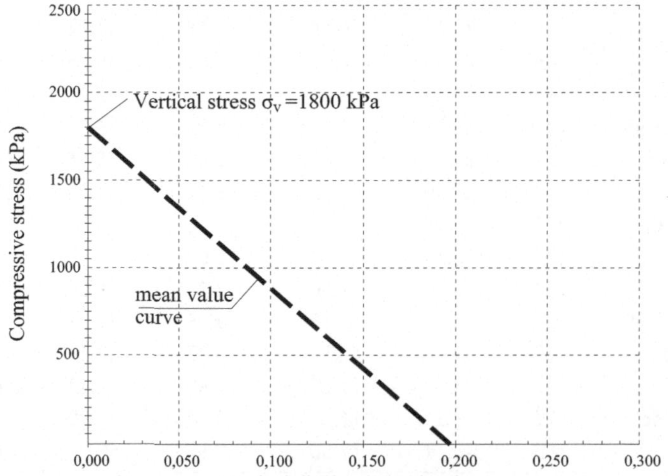

Figure 6.11 Typical pressure–displacement diagram of relaxation test.

- Compressive strength measurement of masonry
  Equipment: two flat jacks and three deformometers (Figures 6.12 and 6.13)
  Procedure: make two cuts.
  Install two flat jacks and three deformometers.
  (Figure 6.12)
  Apply pressure to the flat jacks until the masonry fails.
  Elaborate the relevant P–δ diagram (Figure 6.13).
- Shear strength measurement
  Equipment: one cylindrical jack and one deformometer (Figures 6.14 and 6.15)
  Procedure: Install the cylindrical jack.
  Install the deformometer.
  Estimate the vertical compressive stress on the wall (Figure 6.14).
  Apply pressure into the jack until masonry failure.
  Elaborate the P–δ diagram (Figure 6.15).

It may be easily concluded that the above procedure causes only small local defects and, in this respect, it is characterised as an NDT or a semidestructive *in-situ* test.

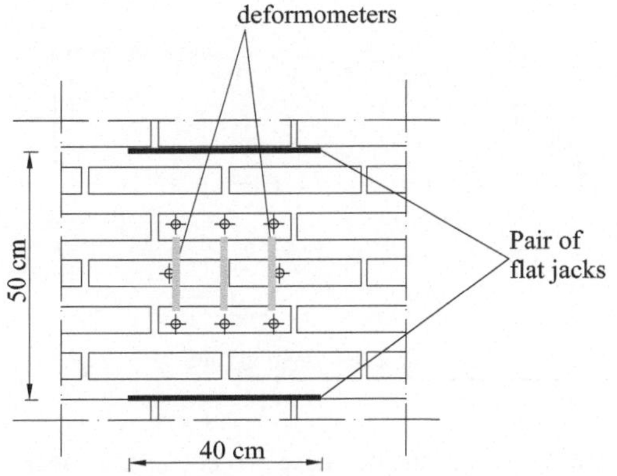

Figure 6.12 Typical compressive test. Estimate of the *in-situ* masonry compressive strength.

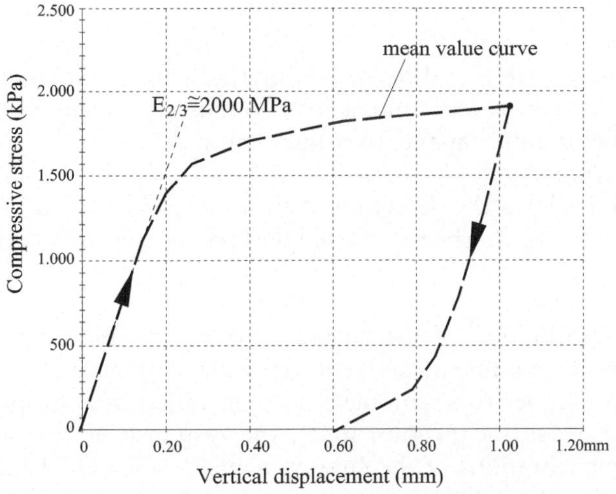

Figure 6.13 Typical pressure–displacement diagram of an *in-situ* masonry flat-jack test.

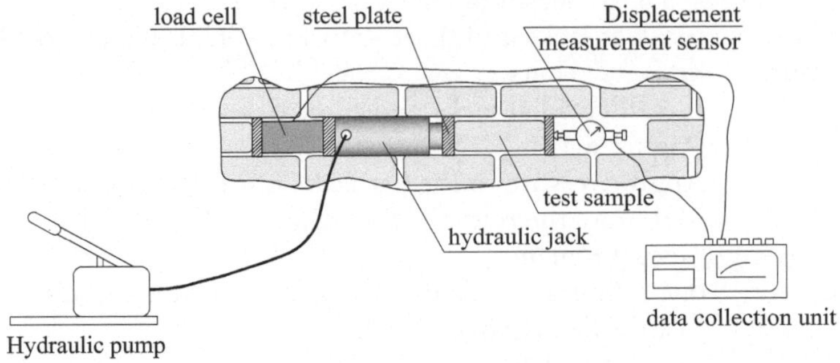

Figure 6.14 Galata Greek School renovation and reuse project: masonry shear setup (Courtesy of Omer, U. (2019). Ulker Consulting Engineers. Archives of Galata Greek School renovation and reuse in Istanbul.)

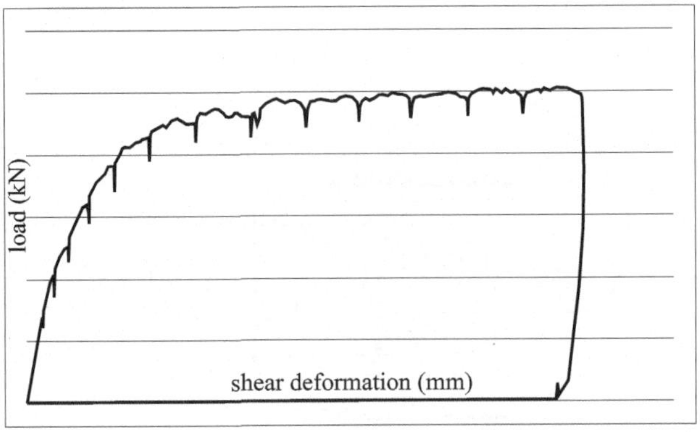

Figure 6.15 Galata Greek School renovation and reuse project: *in-situ* shear test load–displacement plot. (Courtesy of Omer, U. (2019). Ulker Consulting Engineers. Archives of Galata Greek School renovation and reuse in Istanbul.)

iv. *Thermography*

Thermal images are useful for providing the weak points in buildings. They can be used to check construction integrity in buildings, as well as to provide valuable information before, during and after interventions.

v. *Electromagnetometry*

This method allows the detection of iron or steel items in masonry of concrete up to a depth of 5.0 cm. The most usual devices for this measurement are known as 'pachometers'.

vi. *Dynamic tests*

For the determination of the natural period, the dumping coefficient and the eigenmodes of a monument, ambient vibration measurements are facilitated. For this reason, a number of seismometers are installed in appropriate positions, and using electronic data acquisition units, the response of the building is recorded due to ambient vibrations of the environment (Penelis, G., 1982). The records are then evaluated using a fast Fourier transform device, and the output is used for comparisons with the analytical calculation of the dynamic characteristics of the

Figure 6.16 Rotunda of Thessaloniki: seismometer for ambient vibration measurements. (From Penelis, G. (1982). *The Masonry Roman and Byzantine Monuments in Greece. UNDP/UNIDO, PROJECT (RER/79/015) National Report of Greece.*)

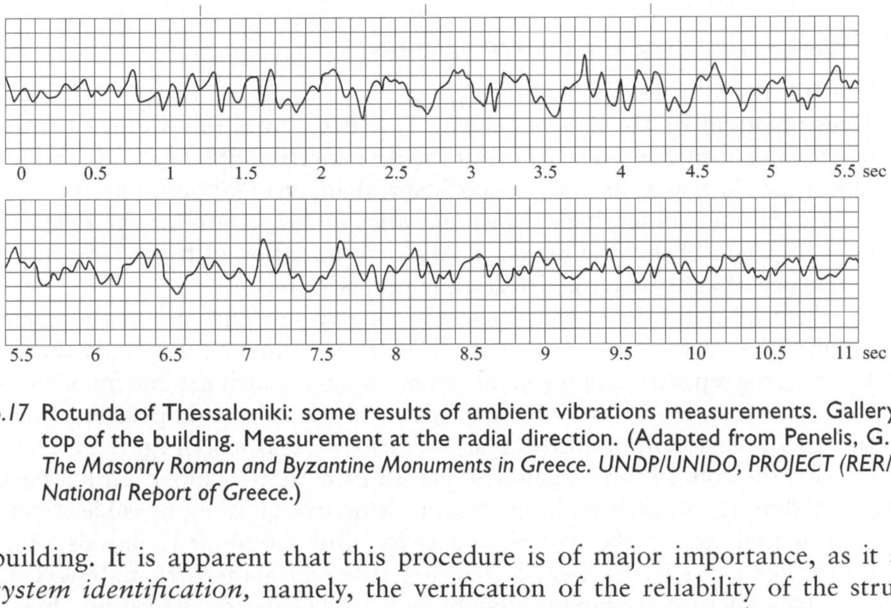

Figure 6.17 Rotunda of Thessaloniki: some results of ambient vibrations measurements. Gallery at the top of the building. Measurement at the radial direction. (Adapted from Penelis, G. (1982). *The Masonry Roman and Byzantine Monuments in Greece. UNDP/UNIDO, PROJECT (RER/79/015) National Report of Greece.*)

building. It is apparent that this procedure is of major importance, as it allows *system identification,* namely, the verification of the reliability of the structural model that has been introduced in the analysis (Figures 6.16 and 6.17).

Forced vibration measurements are of greater significance, but they could only be performed if the monument is not liable to collapse.

6.3.5 Laboratory tests of original masonry materials

6.3.5.1 Introduction

The assessment of strength and response to dead loads or seismic excitation calls for knowledge of the density, unit strength, moduli of elasticity etc. of the materials. On the other hand, for the purpose of assessing the long-term suitability of materials, such as grouts used for strengthening and repair, knowledge of the physical, mineralogical and chemical composition of the original materials is also required.

The engineer responsible for strengthening or repairing of an old existing structure is confronted with three difficulties:

- There may be difficulties in extracting samples suitable for testing, in case the archaeological authorities don't approve such extractions.
- There is likely to be less homogeneity and uniformity than those in conventional constructions, so the overall strength, related to the unit strength of the constituent materials, will be less predictable.
- There may be considerable variation in the properties of a particular type of material.

The extraction of samples for physical, mineralogical and chemical tests is essential, but the samples for this purpose can be small enough to present problems with the archaeological authorities. For reliable values of mechanical properties, however, larger specimens are essential for standard laboratory DTs specified by codes. It may, therefore, be necessary (either in place of these tests or alongside them) to carry out *in-situ* NDTs, the details of which are provided in Subsection 6.3.4.2.

6.3.5.2 Physical, chemical and mineralogical tests

Physical, chemical and mineralogical tests *for stone or marble* aim at the determination of the places of origin of the raw materials. This is of great importance for the local rebuilding of walls or facades. The grade of the deterioration of stone and marble due to atmospheric attack is also a task of physicochemical analyses (Skoulikidis, 1977).

Physicochemical and mineralogical tests on *ancient bricks* aim at the determination of the places of origin of the raw materials, the determination of firing temperature, the eventual existence of additives, the density, the porosity etc.

Physical, chemical and mineralogical tests on *ancient mortars* aim at the quantitative determination of the constituents of the mortars. This determination is not easy at all. Several problems arise in the separation of the binder from the aggregate if the fine fragment consists of the same main minerals as those of binding agent. It is also almost impossible to determine the constituents of the organic additives that were used such as blood, eggs, milk of figs etc.

Furthermore, the conventional chemical quantitative analysis alone cannot ensure the existence of hydraulic components in the mortar. Moreover, it is not possible to determine the component that caused the hydraulic activity. Only combined analysis of mortars using different methods such as X-ray diffraction, differential thermal analysis (DTA) and scanning electron microscopy can give reliable results. The mechanical strength must also be taken into account. From the above discussion, it is concluded that a great number of tests are required either for stones and bricks or for ancient mortars:

- Conventional chemical analyses for quantitative determination of constituents (Tables 6.2 and 6.3)
- Tests for determination of physical properties, e.g. density, water absorption, apparent porosity, suction rate, efflorescence etc.
- Test for mineralogical composition:
  - X-ray diffraction analysis for the determination of crystal phase of existing minerals (Figure 2.6)

Table 6.2 Rotunda of Thessaloniki: Chemical analysis of the Roman period bricks

	Determined value		
Composition	*(gr)*	%	%
Burning loss	1.144	7.00	
Insoluble compounds	11.108	67.97	
Soluble compounds	(4.091)	(25.03)	Composition of compounds
SiO_2	1.55	9.48	37.89
R_2O_3	1.40	8.57	34.22
CaO	0.884	5.40	21.61
MgO	0.065	0.40	1.59
SO_3	0.012	0.07	0.29
Na_2O	0.021	0.13	0.51
K_2O	0.135	0.83	3.30
Not determined	0.024	0.14	0.59
Total	16.343	100.0	100.0

Source: Adapted from Penelis, G., Karaveziroglou, M., and Papayianni, I. (1989). Materials used in the restoration of historical buildings and monuments. *Seminario per Conservao Recuperaçäo e consolidacäo de monumentos e edificios histöricos*, Conselho Directivo National (ed) Lisbon.

Table 6.3 Rotunda of Thessaloniki: Chemical analysis of the mortar of the Roman period

	First analysis with coarse ceramic material			*Second analysis without coarse ceramic mater*		
	Determined values			*Determined values*		
Composition	*(gr)*	%	%	*(gr)*	%	%
Burning loss	2.672	12.78		4.3045	21.52	
Insoluble compounds	13.852	66.30		8.989	44.94	
Soluble compounds	(4.372)	(20.92)	Composition of compounds	(6.71)	(33.54)	Composition of compounds
SiO_2	0.540	2.58	12.35	0.640	3.20	9.53
R_2O_3	0.673	3.22	15.39	0.775	3.87	11.55
CaO	2.988	14.31	68.36	5.10	25.48	76.00
MgO	0.107	0.51	2.44	0.1175	0.59	1.75
SO_3	0.029	0.14	0.67	0.029	0.14	0.44
Na_2O	-	-	-	0.0060	0.03	0.09
K_2O	-	-	-	0.0400	0.20	0.59
Not determined	0.035	0.16	0.79	0.0025	0.03	0.05
Total	20.896	100.00	100.00	20.003	100.00	100.00

Source: Adapted from Penelis, G., Karaveziroglou, M., and Papayianni, I. (1989). Materials used in the restoration of historical buildings and monuments. *Seminario per Conservao Recuperaçäo e consolidacäo de monumentos e edificios histöricos*, Conselho Directivo National (ed) Lisbon.

- Scanning electron microscopy for the study of microstructure of the material and identification of the crystal ingredients found by X-ray diffraction analysis (Figure 6.18)
- DTA analysis which helps to recognise the mineralogical phases from their changes during heating (Figure 6.19).

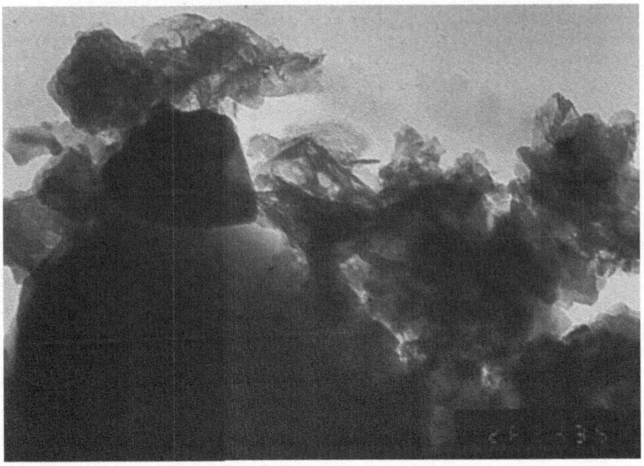

Figure 6.18 Rotunda of Thessaloniki: scanning electron microscopy analysis. (Penelis, G., Karaveziroglou, M., and Papayianni, I. (1989). Materials used in the restoration of historical buildings and monuments. *Seminario per Conservao Recuperaçäo e consolidacäo de monumentos e edificios histöricos*, Conselho Directivo National (ed) Lisbon.)

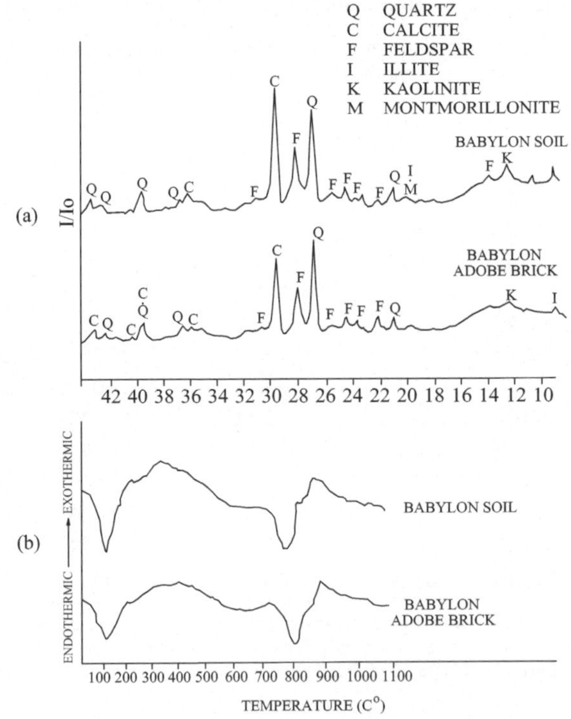

Figure 6.19 (a) X-ray diffraction patterns and (b) DTA curves for Babylon soil, adobe brick. (Adapted from Al-Kass, R., Hadi, M., Khalil, N., and Ali, F. (1989). Ancient Babylonian bricks: Their technological characterisation and simulation. *Proceedings of International Conference on Structural Repairs and Maintenance of Historical Buildings*. Brebbia, C., (ed.). Birkhäuser Verlag, Berlin. pp. 71–80.)

Table 6.4 Chemical, physical and mechanical properties of Babylon bricks

			Chemical analysis	
Physical properties	*Group 1*	*Group 2*	*Oxides*	*(%)*
Density (kg/m^3)	1,634	1,343	SiO_2	50.38
Water absorption (%)	19.7	29.6	Al_2O_3	13.79
			Fe_2O_3	4.14
Apparent porosity (%)	32.2	39.8	CaO	17.64
			MgO	4.08
Suction rate (kg/m^2. min)	1.2	3.4	SO_3	1.07
			Loss of Ignition (L.O.I)	6.95
Compressive strength (N/mm^2)	9.1	3.2	Total	98.05
			Main minerals	
Efflorescence	Nil to slight	Nil to slight	Quartz	
			Feldspar	
Colour	Pale red brown		Calcite	

Source: Adapted from Al-Kass, R., Hadi, M., Khalil, N., and Ali, F. (1989). Ancient Babylonian bricks: Their technological characterisation and simulation. *Proceedings of International Conference on Structural Repairs and Maintenance of Historical Buildings*. Brebbia, C., (ed.). Birkhäuser Verlag, Berlin. pp. 71–80.

It is interesting to give here the results of some case studies carried out in the laboratories in the University of Thessaloniki.

Tables 6.2 and 6.3 present the quantitative analyses of the bricks and of the mortar of Roman part of Rotunda, respectively (UNDP/UNIDO, 1984; Penelis, G., 1982). Figure 2.6 presents the X-ray diffraction analysis of the mortar of Rotunda of Thessaloniki. Figure 6.19 presents the scanning electron microscopy of the mortar of Rotunda.

It is also interesting to observe from Table 6.4 and Figure 6.19 some results of a combined effort for the determination of the properties of Babylonian bricks carried out by AI-Kass, et al. in the Building Research Centre of Iraq (Al-Kass, et al., 1989).

6.3.5.3 Mechanical tests

The most important mechanical properties of stones, bricks and mortars of a monument are the following:

f_c: compressive strength
f_{ct}: tensile strength in bending
E_o: modulus of elasticity
v: Poisson's ratio
ε_u: ultimate compressive deformation
τ_c: bonding of mortar and brick or stone
μ: coefficient of friction at the joint of mortar and brick or stone.

These properties are 'necessary for the assessment' or calculation of the mechanical properties of the original masonry and also for the choice of the restoration materials.

The methods and the specifications used for the determination of the above-mentioned properties are usually the same as those used for the modern materials, as there are not

Table 6.5 Rotunda of Thessaloniki: Bending tests of ancient mortars

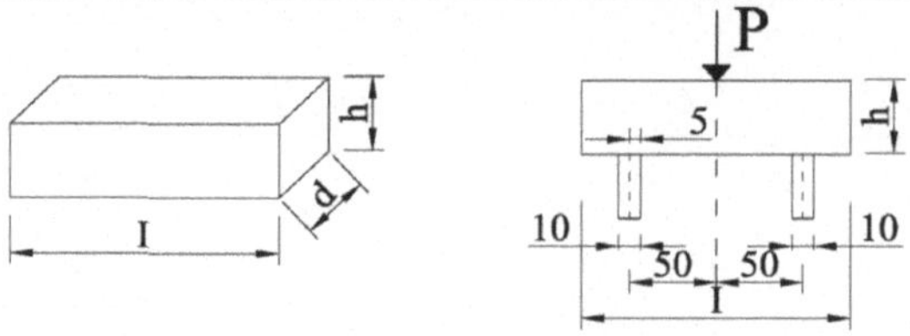

Origin			Dimensions		Failure load	Strength
Part	*Core Nr*	*Specimen Nr*	*n (cm)*	*D (cm)*	*P (kg)*	*σd (kg/cm2)*
Roman	5	1	3.03	4.55	14.60	3.49
	18	2	4.10	4.00	25.20	5.76
	19	3	4.06	3.94	32.10	7.64
	19	4	4.00	3.92	54.80	13.37
Christian	3	1	4.08	4.44	64.00	11.94
	3	2	4.03	4.50	28.10	5.16
	3	3	3.64	4.27	77.20	17.45
	3	4	3.56	4.43	74.00	15.89
	8	5	3.80	3.30	19.80	7.18
	8	6	3.50	4.10	20.20	5.15
	8	7	3.79	4.50	62.30	12.18
	8	8	3.50	4.50	43.50	9.21

Source: Adapted from Penelis, G., Karaveziroglou, M., and Papayianni, I. (1989). Materials used in the restoration of historical buildings and monuments. *Seminario per Conservao Recuperação e consolidacão de monumentos e edificios histöricos*, Conselho Directivo National (ed) Lisbon.

yet specifications for testing of ancient ones. *These methods are modified from case to case to comply with the form and the size of the extracted specimens.*

It is interesting to give here some information from our experience in testing mechanical properties of ancient materials in the laboratories of the University of Thessaloniki. Figure 6.5 depicts the core-taking procedure from Rotunda, Thessaloniki. Tables 6.5 and 6.6 give the results of the bending and compression tests of mortar specimens from the same monument. Finally, Table 6.7 depicts the bonding strength of mortar to the brick, while Figure 6.20 presents the relevant test procedure.

6.3.6 Laboratory tests of mortars and grouts in use for masonry restoration

It is obvious that mortar and grout mixes should be subjected to standard tests in order to confirm their compliance to the requirements of the conservation. These tests usually conform to the standards for cement mortars and grouts. Anyhow, most of the following tests should be carried out for mortars and grouts in use for masonry restoration:

- Quality control of materials
- Workability (for mortars only)
- Setting time

Table 6.6 Rotunda of Thessaloniki: Compression tests of ancient mortars

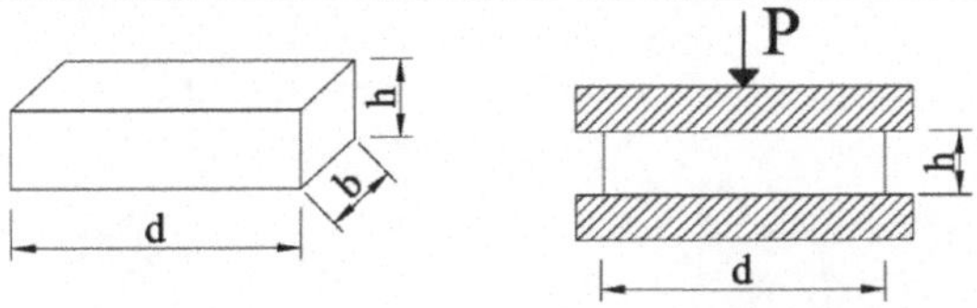

Origin			*Dimensions*		*Failure load*	*Strength*
Part	*Core Nr*	*Specimen Nr*	*b (cm)*	*d (cm)*	*P (kg)*	*σd (kg/cm²)*
Roman	5	1	3.03	7.40	510.00	22.75
	9	2	3.00	3.80	170.00	14.91
	12	3	4.40	9.35	570.00	13.86
	14	4	4.93	9.18	890.00	19.66
	16	5	4.10	4.80	330.00	16.77
	18	6	4.40	8.70	815.00	21.29
	18	7	4.30	6.56	700.00	24.82
	19	8	4.10	9.10	740.00	19.83
	19	9	4.10	6.90	580.00	20.50
Christian	3	1	4.08	6.02	815.00	33.18
	3	2	4.03	6.66	1,085.00	40.43
	3	3	3.90	4.16	835.00	51.47
	3	4	3.64	6.10	1,110.00	50.00
	3	5	3.60	5.57	1,115.00	55.61
	3	6	3.55	4.00	880.00	61.97
	8	7	3.90	7.50	1,650.00	56.41
	8	8	4.15	7.50	1,200.00	38.55
	8	9	3.50	6.90	600.00	24.84
	8	10	3.58	4.55	735.00	45.12
	8	11	3.50	6.09	950.00	44.57
	8	12	3.58	7.38	910.00	34.44

Source: Adapted from Penelis, G., Karaveziroglou, M., and Papayianni, I. (1989). Materials used in the restoration of historical buildings and monuments. *Seminario per Conservao Recuperação e consolidacão de monumentos e edifícios histöricos*, Conselho Directivo National (ed) Lisbon.

- Shrinkage
- Mechanical properties
- Adhesion to masonry units
- Thermal expansion coefficient
- Soluble salts
- Weathering resistance
- Water absorption and porosity
- Pore size distribution
- Fluidity (for grouts only)
- Bleeding (for grouts only)
- Segregation (for grouts only).

Table 6.7 Rotunda of Thessaloniki: Bonding strength between mortar and brick

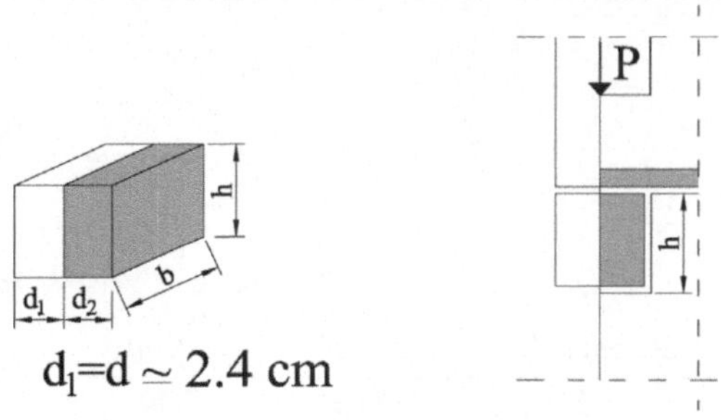

Origin			Dimensions		Failure load	Bond strength
Part	*Core Nr*	*Specimen Nr*	*b (cm)*	*h (cm)*	*P (kg)*	*σl (kg/cm²)*
Roman	1	1	5.20	6.00	54.0	1.73
	9	2	4.93	4.95	55.0	2.25
	9	3	4.91	4.50	40.0	1.81
Christian	2	1	5.35	5.46	78.0	2.67
	8	2	5.00	4.80	55.0	2.29

Source: Adapted from Penelis, G., Karaveziroglou, M., and Papayianni, I. (1989). Materials used in the restoration of historical buildings and monuments. *Seminario per Conservao Recuperaçäo e consolidacäo de monumentos e edificios histöricos*, Conselho Directivo National (ed) Lisbon.

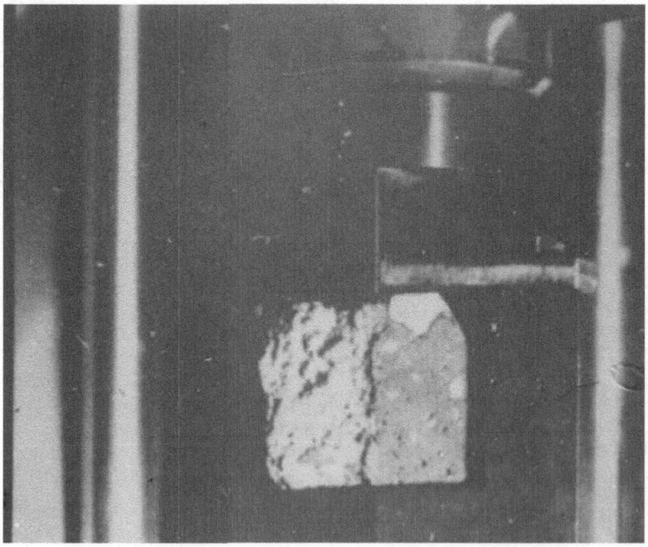

Figure 6.20 Rotunda of Thessaloniki: bonding test. (Penelis, G., Karaveziroglou, M., and Papayianni, I. (1989). Materials used in the restoration of historical buildings and monuments. *Seminario per Conservao Recuperaçäo e consolidacäo de monumentos e edificios histöricos*, Conselho Directivo National (ed) Lisbon.)

It is self-understood that, in the case of monuments of minor importance, this wide range of tests may be limited to the most important ones. It should also be mentioned that, in the event that a wide range of mixes are designed, the results can be useful for more than one monument, since it is well known that, in monuments of the same period, similar techniques and materials were used.

6.3.7 Laboratory tests on models

Sometimes thoughts may prevail of the construction of scaled walls or piers (usual scale 1:3) in the laboratory for loading a model with axial load or axial load combined with shear up to failure. In this way, we have the opportunity of combining the failure load with the strength of the mortar and the bricks for both the model and the actual monument in order to determine more precisely the strength of the compound masonry. This approach is accompanied by strong drawbacks due to the high cost and the longer time required for the program. Therefore, the assessment of the masonry strength is usually limited to indirect analytical methods as those presented in Chapter 3 which combine the strength of the mortar joint and units with their geometrical properties.

In case of very important monuments (e.g. Hagia Irene of Istanbul), physical models are tested on a seismic table for the verification of analysis and design calculations. The scale of the model depends on the capacity of the seismic table and could be about one-tenth of the actual monument. It should be noted that such a procedure presupposes high technical expertise for the model preparation, the estimate of model masses and their proper application, the seismic excitation and, finally, the interpretation of the results.

6.3.8 Soil tests

Where existing information for the soil properties is insufficient to take local soil conditions into account, particularly in estimating seismic loading, boreholes should be drilled to determine the stratigraphy and the water table level. Samples from each soil layer should be analysed in the laboratory for the determination of grain content, mass density, void ratio, moisture content, triaxial strength etc. Additionally, standard penetration test (SPT) or other appropriate *in-situ* tests could be executed for the assessment of the soil's bearing capacity.

In case of serious monumental buildings, additional tests might be executed, essentially useful for the assessment of seismic properties of the ground under consideration. The most usual are the following:

a. *Cross-hole seismic test*
   It is implemented for the determination of the velocity of seismic waves between boreholes at the successive soil layers. The final product of a cross-hole survey is a model of S-wave velocity (Figures 6.21 and 6.22) (Geometrics, 2018).
b. *Multichannel analysis of surface waves (MASW)*
   This method, based on Rayleigh waves (surface waves) analysis, allows the construction of a V_s 'shear wave' vertical profile (Geometrics, 2018). In parallel, this method, using the same equipment, enables the stratigraphic survey of the soil under investigation.

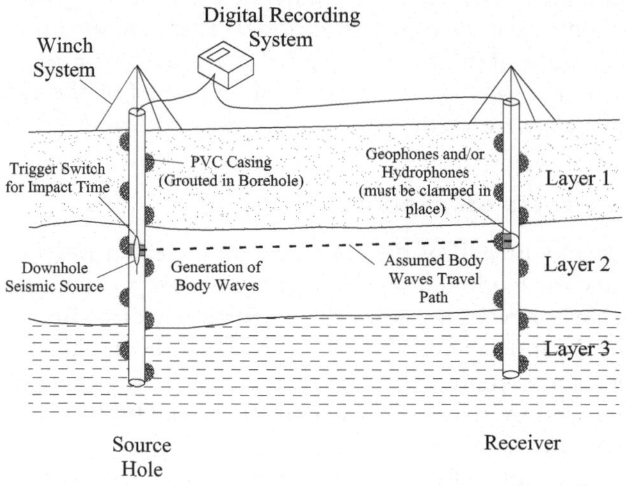

Figure 6.21 General layout of cross-hole testing. (Adapted from GEOMETRICS (2018). Crosshole seismic testing. www.geometrics.com/applications/geophysical-methods/cro. pp. 1–5.)

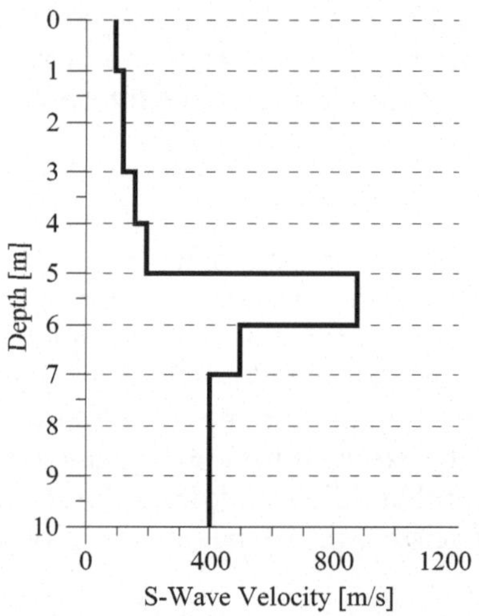

Figure 6.22 Cross-hole testing: S-wave velocity diagram (m/s). (Adapted from GEOMETRICS (2018). Crosshole seismic testing. www.geometrics.com/applications/geophysical-methods/cro. pp. 1–5.)

Chapter 7

Structural analysis and design

7.1 TASK OF ANALYSIS AND DESIGN

Analysis and design are the basic tools for the safety evaluation of new or existing structures. Particularly, for monumental buildings, analysis and design are the basic tools in two successive steps of structural restoration:

i. *Safety evaluation* of the monumental building as a whole and of its structural elements, *before any intervention.* This step is also called *structural verification.*
ii. *Safety evaluation* of the monument and its structural elements, taking into account the intervention scheme under consideration for strengthening. This step is also called *reanalysis and redesign.*

It is apparent that the same procedure is also followed in repair and strengthening of the structural system of existing conventional buildings. The main difference between monumental buildings and conventional ones is attributed to the following issues:

- Masonry being the main building material of historical buildings presents low and *unreliable tensile and shear strength.*
- It presents a highly orthotropic behaviour defined by the bed joint orientation.
- The spatial structural form of the majority of the historical buildings with curved structural elements such as domes, vaults, arches and thick walls or piers, together with the above two issues, make the structural modelling and the consecutive analysis and design difficult tasks.

These characteristics, and mainly the low tensile strength, result in the development of failure zones (cracking) for a very low percentage of the loading. The redistribution of the stresses, as a result of the crack evolution, leads to an early rearrangement of the load paths step by step up to final collapse. All the above, together with the requirements for soft interventions and, therefore, analysis and design of high accuracy have led to radical reconsideration of the concept of linear analysis and design which constitutes the basis for structural verification or reanalysis and redesign of all conventional structures, even those of masonry. So, since early 1970s, *inelastic methods*, either in the form of a *limit-state method* or in the form of a *step-by-step method* have been developed aiming at incorporating the early cracking in the analytical model.

Additionally, it should be noted that, even for structural evaluation of existing conventional structures subjected to seismic actions, *the displacement-based design* has been recently introduced *as reference method* which has as a prerequisite the elaboration of the *capacity curve* of the structure obtained as a result from a *pushover analysis*, namely from an *inelastic static step-by-step analysis* till failure, for the combination of dead loads and the seismic actions simulated by horizontal forces distributed along the height of the building.

Taking the above into account, it seems to be inevitable in seismic regions for the structural engineer to be in position to produce the relevant capacity curve, also for monumental buildings, so that a displacement-based design be implemented in compliance with the code requirements for all buildings, and, in this respect, the inelastic analysis for monumental buildings has become indispensable.

Closing, we may summarise below the methods of analysis in use on their own or in combination one with another for the verification and redesign of monumental buildings:

- Static elastic analysis for systems of frames or of 2D elements (walls–shells)
- Dynamic elastic analysis for the above type of structures
- Static inelastic analysis using the step-by-step method for the determination of the collapse mechanism and loads for frame or frame equivalent systems and systems of 2D elements (walls–shells)
- Limit-state methods using the upper- and lower-bound theorems (special cases)
- Dynamic inelastic analysis (time history analysis) in special cases of limited structural complication.

A special reference should be made here to the paper of Roca et al. (2010), where an extended review of methods applicable to the study of masonry in historical buildings, encompassing both classical and advanced ones, is presented.

7.2 HISTORICAL NOTES

The year 1743 may be characterised as the *turning point* in the design of structures. In fact, it was the first time in human history that mechanics was called for a quantitative scientific support of an assessment and redesign of a structure namely Saint Peter's dome in Rome that suffered due to extended and dangerous cracks (Figure 7.1). According to Poleni (1748), '...the entire wall of the drum and the attic, together with the columns and the buttresses, has rotated outwards, dilating the dome and lowering the lantern...'. So, Pope Benedict XIV appointed a committee of three scientists known thereafter as the 'Three Mathematicians' to report on the condition of the dome. The assessment of the Three Mathematicians, published as *opinion* (1742), was that the dome was seriously damaged and in danger of collapse and that its repair required extensive reinforcement operations and architectural alterations (Le Seur, et al., 1742). This *opinion* caused extended disputations by other scholars, so Pope Benedict XIV sought the advice of Giovanni Poleni, Professor of Physics in the University of Padua. The report of Poleni published as *his memories* (Poleni, 1748) instead assessed that the dome's state was less threatening than the assessment of the Three Mathematicians and that the defects could be repaired without major modifications.

Finally, Poleni's proposal was adopted, and the restoration of the dome was carried out by Poleni with the cooperation of L. Vanvitelli (Como, 2019). The intervention included the following:

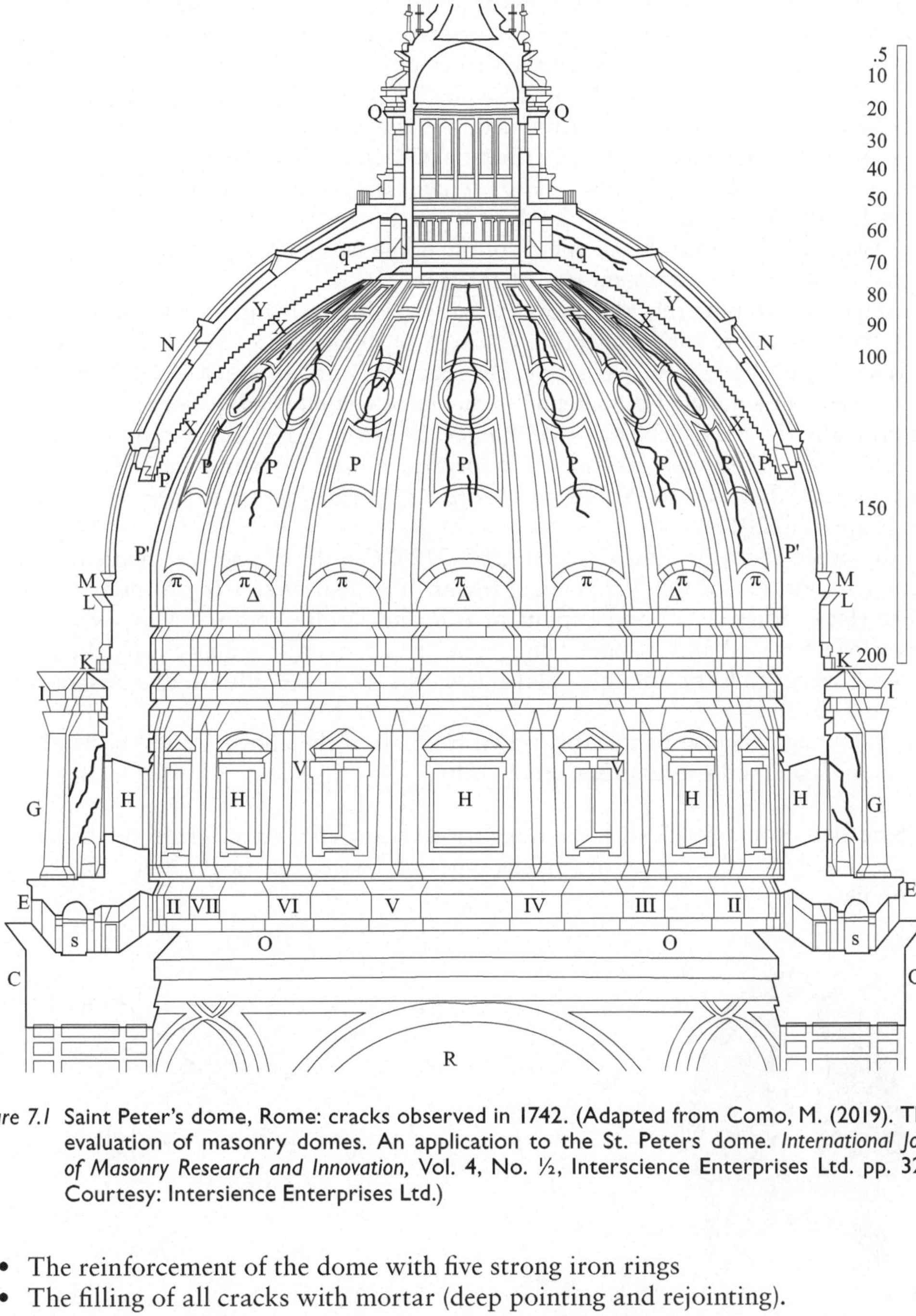

Figure 7.1 Saint Peter's dome, Rome: cracks observed in 1742. (Adapted from Como, M. (2019). Thrust evaluation of masonry domes. An application to the St. Peters dome. *International Journal of Masonry Research and Innovation,* Vol. 4, No. ½, Interscience Enterprises Ltd. pp. 32–49. Courtesy: Intersience Enterprises Ltd.)

- The reinforcement of the dome with five strong iron rings
- The filling of all cracks with mortar (deep pointing and rejointing).

Despite the serious differences in their conclusions, these two studies both were based on a sound scientific concept expressing the two main approaches of mechanics for the determination of the failure load of a structure, namely *the static equilibrium concept* on one hand and *the kinematic theorem* (theorem of virtual work) on the other.

The Three Mathematicians implemented the kinematic theory of virtual work (Figure 7.2) for the determination of the required ring reinforcement for stopping the further opening of the meridional cracks. Poleni used the theory of static equilibrium for arches that had been developed some years ago by Robert Hooke (1676) who discovered that the ideal shape of a masonry arch in equilibrium is that of an *inverted catenary curve* drawn by a chain subjected to the same weight distribution (Croci, 2007). Therefore, an *arch is stable when the catenary can be fitted within its thickness*. The implementation of this theory to Saint Peter's dome (Figure 7.3), analytically and experimentally in the laboratory of Padua, assured Poleni that the geometry of the dome was actually admissible and, therefore, that there was not an urgent risk of collapse.

It is important to note that the above two approaches to the game of forces in a structural system are the result of scientific research of thousands of years. Aristotle first in ΦΥΣΙΚΑ (Physic VIII 5, 249b) (Aristotle, Physica, 1955), and in ΜΗΧΑΝΙΚΑ (Mechanics 3,850b), stated in a nutshell the kinematic theory of virtual work (Capecchi, 2012).

Additionally, in ΜΗΧΑΝΙΚΑ (Mechanics 1, 848b), Aristotle stated the theory of parallelogram for the *composition of velocities*, almost two millennia before Stevin (16th Century A.D.) who stipulated *the theorem of parallelogram for the composition of forces* (Capecchi, 2012).

On the other hand, Archimedes first (287–212 B.C.), developed the theorem of levers and *the composition of parallel vertical forces*. It is also interesting to note the efforts of Hero (1st century A.D.) and Pappus of Alexandria (4th century A.D.) who in their treatises 'Mechanical' and 'Pappus Synagogue', respectively, tried to clarify the theorem of virtual work and the inclined plane law, the last one admittedly unsuccessfully (Capecchi, 2012).

The following conclusions may be drawn from all of the above, on the basis of sound facts on one hand and speculations on the other:

i. Since late Roman period, there was serious knowledge on mechanics which was in use mainly for the design of various types of devices, particularly for lifting heavy bodies.

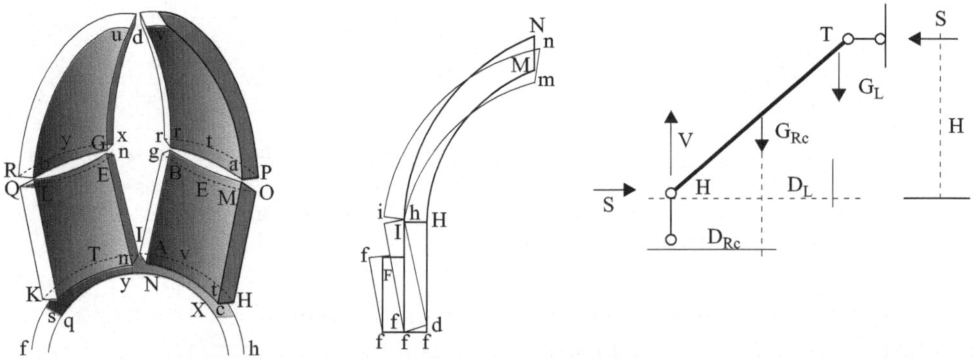

Figure 7.2 The Three Mathematicians model and the corresponding failure mechanism based on the kinematic theorem of virtual work. (Adapted from Como, M. (2019). Thrust evaluation of masonry domes. An application to the St. Peters dome. *International Journal of Masonry Research and Innovation,* Vol. 4, No. ½, Interscience Enterprises Ltd. pp. 32–49. Courtesy: Intersience Enterprises Ltd.)

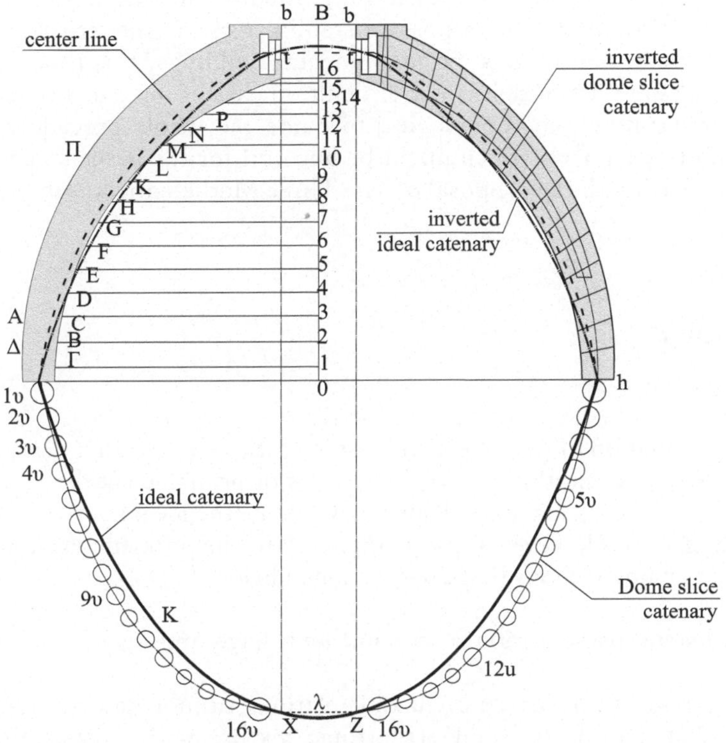

Figure 7.3 Poleni's static model. (Adapted from Como, M. (2019). Thrust evaluation of masonry domes. An application to the St. Peters dome. *International Journal of Masonry Research and Innovation*, Vol. 4, No. ½, Interscience Enterprises Ltd. pp. 32–49. Courtesy: Intersience Enterprises Ltd.)

ii. In this period, serious domed buildings (e.g. Pantheon of Rome, Rotunda of Thessaloniki, Hagia Sophia of Istanbul etc.) have been designed in such a way that their aftermath (contemporary) structural analysis proves that they were designed as if there was sound knowledge of mechanics as background.
iii. For the design of serious projects, scholars in mechanics and mathematics were employed, e.g. for Hagia Sophia, Anthemius and Isidorus. Undoubtedly these scientists were employed to offer their scientific knowledge and not *honoris gratias*. However, until 1743 there had not been found any written evidence of such a scientific approach for structural analysis and design, even for serious buildings.

It is our opinion, being based on the above considerations, that the existing knowledge of mechanics – at least the rule of parallelogram for forces composition and the funicular theory – was used for the safe design of big domes and arches *without written justification*, this knowledge being, perhaps, a hidden secret *of the* guild of 'designers'.

The development of the thermal machine and, particularly, the railway in subsequent years after 1743 resulted in the explosive development of science and technology and, consequently, in the gradual development of analysis and design of structures to the contemporary level. At the same time, the issue of a building permit by the authorities for the construction or repair of any building required the justification of its safety by a static calculation based on mechanics and on gradually established relevant codes. The above evolution was undoubtedly a positive step in ensuring the structural safety. Even so, in case of intervention to existing monumental buildings, this procedure should be implemented with care, because it is not difficult, in case of introduction of conservative assumptions or disputable methodologies, to guide the whole procedure to erroneous conclusions and to proposals which might be harmful for the authenticity of the monument, as it happened with the proposal of The Three Mathematicians in the case of Saint Peter's dome.

7.3 MODELLING

7.3.1 General

An analytical calculation of the time history of inelastic response of a monument for dead loads and for seismic excitation caused by a series of probable accelerograms, using the finite element method on solid-state elements, would be the ideal approach to the problem of analysis and design. However, such an approach to the problem, even today, seems to be unrealistic for many reasons that are mentioned below:

- The solid-state discretisation of complicated, huge and massive structures is very difficult.
- Reliable stress–strain inelastic triaxial relations cannot be easily adopted, as the masonry has a unilateral and orthotropic texture with varying orthotropy and material properties almost from point to point. Additionally, the descending branch in the stress–strain constitutive law, necessary for cycling loadings, cannot be estimated in a justified way.
- The adoption of a reliable strength envelope varying from area to area of the structure cannot be supported by experimental evidence, moreover, when the independent variables are not only the stresses σ_1, σ_2, σ_3 but also the direction cosines of the masonry orthotropy (direction of the bed joints).
- The loading path, which plays an important role in inelastic analysis, cannot be easily guessed for a monument with a history of many centuries.
- The existing cracks due to foundation settlements, various restraints or accidental actions in the past (i.e. earthquakes) cannot be easily taken into account in the analysis of the existing structure, and even more, they cannot be analytically reproduced in the successive loading steps up to failure.
- Last but not least, it is very difficult to adopt some reliable accelerograms for the dynamic analysis of the monument when the existing information on this matter is limited to the last 40–50 years.

Taking all these uncertainties into account, it appears fairly meaningless for practical purposes to try a sophisticated inelastic analysis using modern general-purpose FEM programs or even problem-oriented FEM programs. In fact, when so many parameters of the problem are uncertain and the outcoming results are so sensitive and highly dependent on these parameters, these results cannot be trustworthy. Therefore, for the analysis and design of monumental or historical buildings, simplified models should rather be introduced.

The procedure for analysis and design presents various difficulties, the most important of which are the following:

- Geometry of the construction
- Internal morphology of the structural elements (e.g. three-leaf walls)
- Mechanical properties of masonry
- Modification of the structural system in the past
- Cracks and damage of the past
- Actions (gravity–wind–seismic)
- The behaviour factor (q-factor)
- M–θ diagrams.

These issues will be analysed in short in the subsequent subsections.

7.3.2 Geometry

Monumental buildings have usually a very complicated geometry. As mentioned before (see Chapter 4), their structural capacity is based mainly on their form which ensures multiple load path possibilities, and, in this context, the crack development for various events is balanced by an internal redistribution of forces and, therefore, new load paths. It is not curious that after strong earthquakes, big monumental Roman, Byzantine or Islamic buildings have survived despite the general calamity. In this context, 1D curved elements, like arches or semiarched buttresses, combined with 2D ones, like walls, vaults or domes, or combined even with 3D elements, like massive piers and pendentives, constitute a usual structural form. The geometric modelling of such a structure may be easily formulated using 3D finite elements (Figure 7.4). However, the required information for the mechanical behaviour of these elements, which is not available, makes such an approach very complicated and results in limited reliability. So, the solid-state formulation is implemented only in special cases or for the verification of the achieved accuracy in case of reduction of a 3D element to a 1D or a 2D model which constitutes the usual form. For example, the thick dome of Rotunda of Thessaloniki was analysed first, as an independent axisymmetric homogenous thick shell with 3D finite elements, and, then, after the verification that there was not a substantial difference from the concept of a membrane behaviour (2D elements), the whole structure including the supports was reduced to 2D shell elements and 1D elements for columns, beams and arches (Figures 7.5 and 7.6) (UNDP/UNIDO, 1984).

7.3.3 Internal morphology of structural elements

It is known that the massive structural elements are not homogeneous and show complex internal structure including, usually, three leaves connected with stone or brick layers in successive levels. It should be noted here again that endoscopic investigation constitutes a valuable method providing visual and diagnostic information at nonvisible points inside the masonry body and, in this respect, helps in overcoming the above-mentioned difficulties. Consequently, massive walls or piers themselves constitute a space solid-state structure, whose mechanical properties are evaluated using approximate and limited-reliability methods (see Chapter 3). Even so, the mechanical properties of the massive elements are uncertain as the loading in the past, due to various events like strong earthquakes,

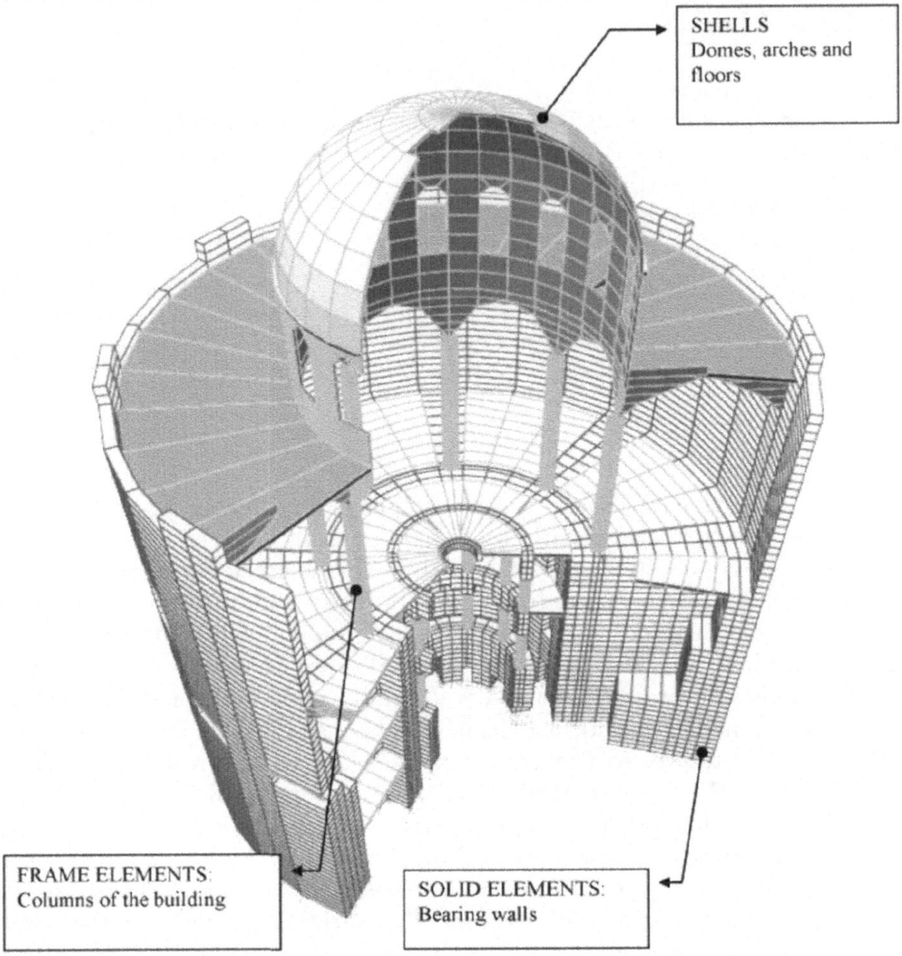

Figure 7.4 Mar Girgis, Cairo: 3D solid FEM analysis. (Adapted from Penelis, G. and Penelis, Gr. (2018). Restoration of the MarGirgis Church and Roman tower in Cairo. *Proceedings of the 4th Architectural and Urban Restorations BRAU4,* CI COP Italia. pp. 163–174.)

thermal changes due to fire events etc., may have led to a relaxation between the external leaves and the core.

7.3.4 Mechanical properties of masonry

Extensive reference has been made already in Chapters 2, 3 and 6 to the materials of the original structure, their mechanical properties and the NDT and DT investigations on site and in the laboratory for an acceptable estimate of the mechanical properties that should be introduced in the analysis and design of the building. Additionally, it should be reminded that recent extensive research (see Section 3.4) has been developed in this direction by means of FEM (Stavrakakis, et al., 1995) using microelements of mortar and bricks (Figure 7.7). It is the authors' opinion that this procedure is the most promising

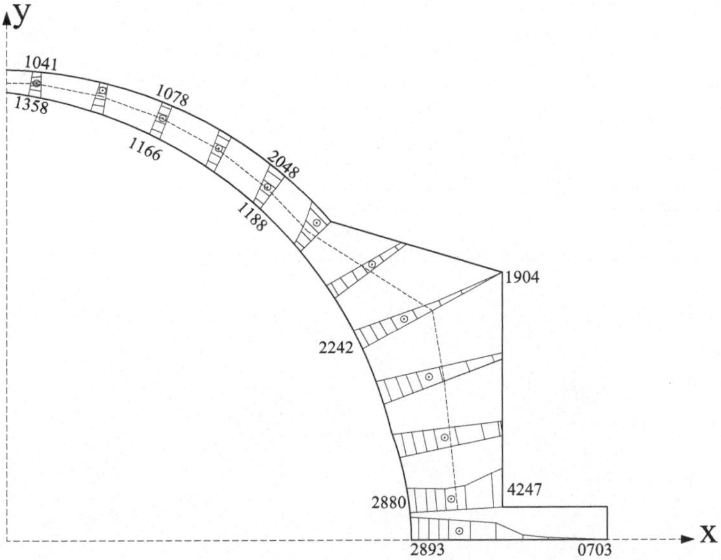

Figure 7.5 Dome of Rotunda of Thessaloniki: modelled as a thick shell of revolution. Distribution of meridian stresses for dead loads. (Adapted from Penelis, G. (1982). *The Masonry Roman and Byzantine Monuments in Greece. UNDP/UNIDO, PROJECT (RER/79/015) National Report of Greece.*)

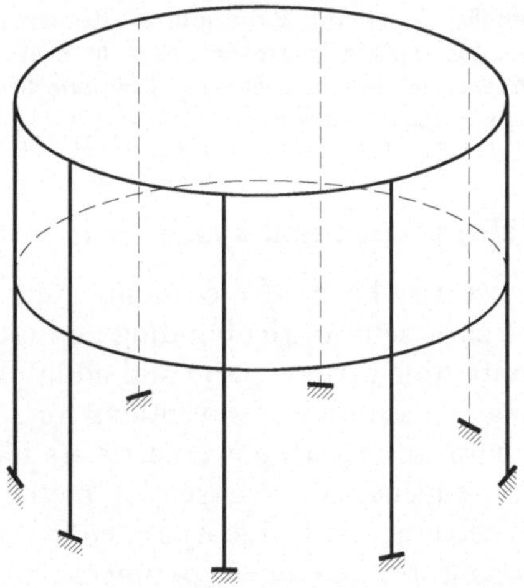

Figure 7.6 Rotunda of Thessaloniki: supporting system of the dome modelled as a 1D frame. (Adapted from Penelis, G. (1982). *The Masonry Roman and Byzantine Monuments in Greece. UNDP/UNIDO, PROJECT (RER/79/015) National Report of Greece.*)

for the determination of the failure envelope (σ_1–σ_3) and the constitutive law σ–ε of the original masonry.

Even so, the massive dimensions of the structural elements, their heterogeneous structure, their inherent low and unreliable tensile and shear strength make the whole procedure for the estimation of the mechanical properties of the various regions of the monumental building quite problematic and questionable, leading to the introduction of conservative values.

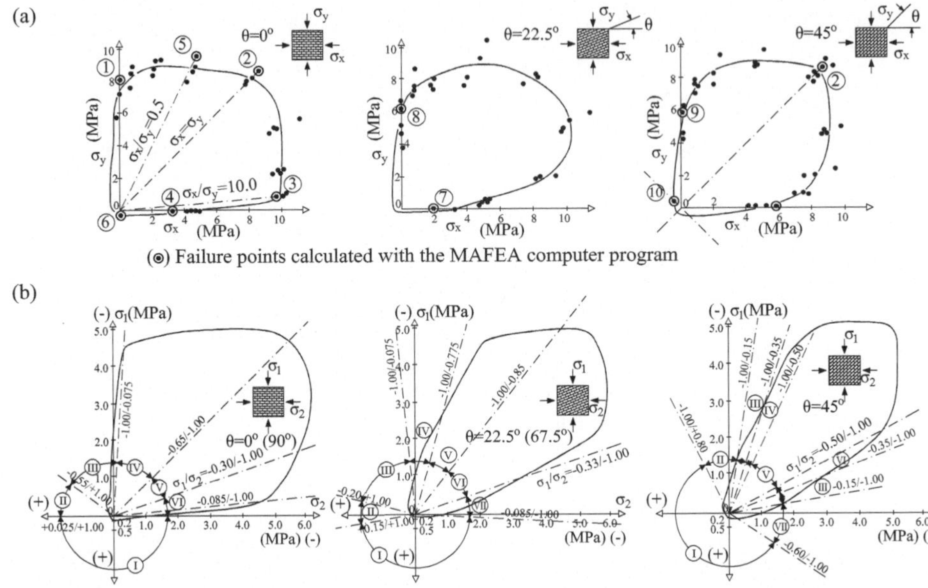

Figure 7.7 Analytically determined failure envelopes of masonry: (a) evaluation of MAFEA FEM code using experimental failure envelopes; (b) analytically determined failure envelopes of Roman masonry (Rotunda of Thessaloniki) using the MAFEA FEM code. (Adapted from Stavrakakis, E., Ignatakis, C., and Penelis, G. (1995). Behaviour of Roman masonry under biaxial loading analytical assessment of complete failure criterion. In Brebia, C.A. (ed.). *Proceeding of the 4th International Conference on Structural Studies of Historical Buildings,* Southampton, Boston; computational mechanics. pp. 339–347.)

7.3.5 Modification of the structural systems in the past

General alterations in the past may have affected significantly the response of the structure. History is an essential parameter of the building and must be considered and integrated in the model. Demolition of various parts and additions of others tending to the improvement of the building functioning; interventions for repair and strengthening to tackle damages caused by natural or man-powered events like earthquakes, soil settlements, fires, war and the consequences of construction methods that may have been followed, should be taken into account and be incorporated in the model under elaboration. The history of a building constitutes a unique experiment that occurred in a true scale of space and time that should be taken into account (Croci, 2007).

7.3.6 Cracks and damage of the past

Cracks, local masonry splitting due to compression overload, local collapse and repair should be surveyed and taken into account in structural modelling. The analysis of the model with no damage at all should prove that, for a rational load combination in the past, the building has cracked in the existing crack pattern. For example, in a cracked dome with cracks in meridional direction, the tensile strength of masonry along the peripheral direction should be considered zero (Figure 7.8) for the estimate of the response of the existing situation (Penelis, et al., 1983).

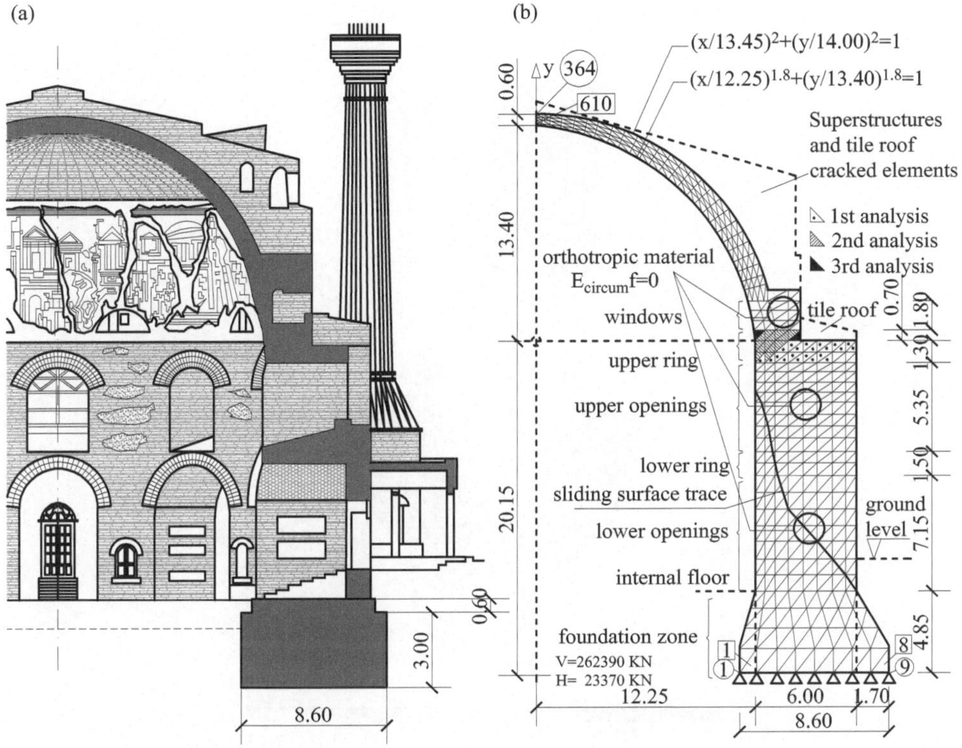

Figure 7.8 Rotunda of Thessaloniki: (a) meridional cross section; (b) axisymmetric finite element model, predicted cracking. (Adapted from Penelis, G., Karaveziroglou, M., Sytlianides, K., and Leontarridis, D. (1983). Strengthening of the Rotunda Monument in Salonica. *Proceedings of the IABSE Symposium Strengthening of Building Structures – Diagnosis and Therapy*. Venezia, Italy.)

7.3.7 Actions

The actions that should be taken into account for the analysis are the following.

7.3.7.1 Dead and live loads

The determination of dead loads should take into account the real situation resulting from the survey and material investigation (densities of the structural system, fillings, roofing, flooring system etc.).

The definition of live loads should take into account the relevant codes and the eventual new use of the building after restoration.

7.3.7.2 Imposed deformations

They include *thermal effects, soil settlements* etc.

Thermal effects are of major importance in case that external posttensioned steel rings are introduced for structural strengthening, e.g. in the form of rings. It should be noted that while summer–winter temperature difference causes axisymmetric dilatation or subtraction, noon–midnight differences cause local temperature influence (Figure 7.9) (Stylianidis, K., 2000; Penelis, G., 2014).

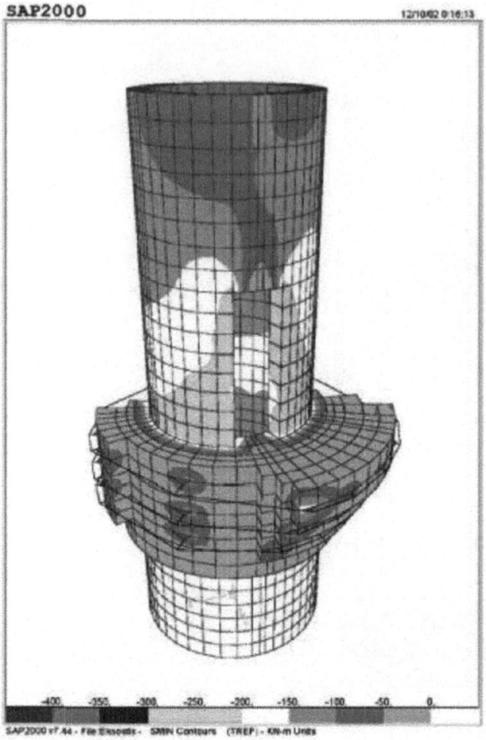

Figure 7.9 Minaret of Rotunda of Thessaloniki: Isostress curves $[\sigma_{\min}]$ for dead load and confinement using externally posttensioned tendons under minimum winter temperature. (Adapted from Stylianidis, K. (2000). Strengthening of the minaret of Rotunda in Thessaloniki. G. Penelis. *International Symposium of Concrete and Masonry Structures*, Stylianidis, K. and Kappos, A. (eds.). Thessaloniki, Greece pp. 335–348; and Penelis, G. (2014). Structural restoration of monumental buildings in seismic areas, *MARE.14 materials and techniques for the restoration of monuments lifelong learning unit. AUTH*, Thessaloniki.)

Soil settlements are introduced in the form *of constraints* in case that some areas on the foundation present well-defined settlements. Otherwise, the foundation restraints are introduced in the form of springs. In this case, the soil subgrade modulus must be known. In case of posttensioning of steel elements, creep phenomena of masonry cannot influence significantly the stress state. In fact, masonry in a historical building has lost almost all its tendency to creep in the past years; even so, groutings for repair that have been injected in its body cannot creep alone due to the posttensioning forces, since the main masonry matrix where grout has been injected cannot creep.

7.3.7.3 Seismic action

As known, seismic action is a vibration for a short period of time imposed on foundations of buildings of a region affected by an earthquake (Figure 7.10) (Penelis, G. and Penelis, Gr., 2019). Seismic action is introduced in the analysis in one of the three following forms, depending on the method of analysis that will be used:

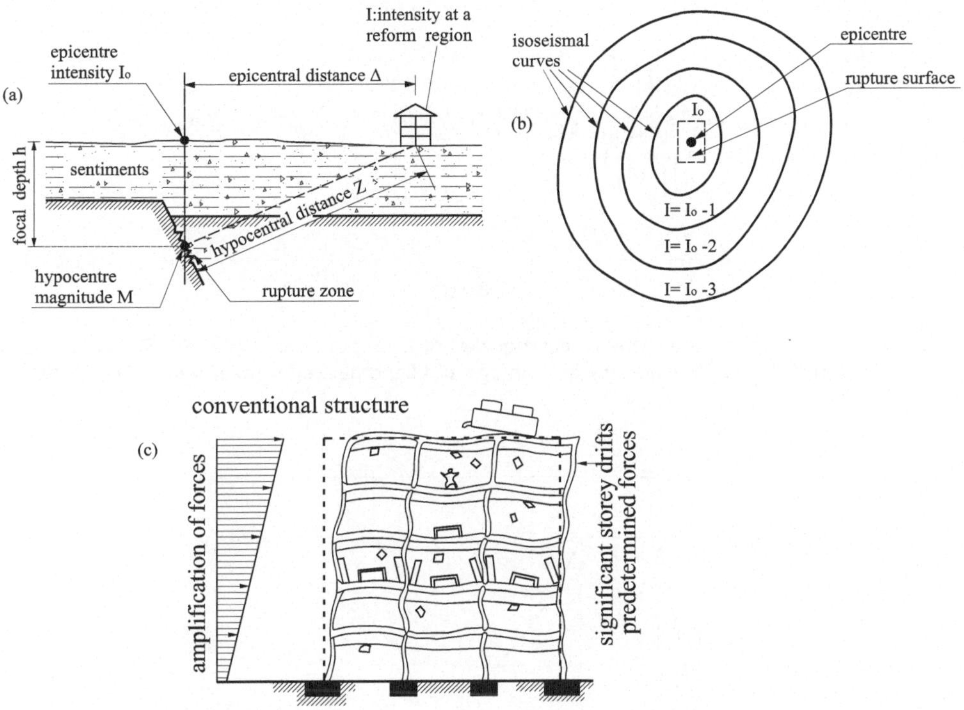

Figure 7.10 Terminology related to earthquake: (a) generation of earthquakes; (b) isoseismic curves; (c) building response to seismic action. (Adapted from Penelis, G. and Penelis, Gr. (2019). *Concrete buildings in seismic regions*. CRC Press (A SPON Press Book), London.)

- A set of seismic accelerograms (Figure 7.11)
- An elastic response spectrum (Figure 7.12)
- A set of equivalent static forces (Figure 7.13).

The first form is seldom used, used particularly for time-history elastic or inelastic analysis. The suite of three accelerograms which are introduced in the analysis should comply with the codified elastic response spectrum (Figure 7.14) (Penelis, G. and Penelis, Gr., 2019).

The second form constitutes the basis for the determination of the seismic action in most of the countries all over the world. It represents in a diagram the spectrum of maximum accelerations of a series of one-degree-of-freedom systems with a natural period $\left(T_o = 2\pi\sqrt{\frac{m}{k}}\right)$ ranging from zero to infinity (Figure 7.15) under the action of a codified earthquake.

This spectrum is codified to a generic diagram (Figure 7.12) for the quantification of which the following parameters should be defined in advance:

- Peak ground acceleration (PGA) of the region (a_{gr})
- Importance factor of the structure (γ_1)
- Soil factor (s)
- Mass of the building (m)

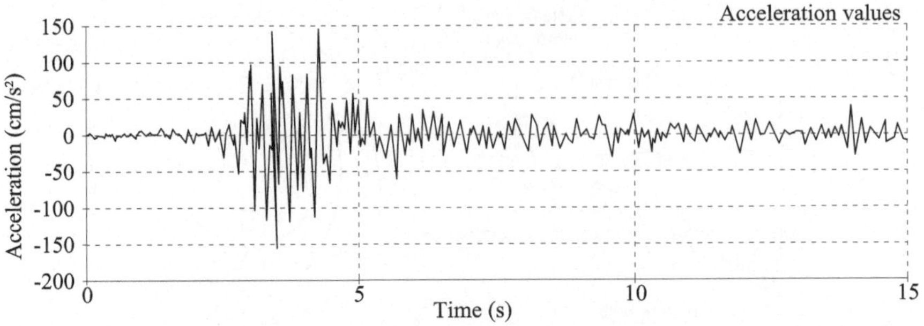

Figure 7.11 Accelerogram of the Athens earthquake of 7 September 1999 (N–S). (Adapted from IESEE (Institute of Engineering Seismology and Earthquake Engineering Greece), 2003.)

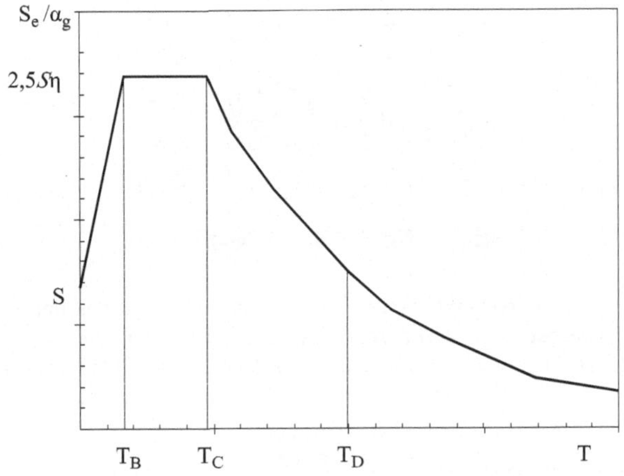

Figure 7.12 Generic shape of the elastic response spectrum. (Adapted from E.C.8-1/EN 1998-1 (2004). *Design of structures for earthquake resistance: General rules, seismic actions and rules for buildings.* CEN, Brussels, Belgium.)

- Stiffness of the building (k)
- Behaviour factor (q)
- Damping correction factor (n).

This quantified spectrum called *design spectrum* is the input of seismic action for a *linear dynamic multimodal response spectrum analysis* which is the reference method of Eurocode 8 (E.C.8-1/EN1998-1, 2004).

In the third form, the dynamic response spectrum analysis is replaced by an equivalent static approach, based approximately on the determination of the base shear force of the system, making use of its fundamental period T_o and of the corresponding design spectrum. Then the base shear is distributed along the height of the building (Figure 7.13d) proportionally to the ordinates of the fundamental eigenmode at the level of the concentrated masses and in proportion to the mass values (Figure 7.16).

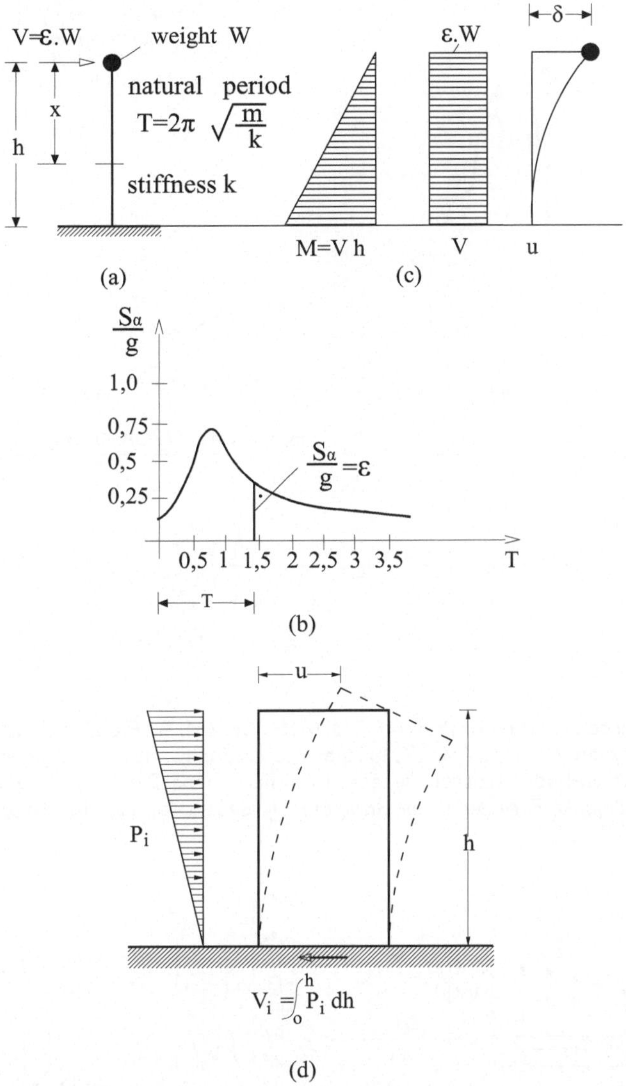

Figure 7.13 Transition from the dynamic to static response: (a) an SDOF system; (b) acceleration response spectrum; (c) response of the structure; (d) lateral load P_i of a building developing base shear V_B.

This loading form is proper for regular multistorey buildings with low torsional flexibility. It should be added that this type of static simulation of seismic actions is used also for the elaboration of a *pushover inelastic analysis with a step-by-step static loading* up to failure for the seismic resistance assessment of historical buildings (see Section 7.7)

Crucial point for all above procedures is the value of the PGA a_{gr} (Figure 7.17). As known, all contemporary buildings are analysed and designed for a seismic action with a PGA a_{gr} which corresponds to a probability of 10% of exceedance in 50-year building life or 475-year return period. For new, important or monumental buildings, according to existing seismic codes, this value is increased by an *importance factor* $\gamma_1 = 1.20$ which

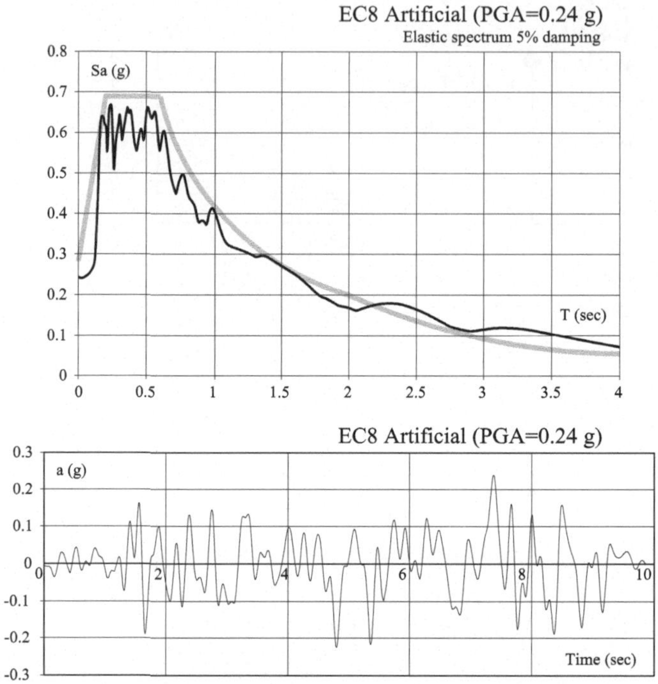

Figure 7.14 Artificial accelerogram for PGA = 0.24g. (After Sextos, A., Pitilakis, K., and Kappos, A. (2003). Inelastic dynamic analysis of R/C bridges accounting for spatial variability of ground motion, site effects and soil-structure interaction phenomena Part I: Methodology and Analytical tools. *Earthquake Engineering and Structural Dynamics*, Vol. 32, No. 4. pp. 607–627.)

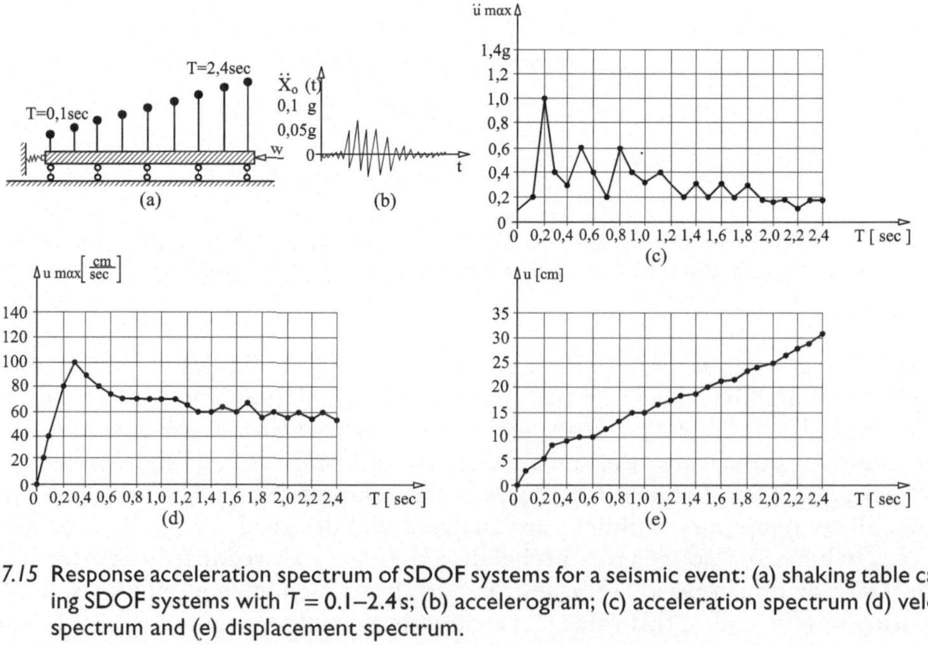

Figure 7.15 Response acceleration spectrum of SDOF systems for a seismic event: (a) shaking table carrying SDOF systems with $T = 0.1–2.4$ s; (b) accelerogram; (c) acceleration spectrum (d) velocity spectrum and (e) displacement spectrum.

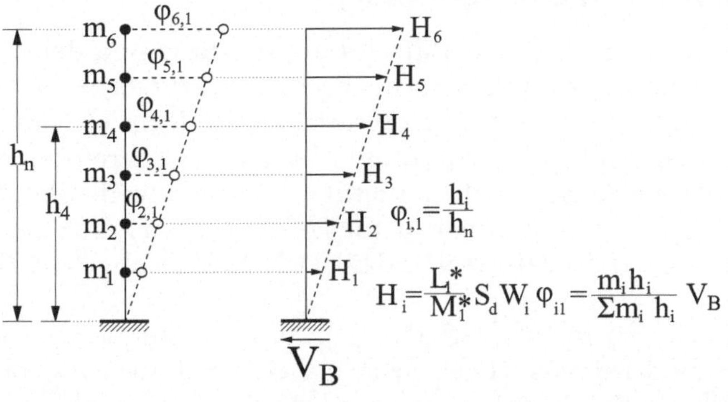

Figure 7.16 The fundamental mode of a multistorey system and the corresponding inertial forces.

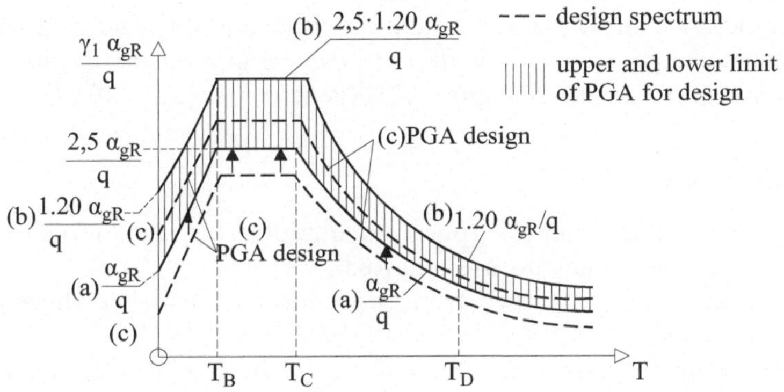

Figure 7.17 Demand design spectrum: (a) $\gamma_1 = 1.0$; (b) $\gamma_1 = 1.20$; (c) γ_1 defined from failure of the existing structure at critical zones, in the past.

corresponds to a probability of 10% of exceedance in 100-year building life or 950-year return period. Such a probabilistic value of $\gamma_1\, a_{gr}$ may easily be managed in case of a new building. But for an existing monumental building, such values may drive the whole procedure to unacceptable interventions from the point of conservation.

Having in mind that a monumental building has survived for centuries, its behaviour is an irresistible criterion of high structural capacity. Thanks to this capacity, the monument has overcome many destructive earthquakes without serious damage or collapse. Therefore, based on this fact, the authors believe that it is rational that, for the design value $\gamma_1\, a_{gr}$, the seismic action that corresponds to the failure of critical points of the building should be used. However, this value should not be lower than a_{gr} (PGA for conventional buildings) and not higher than 1.20 α_{gr} (PGA for very important contemporary buildings) (Figure 7.17). In this respect, the design spectrum on one hand should lie between the boundary requirements of the code and on the other is validated to a degree from the existing capacity of the structure.

The scope of this book doesn't allow more details on the subject. However, some issues arising, particularly for monumental buildings, will be analysed below.

7.3.8 The behaviour factor (*q*-factor)

The behaviour factor, which is necessary for a quasistatic or a dynamic elastic analysis of a structure, cannot be taken from the codes of practice, which are valid for codified masonry structures, since the texture and the form of conventional contemporary buildings defer radically from those of the past. Therefore, an inelastic static approach of the main elements of the structure (i.e. piers, walls) should be carried out (pushover analysis) for the determination the H–δ curve, and the relevant ductility factor (Figures 7.18, 7.19 and 7.20) (UNDP/UNIDO, 1984; Kappos et al., 2002; Penelis, Gr., 2006b).

It should be noted that for the determination of the flexural inelastic behaviour of a masonry element modeled as a 1D element, a theoretical phenomenological procedure may be used, while for the shear behaviour, a hybrid procedure should be implemented. Then both of these models are inserted in a software code, which allows the prediction of the moment–displacement curve of the 1D element and therefore allows the determination of the ductility factor μ and, then, the corresponding q-factor. A more detailed reference to this issue may be found in Subsection 7.6.2.

More recent methods allow the determination of the ductility factor of basic structural masonry elements using *a macroelement analysis* or a *microelement analysis* combined with *homogenisation procedures* (Figure 7.21) (Penelis, G., et al., 2010).

7.3.9 M–θ diagrams

Even for the inelastic static analysis (pushover analysis) of frame models (1D systems), M–θ diagrams of all basic elements are required.

This information may be obtained from procedures analogue to those of Subsection 7.3.8 above.

The problem becomes almost insuperable in case of surface or solid elements where *the stress–strain tensor relations* of orthotropic and inelastic material properties are needed.

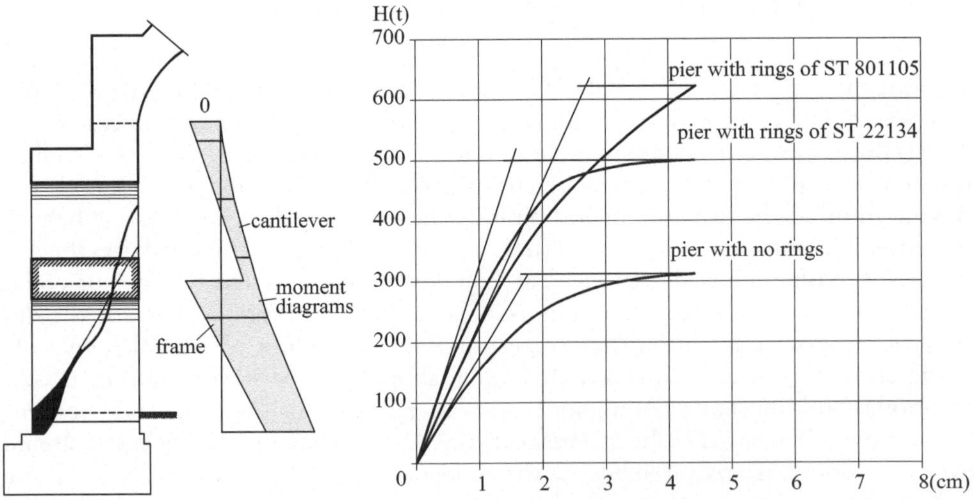

Figure 7.18 Analytical H–δ diagrams of Rotunda of Thessaloniki: piers. (Adapted from UNDP/UNIDO (1984). *Repair and Strengthening of Historical Monuments and Buildings in Urban Nuclei*, 6, Vienna.)

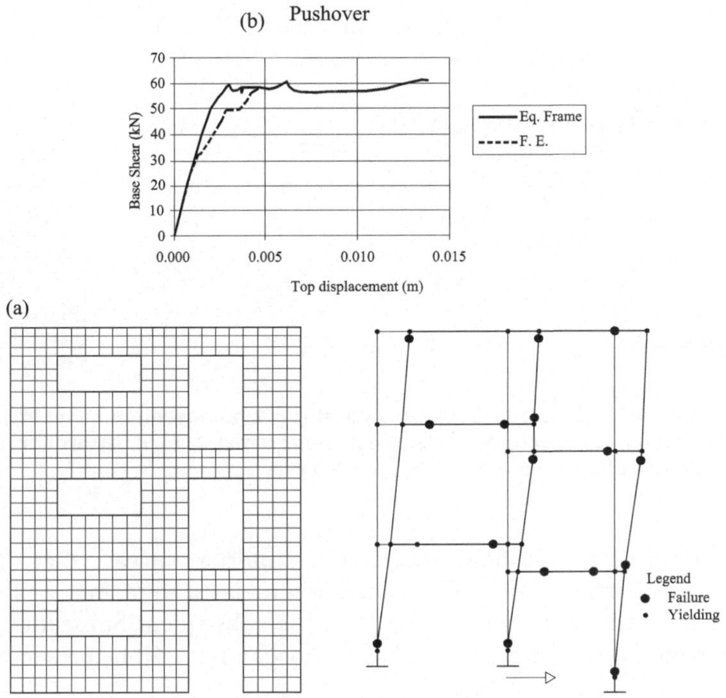

Figure 7.19 Pushover curves for equivalent frame and FEM model: (a) geometry of the perforated masonry and the corresponding equivalent frame model; (b) pushover curves. (Adapted from Kappos, A., Penelis, Gr. and Drakopoulos, C. (2002). Evaluation of simplified models for lateral load analysis of Unreinforced Masonry Buildings. *Journal of Structural Engineering*, Vol. 128, No.7, ASCE. pp. 1–8.)

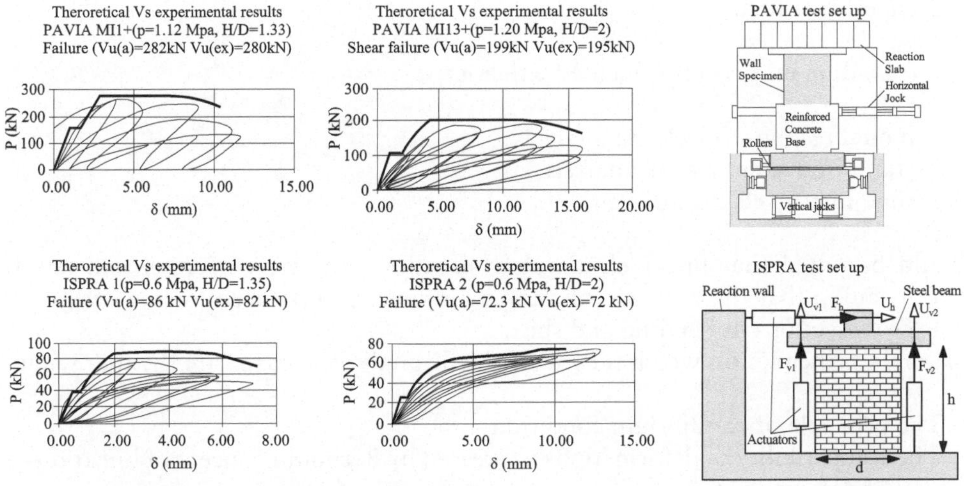

Figure 7.20 P–δ diagrams. Analytical versus experimental results (equivalent frame models). (Adapted from Penelis, Gr., 2006b.)

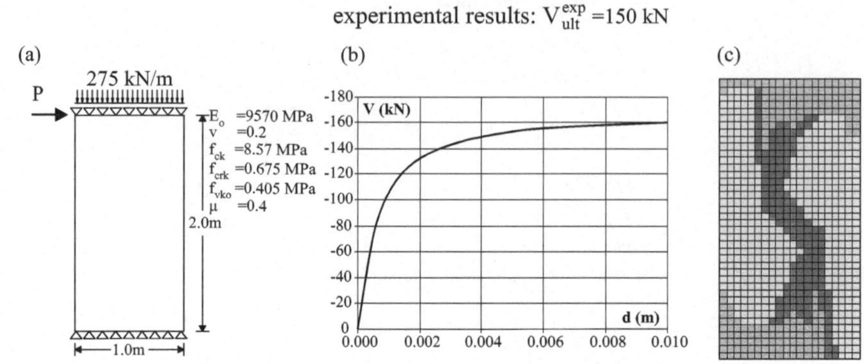

Figure 7.21 Panel test of Magenes and Calvi. Analytical (FEM homogenised elements) versus experimental results: (a) structural model; (b) *P*–δ diagram; (c) crack pattern for $\delta = 0.01$ m (drift: 0.5%). (Adapted from Penelis, G., et al., 2010.)

It is self-understood that additional uncertainties interfere in case that the descending branch of the stress–strain 'curve' is needed for inelastic dynamic analyses. A more detailed reference to this issue will be made in the subsequent Subsection 7.6.2 of this chapter where contemporary trends on masonry mechanics will be presented.

7.4 LINEAR ELASTIC ANALYSIS AND DESIGN

Linear elastic analysis and design is the reference method for almost all design codes in effect for conventional new buildings and to a degree for existing ones too. It may be applied to 1D, 2D and 3D structures for

- Static loading and
- Dynamic loading.

Dynamic loading may be implemented either as

- A modal response spectrum analysis
- A static equivalent elastic analysis
- A time-history elastic analysis.

It should be noted that linear elastic analysis allows the separation of *analysis* from *design,* meaning that the outcome of analysis is subsequently used for the design of the various cross sections in bending and shear.

The main principles on which the linear elastic analysis is based are the following:

- The principle of equilibrium condition
- The compatibility of deformations expressed by Bernoulli concept of linear distribution of deformations on any cross section in case of 1D systems
- The assumption of linear constitutive law of material, either in tension or in compression.

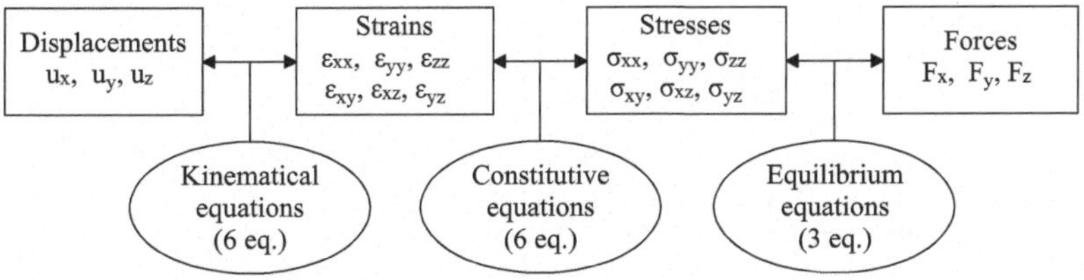

Figure 7.22 Relations between the basic quantities in structural mechanics. (Courtesy: Vrouwenvelder, A. (2003). *The plastic behaviour and the calculations of beams and frames subjected to bending.* Technical University Delft, faculty of civil engineering and geosciences Delft, Holland.)

All the above principles are expressed by linear equations (Figure 7.22), and as a result, the principles of

- Analogy
- Superposition and
- Separation of design from the analysis

are the main characteristics of the method. All the above advantages have established this method as the basic tool in the hands of the structural engineer for the analysis and design of any type of structure and for all types of materials. Even conventional contemporary masonry buildings, according to Eurocodes, are analysed using linear elastic methods, while the design or resistance verification of the structural elements which are masonry walls, is carried out with the assumption of cracked cross sections in the region under tension (see Chapter 3).

However, elastic analysis – either static or dynamic – presents many inadequacies for monumental buildings with arches and carved surfaces, due to the very low and unreliable tensile strength of masonry together with its orthotropy. Because of these reasons and of the high degree of redundancy of masonry structural systems, extended stress redistribution takes place, so that the original picture of stress distribution of the elastic analysis before cracking defers seriously from the final one at collapse limit. For example, a masonry arch supported by two piers according to an elastic analysis presents cracks at the tension region of its crown for 25% of the dead load of the system (Figure 7.23) without any risk of collapse. In fact, if, on the other hand, a step-by-step inelastic analysis is adopted up to collapse, the ultimate load in bending goes up to 350% of its dead load, while in the meantime, for almost 150%–200% of this load, a shear compression failure takes place at the supports.

So, as these characteristics had been observed very early, extended efforts had been made for the development of more reliable methods; reference to these will be made in subsequent sections.

Even so, the linear elastic analysis continues to be a valuable tool, even for monumental buildings for the following reasons:

- It identifies zones where cracks or serious displacements according to the analysis are expected and therefore allows a cross reference to the cracks and displacements that have been surveyed. In case that the analytical results and the survey do not correlate with each other, the model of analysis should be reconsidered and revised.

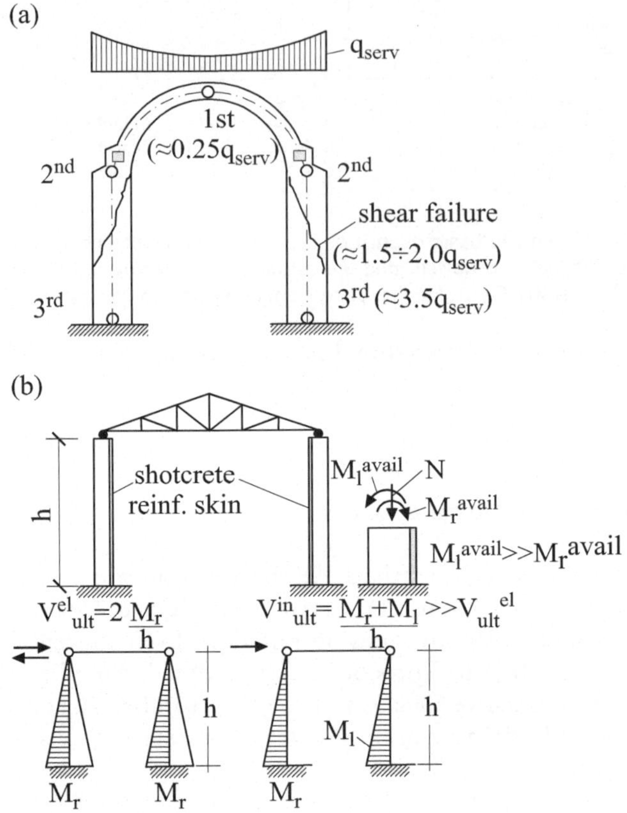

Figure 7.23 Successive steps of plastic hinges development: (a) an arch on piers; (b) a roofed hall, strengthened internally with shotcreting. (Adapted from Penelis, G. (2014). Structural restoration of monumental buildings in seismic areas, *MARE.14 materials and techniques for the restoration of monuments lifelong learning unit. AUTH*, Thessaloniki.)

- It is useful for the determination of the distribution and reinforcement demand for carrying the tensile stresses.
- For the time being, it is the only feasible method for the dynamic analysis of real monument buildings with complicated configuration. As known, this analysis is very useful for the determination of the eigenperiods and the eigenmodes of the structure; they are used for the horizontal seismic action distribution for the development of a further realistic static inelastic analysis (pushover analysis, capacity curves).

Closing, it should be noted that very important and significant monuments, such as the Colosseum of Rome, Pantheon, San Marco of Venice, the Metropolitan Cathedral of Mexico, the Rotunda of Thessaloniki, Hagia Sophia of Istanbul, Acheiropoietos of Thessaloniki etc., have been analysed first using linear elastic analysis. Using these results as a guide, more detailed analyses have been carried out then at critical regions of the structure using *limit-state theory* or *inelastic step-by-step analysis*. The cross reference of these analyses with the survey of the deficiencies of the structure and the mechanical properties of the materials allow the formation of a safe assessment.

7.5 LIMIT-STATE ANALYSIS

Limit-state analysis, known also as *plastic analysis*, was developed in about *1950* for the analysis of steel structures (Baker, et al., 1956) based on the general theory of plasticity, developed by Drucker and Prager, 1950 (Vrouwenvelder, 2003). Later on, the method was extended for use in R/C structures, and in 1966, Heyman (1966) implemented the method to masonry arches and vaults.

The basic principles of this method are summarised in the following three theorems (limit theorems).

- *Lower-bound or static theorem*

  All arrangements that accomplish for a given load the static equilibrium conditions, together with the condition of not exceeding the ultimate strength of the structure (plastic moment M_u) along its axis, constitute *admissible states of equilibrium with values of load P_{st} below or equal to collapse load*, namely

  $$P_{st} \leq P_u, \tag{7.1}$$

  where
  P_{st} is the load under consideration complying with the lower-bound theorem
  P_u is the load at collapse *(safe approach)*.
- *Upper-bound or kinematic theorem*

  All arrangements that accomplish a kinematical admissible mechanism, in which the work developed by external forces P_{kin} is greater than or equal to the work developed by the internal plastic moments M_u *at plastic hinges, constitute loading states above or equal to collapse*, namely

  $$P_{kin} \geq P_u, \tag{7.2}$$

  where
  P_{kin} is load under considerations complying with the upper-bound theorem
  P_u is the load at collapse *(unsafe approach)*.
- *The uniqueness theorem*

  A load P for which the value of the bending moments along the axis of the structure is statically acceptable and is equal to the plastic moment M_u in a number of sections, sufficient for the formation of a collapse mechanism is the *collapse load*.

The above three theorems known also as Drucker–Prager theorems constitute the basis for the limit-state analysis.

In Figure 7.24 on the left, three cases are given with the first two complying with the lower-bound theorem and the third one violating it.

On the right in Figure 7.24, three other cases are given with the first two complying with the upper-bound theorem and the third one violating it.

Finally, in Figure 7.25, a loading scheme is given, for which both theorems are valid and, in this respect, the resulting load is the *collapse load*.

Heyman formulated the plastic analysis for masonry arches and vaults based on the following realistic assumptions (Chapter 4) (Heyman, 1977):

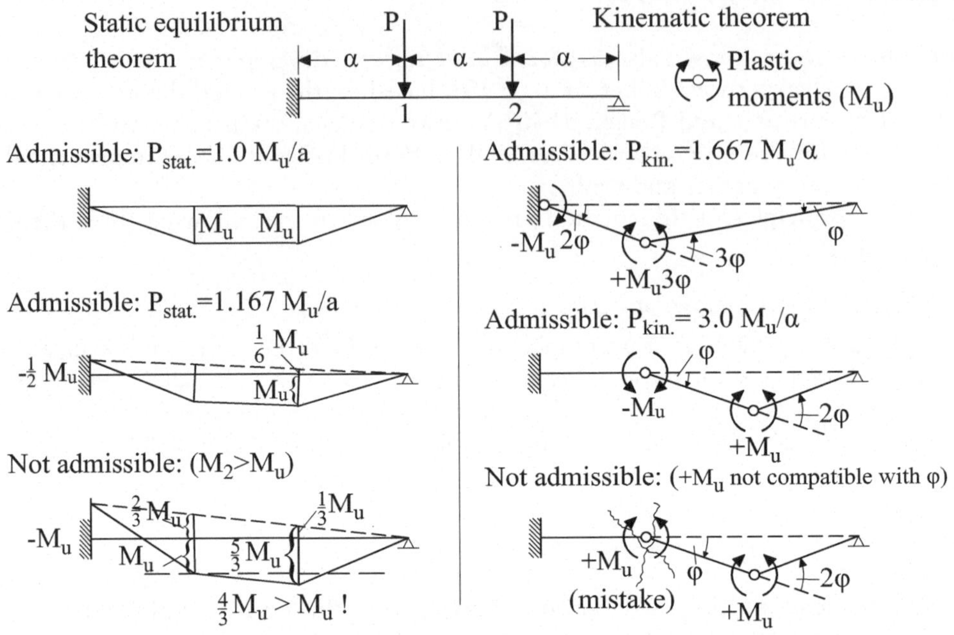

Figure 7.24 Examples of static and kinematic admissible and inadmissible arrangements for the determination of P_u. (Adapted from Penelis, G. (1976). Load bearing capacity of R/C Linear structures. *Seminar on modern trends in analysis and design of R/C structures.* In Greek. Penelis, G., (ed.). Thessaloniki, Greece. pp. 1–41.)

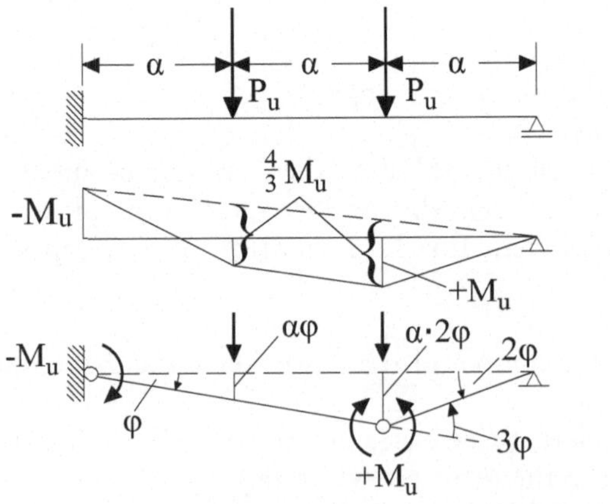

$$P_u(\alpha\varphi+\alpha 2\varphi)=M_u(\varphi+3\varphi) \qquad P_u=\tfrac{4}{3}\,M_u/\alpha$$

Figure 7.25 Arrangement complying with both theorems (static and kinematic). (Adapted from Penelis, G. (1976). Load bearing capacity of R/C Linear structures. *Seminar on modern trends in analysis and design of R/C structures.* In Greek. Penelis, G., (ed.). Thessaloniki, Greece. pp. 1–41.)

- *The compressive strength of the masonry material is unlimited.*
  In fact, particularly in cases of stone arches with dry joints, this condition must be considered valid, since the compressive stresses that develop at failure state are very low in relation to the compressive strength of the stone.
- *Sliding between units is impossible* due to the friction resistance between the voussoirs (stones).
- The tensile strength of masonry is null.

These conditions enable the application of the limit theorems presented above. In arches and vaults, in fact, in case of eccentric position of the thrust line in a region of the arch, but within its boundaries, the boundary under tension presents successive cracks which form a kind of plastic hinge. So, *the lower-bound (or safe) theorem* is formulated as follows: *The structure is safe if a thrust line can be determined, in equilibrium with the external load, which falls within the boundaries of the structure* (Figure 7.26). The load applied is a *lower bound* of the actual ultimate load causing collapse.

On the other hand, the *upper-bound theorem* is formulated as follows: *If a kinematic admissible mechanism is formed by inserting plastic hinges so that the system is transformed to a kinematic chain with one degree of freedom and the virtual work of the external loading is positive or at least equal to the work developed in the hinges due to the plastic moments, then the system is unsafe (collapses).* In this respect, the external load is an *upper bound* of the actual load causing collapse.

Heyman (1966) based his limit-state analysis for arches and vaults on the lower-bound (safe) theorem (diagram of Figure 4.10) (Roca, et al., 2010).

In concept, for more than one century, the lower-bound theorem was implemented intuitively in the procedure followed in the analysis of multistorey buildings by separating them in plane subassemblies (plane frames) considering the carrying capacity of their joints to the transverse frames of null strength. Undoubtedly, all these systems were designed *according to the lower-bound theorem* for a safety factor higher than that anticipated in the analysis, since the implementation of this theorem results in conclusions being on the safe side.

It should also be noted that the methods followed by Poleni and the Three Mathematicians for the residual capacity of 'St. Peter's' dome were in practice an early implementation of the above limit-state theorems.

The limit-state method entails, on the other hand, some disadvantages, most important of which are given below:

- It cannot be used to describe the response and predict damage for moderate or service load levels.

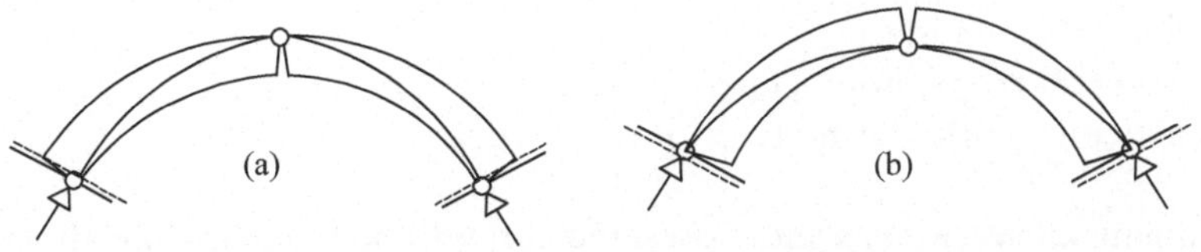

Figure 7.26 Static equilibrium of the arch. Continuous lines drawn within the arches (a) and (b) are possible admissible thrust lines. (Adapted from Mainstone, R. (1975). *Developments in structural form*. Penguin Books Ltd, England.; UNDP/UNIDO (1984). *Repair and Strengthening of Historical Monuments and Buildings in Urban Nuclei*, 6, Vienna.)

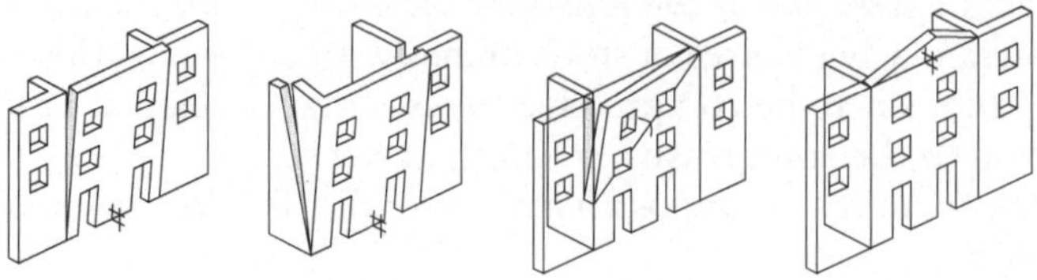

Figure 7.27 Some failure mechanisms for buildings embedded within urban texture. (Adapted from D. Ayala and Speranza, 2003.)

- It cannot predict early failure due to shear or buckling.
- It cannot provide rotation demands θ_u at the plastic hinges for comparison with the rotation capacities of the structural elements at their ends.

In spite of its origin in the past, limit-state analysis is regarded even today as a powerful tool for the safety verification of structures composed of not only arches but also towers, facades and entire buildings. The works of Fajfar (1999) and Lagomarsino et al. (2003) constitute interesting methods applicable to buildings, churches and towers. These methods are based on the observation of real seismic failure modes of historical buildings in Italy, where a kinematic limit-state analysis is implemented (Figure 7.27). Furthermore, the above method has been adopted by the seismic Italian code OPCM 3274 (Ordinanza OPCM3274, 2003) (Chiozzi, et al., 2018).

7.6 NONLINEAR ANALYSIS

7.6.1 General

The limit-state method presented in the previous section was developed in the framework of an effort to overcome the computational difficulties that a step-by-step procedure posed, due to the successive changes of stiffness, from step to step, till collapse.

The storming development of computer technology in the last 50 years has given the ability to structural mechanics for an insight approach to various special mechanical problems of materials like anisotropy, asymmetry, geometrical and material nonlinearities, zero tension or compression strength etc. The basic procedure for approaching the above issues is a step-by-step method combined with the discretisation of the structure to finite elements.

The above procedure covers a very extensive research zone for masonry, laminates, soil etc. which is out of the scope of this book.

Particularly for masonry structures, the following approaches have been developed:

- Equivalent frame models
- Isotropic–orthotropic homogeneous nonlinear material macromodels
- Heterogeneous nonlinear material micromodels.

In the subsequent subsections, a short presentation will be made for the above three main approaches.

Closing, it should be noted that, in the last two decades, the nonlinear static analysis has gained significant impact for the displacement-based analysis of earthquake. In fact, the combination of the *capacity curve,* which is the main outcome of the inelastic

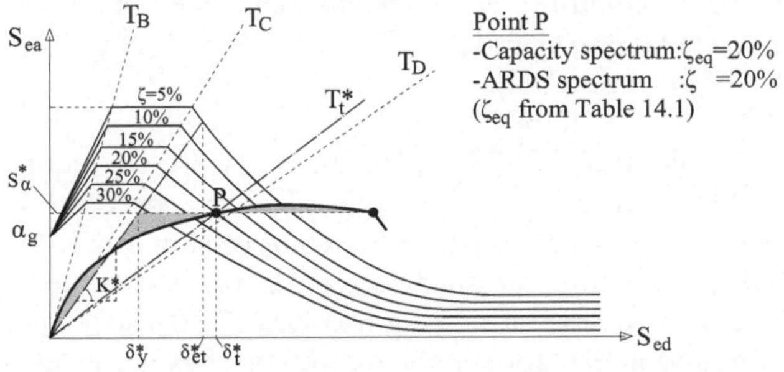

Figure 7.28 Determination of the target displacement implementing an iterative procedure. (Adapted from Penelis, G. and Penelis, Gr. (2019). *Concrete buildings in seismic regions*. CRC Press (A SPON Press Book), London.)

static analysis, transformed to a *capacity spectrum*, with the corresponding acceleration demand response spectra (ADRS) gives a reliable evaluation of the structural safety of the building (Figure 7.28) (Penelis G. and Penelis Gr., 2019) in terms of displacement (ratio of displacement demand to predicted capacity).

The scope of this book does not allow a more detailed discussion on this issue which constitutes one of the main objects of modern seismic design of structures (Penelis G and Penelis Gr., 2019).

7.6.2 Nonlinear analysis of an equivalent frame model

Despite the extended research so far in modelling of URM walls using FEM for inelastic behaviour (see Subsection 7.6.4), the application of this procedure of microelement analysis to real structures has been limited so far only to very few cases. The main reasons are as follows:

a. The required very refined discretisation (bricks, mortar and their interface) that may exceed the capacity of commonly available hardware
b. The required numerous data (mechanical properties of bricks, mortar and their interface), which are not easily available for everyday practice.

So, there has been a lot of work on modelling URM buildings with equivalent frames, and the reports mentioned here are only the ones of greater interest and importance.

Tomazevic (1997) developed and later improved the storey mechanism approach, which assumes a simple bilinear nonlinear relation for every pier of each storey, and predicted the seismic behaviour of multistorey masonry buildings with satisfactory accuracy in comparison to experimental results provided by literature.

Penelis et al. (1984) used a nonlinear static step-by-step analysis of plane frames to simulate the church of Hagios Andreas (Figure 7.23) and the Christian part of the Rotunda. This was probably one of the first attempts to perform pushover analysis on masonry structures.

Magenes and Della Fontana (1998) presented an approach for the nonlinear analysis of URM buildings using *an equivalent frame point hinge model* which uses an elastic–plastic bilinear diagram for both flexure and shear. The behaviour is considered elastic until either moment or shear exceeds the element's capacity, and the elastic stiffness is controlled with the extent of the rigid offsets. The moment capacity is defined using section analysis with rectangular compression stress block, while the shear capacity is defined using experimentally calibrated equations mentioned in a paper by Magenes and Calvi (1997). The ultimate flexural and shear deformation is defined based on experimental data as 1% and 0.5%, respectively.

Penelis Gr. (2006b) has developed a methodology similar to the method presented by Magenes and Della Fontana (1998). Both methods use the equivalent frame point hinge model for URM structures with openings. The proposed model uses an analytical approach for flexure utilising the parabolic compression stress block assumption, a Mohr–Coulomb failure criterion for shear strength and experimentally derived shear stiffness and ultimate shear deformation (Figures 7.29, 7.30 and 7.31).

The use of these methodologies for practical purposes has many advantages, since they allow the application of the explicit static nonlinear methods, using commercial software codes, on URM buildings, something that has been extensively used for R/C and steel buildings over the past few years, and it is foreseen in future rehabilitation guidelines. It allows also the prediction of the failure mechanism by observing the development of plastic hinges at different locations of the building. It should, however, be noted that these plastic hinges are not the actual failure mechanism of URM (being a brittle material) but a computational simulation of the nonlinear behaviour of the whole member they correspond to.

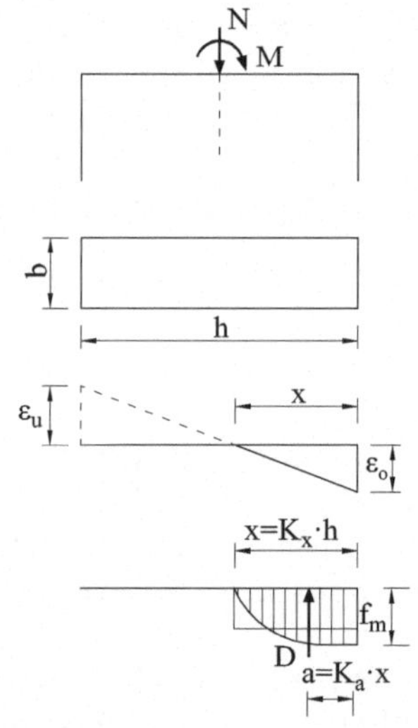

Figure 7.29 Flexural inelastic deformation of a masonry pier. (Adapted from Penelis, Gr., 2006b.)

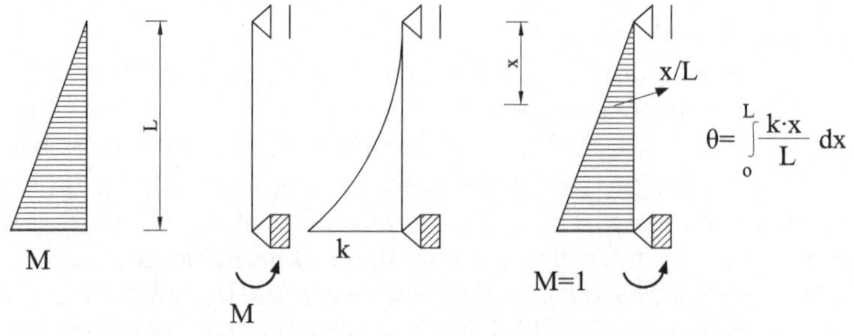

Figure 7.30 Determination of a M–θ diagram. (Adapted from Penelis, Gr., 2006b.)

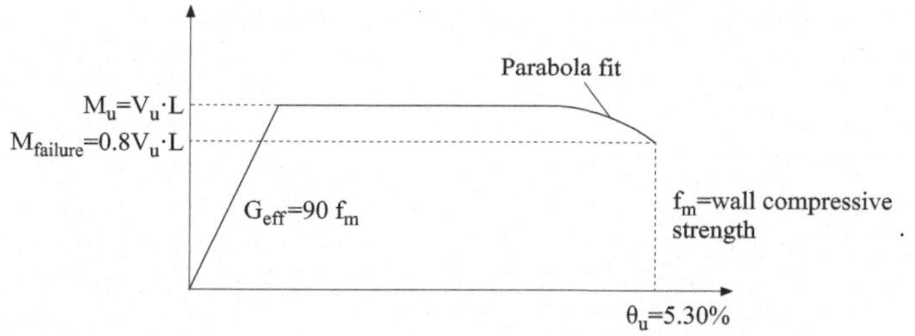

Figure 7.31 Adopted shear mode diagram $V\text{–}\theta_{shear}$. (Adapted from Penelis, Gr., 2006b.)

It is interesting to note that Penelis approach to the equivalent frame models was used for the pushover analysis of more than 40 3D reinforced (URM) prototype buildings for the production of *capacity curves* and *vulnerability curves* (Figures 7.20, 7.32 and 7.33) within the RISK UE project, the so-called European HAZUS (Penelis, Gr. et al., 2003). It was also used in the structural restoration of the National Library of Greece in Athens, extensive reference to which is made at the end of this book in the case studies, (see Chapter 9).

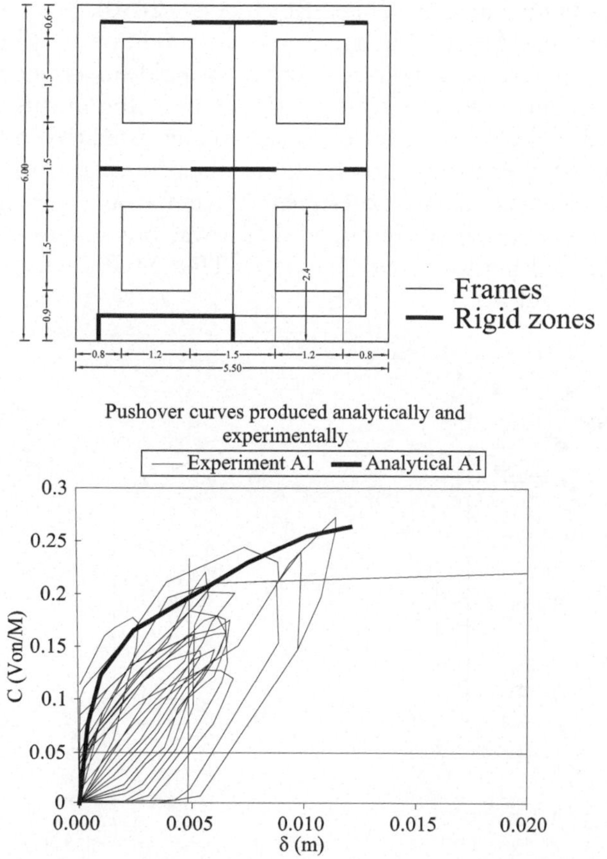

Figure 7.32 ISMES wall model and pushover curve. (Adapted from Penelis, Gr., 2006b.)

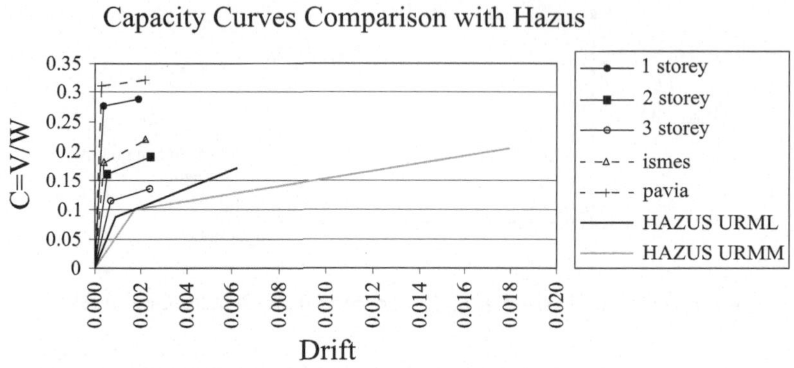

Figure 7.33 Comparison of typical capacity curves to experimental data and HAZUS proposals. (Adapted from Penelis, Gr., 2006b.)

The work of Roca, et al. should also be mentioned here (Roca, et al., 2005). These researchers developed an analysis procedure for 3D systems including masonry structural walls simulated with equivalent frames. The method has been successfully used in the assessment of facades and entire buildings (Roca, et al., 2010).

Finally, it should be noted that in the same category might be classified the development of a computational approach based on rigid and deformable macroelements. Each macroelement represents an entire wall or masonry panel, reducing in this way the number of degrees of freedom drastically. Brencich et al. (1998) have used two-node macroelements, taking into account the overturning and frictional shear mechanisms.

The analytical model was verified experimentally in masonry panels, as in Penelis procedure. The whole procedure is so simplified that it was possible to use it in the development of a computational platform with the name TRE MURI which is very efficient in practice (Figures 7.34 and 7.35).

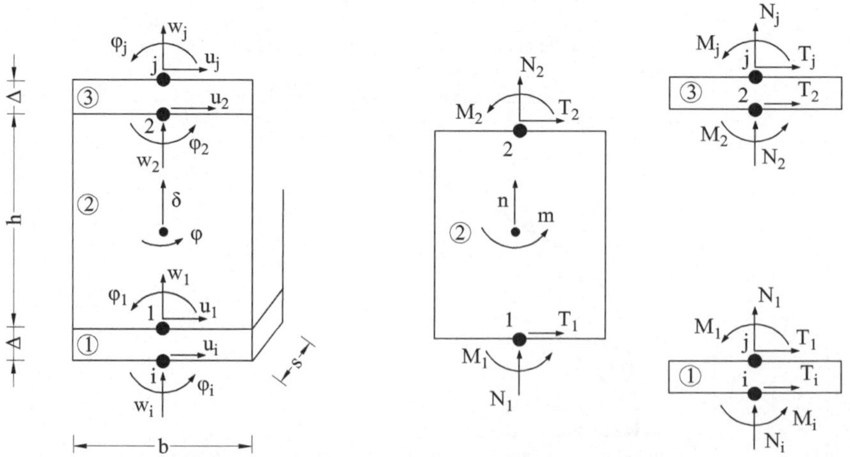

Figure 7.34 Kinematic model of a macroelement. (Adapted from Gambarotta et al., 1996, 13th World Conference on Earthquake Engineering.)

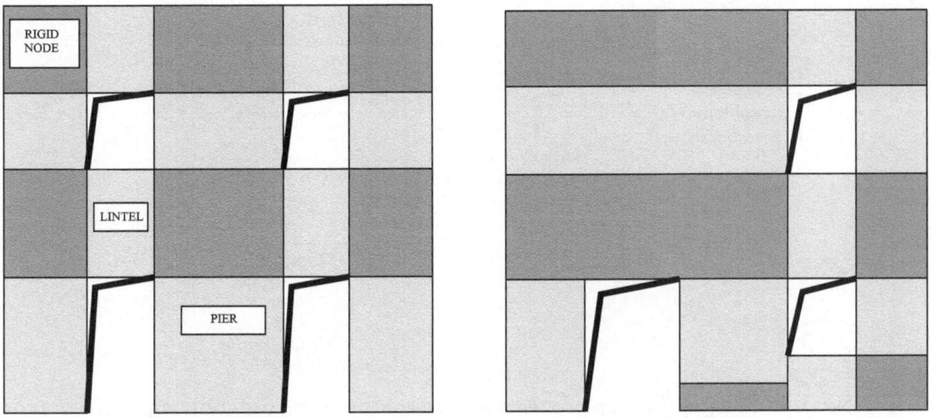

Figure 7.35 Examples of macroelement modelling of masonry walls. (Adapted from Galasco, A., Lagomarsino, S., Penna, A., and Resemini, S., (2004). Non-linear seismic analysis of masonry structures. *13th World Conference in Earthquake Engineering.* Paper No. 843.)

7.6.3 Isotropic–orthotropic homogeneous, nonlinear material macromodels, for 2D structures

The methods based on *isotropic–orthotropic homogeneous nonlinear material assumption* rely mainly on homogenisation techniques through which the properties of the composite material (masonry) are derived from the two individual components (bricks and mortar) and their bond. The most recent applications of this approach allow smeared cracking to simulate the nonlinearity due to cracking of the material (Addessi, et al., 2018). Characteristic applications of this approach are the ones by Page et al. (1985) and Gambarotta et al. (1995). The main advantage of the method is that the nonlinear analysis of masonry structures is possible with the use of only few mechanical properties of the composite material in contradiction to the more advanced and detailed heterogeneous material approaches which use properties of each of the components as well as their interface. So, this methodology leads to significantly less computational time, creating, thus, the ability to analyse complete structures. The drawback of this approach is that, through homogenisation, the local weaknesses of the brick or the mortar are not simulated, leading only to an overall simulation of the nonlinear behaviour of URM structures (Bertolesi, et al., 2018).

In this category might also be classified a method developed by the authors of this book (Penelis, G., et al., 2010). In this method, using the well-known FEM procedures, a widely used and reliable commercial linear analysis software package, SAP2000, is utilised as the core processor of a pure step-by-step loading approach up to collapse. To this task, a properly developed pre-processor and post-processor are used for communication with the built-in API (application programming interface) capabilities of SAP2000. The pre- and postprocessors modify the flexural (EI) and axial stiffness (EA) of each finite element, by properly using the stress data from the previous step via the available in SAP2000 stiffener modifiers, while at every step, the failure mode of each finite element is graphically mapped (crush, crack, shear).

It is obvious that all these are available by including in the postprocessor *failure criteria* for masonry, relations to calculate the *stiffness modifiers* and parameters to take into

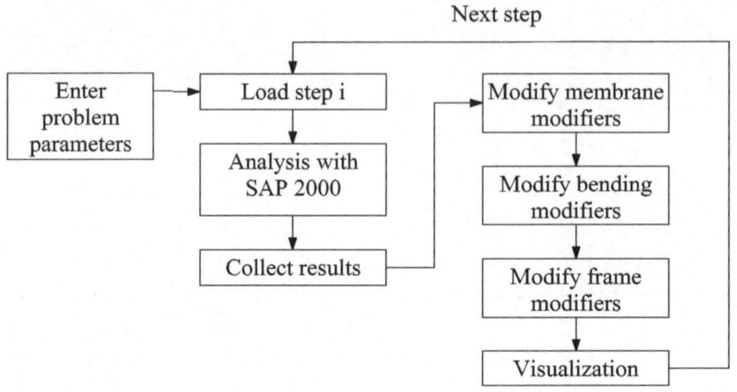

Figure 7.36 Flow chart of a step-by-step procedure up to collapse using pre- or postprocessors together with the main core platform (SAP2000). (Adapted from Penelis, G., et al., 2010.)

account the orthotropy of masonry direction of bed joints (Figure 7.36). The above issues are presented in short below.

i. Stiffness and failure criteria of frame elements

For frame elements, the strength envelope (M–N) introduced in the procedure is shown in (Figure 7.37).

For M–N pairs outside the envelope curve, the element fails, and the stiffness modifier is set to zero, while for pairs representing points within the interaction envelope, the following values are calculated for the stiffness modifiers:

$$\text{for } N_{crk} > N > -[N_3], \quad m = 1.0 \text{ (uncracked section)}, \tag{7.3}$$

$$\text{for } -[N_3] > N > -[N_1], \quad m = m = \frac{k_{eff}}{k_o} \text{ (gradual cracking for compressed section)}, \tag{7.4}$$

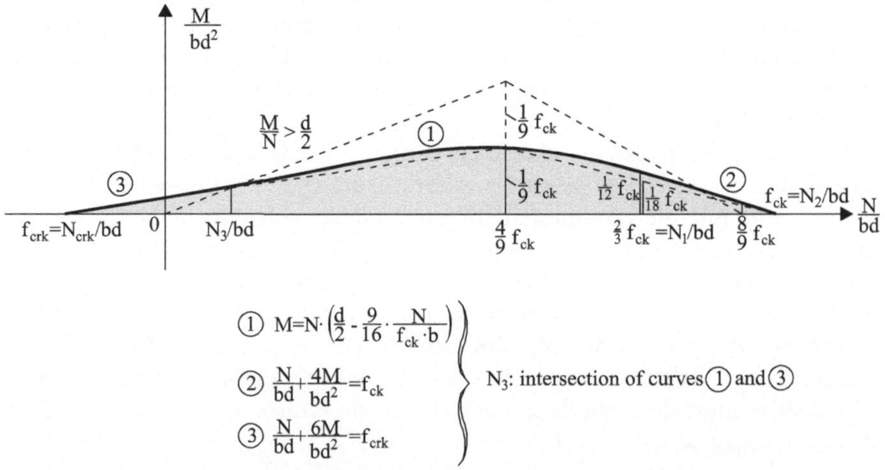

Figure 7.37 M–N interaction curve (strength envelop) for rectangular cross section.

where
k_{eff} is the reduced rigidity of element
k_o is the initial rigidity of element (uncracked).

$$\text{For } -[\mathrm{N}_1] \geq \mathrm{N} > -[\mathrm{N}_2], \qquad m = 1.0(\text{pure compression}).$$

The required input mechanical properties are

- The tensile strength of masonry f_{crk}
- The compressive strength of masonry f_{ck}
- The failure strain $\varepsilon_{\text{lim}} = 2‰$
- The bedjoint cohesion f_{vko}
- The bedjoint coefficient of friction μ.

It should be noted that k_{eff} takes into account the shear deformation, which is significant to squat piers.

ii. Stiffness and failure criteria of membrane elements

It is known that in case of orthotropic materials such as masonry, with orthotropy axes x–x and y–y (the usual case in masonry buildings), the stress–strain relationship for elastic material properties follows the following form (Darwin–Pecknold) (Ignatakis, 2000):

$$\begin{bmatrix} d\sigma_x \\ d\sigma_y \\ z \end{bmatrix} = \frac{1}{1-v^2} \cdot \begin{bmatrix} E_x & v \cdot \sqrt{E_x \cdot E_y} & 0 \\ v \cdot \sqrt{E_x \cdot E_y} & E_y & 0 \\ 0 & 0 & (1-v^2) \cdot G \end{bmatrix} \cdot \begin{bmatrix} d\varepsilon_x \\ d\varepsilon_y \\ d\varepsilon_z \end{bmatrix} \tag{7.5}$$

$$\text{or } \{d\sigma\} = D\{d\varepsilon\},$$

where

$$v = \sqrt{v_x \cdot v_y} \qquad \text{(Poisson's ratio in } x\text{–}x,\ y\text{–}y),$$

$$(1-v^2) \cdot G = \left(E_x + E_y - 2 \cdot v \cdot \sqrt{E_x \cdot E_y / 4}\right).$$

In the above relations, the values of E_x and E_y and ν_x and ν_y represent the modulus of elasticity and Poisson's ratios, respectively, for the $\{d\sigma\}$ load step, while it should be taken into account that for orthotropic material the following relation applies (Ignatakis, 2000; Magenes and Calvi, 1997; Page, 1981, 1983):

$$(E_x) \cdot (e) = (E_y \cdot v_y). \tag{7.6}$$

The strength envelope of masonry is a function of the principal stresses σ_1, σ_2 and the angel θ of the orthotropy angle (Page et al., 1982), which is presented in Figure 7.38 for $\theta = \pi/2$. An algorithm for the calculation of the 3D surface of stresses σ_1, σ_2 and θ has been introduced in the postprocessor using closed-form formulas simulating the

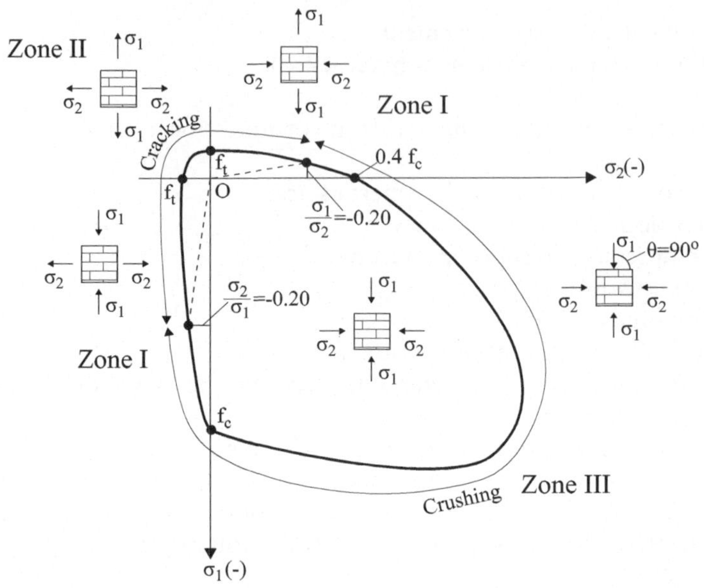

Figure 7.38 σ_1–σ_2 envelope of URM for $\theta = 90°$.

experimental curves produced by Page (1981, 1983). For triads (σ_1, σ_2, θ) representing points outside the 3D surface, the stiffness modifiers m_x, m_y and m_{xy} are set to zero, while in other cases, they have values ranging from 1 to 0.

The required input mechanical properties are

- The prismatic strength of the bricks/stone: f_{bk}
- The limit shear strength of masonry: $f_{vk}^{\lim}$
- The compressive strength of masonry perpendicular to the bedjoints: f_{ck}^{v}
- The cohesion of masonry: f_{vko}
- The tensile strength of masonry for flexure perpendicular to the bedjoints: f_{tk}^{v}
- Poisson's ratio perpendicular to the bedjoints ν^{v}
- The coefficient of friction in the bedjoint (mortar–brick/stone): μ.

iii. Stiffness and failure criteria of shell elements

The shell elements combine the membrane stress with flexural stress. Taking into account that the membrane stress state has been discussed in the previous paragraph, a summary of its combination with flexure will be presented herein.

- The principal moments M_1 and M_2 and their direction angle θ to the orthotropy axes are defined from the flexural stress state, i.e. moments M_x, M_y and M_{xy}.
- The axial forces N_θ and $N_{\theta+s\pi/2}$ are calculated for this direction θ.
- The stress state for the combinations (M_1, N_θ) and (M_2, $N_{\theta+\pi/2}$) is checked, i.e. flexure with axial force.
- The failure criteria and corresponding stiffness for the previous combinations are defined for directions θ and ($\theta + \pi/2$). The formulas given in the paragraph for frame elements are utilised here also since the procedure followed leads to two uncoupled stress states.

The required input mechanical properties are the same as the ones in the membrane case.

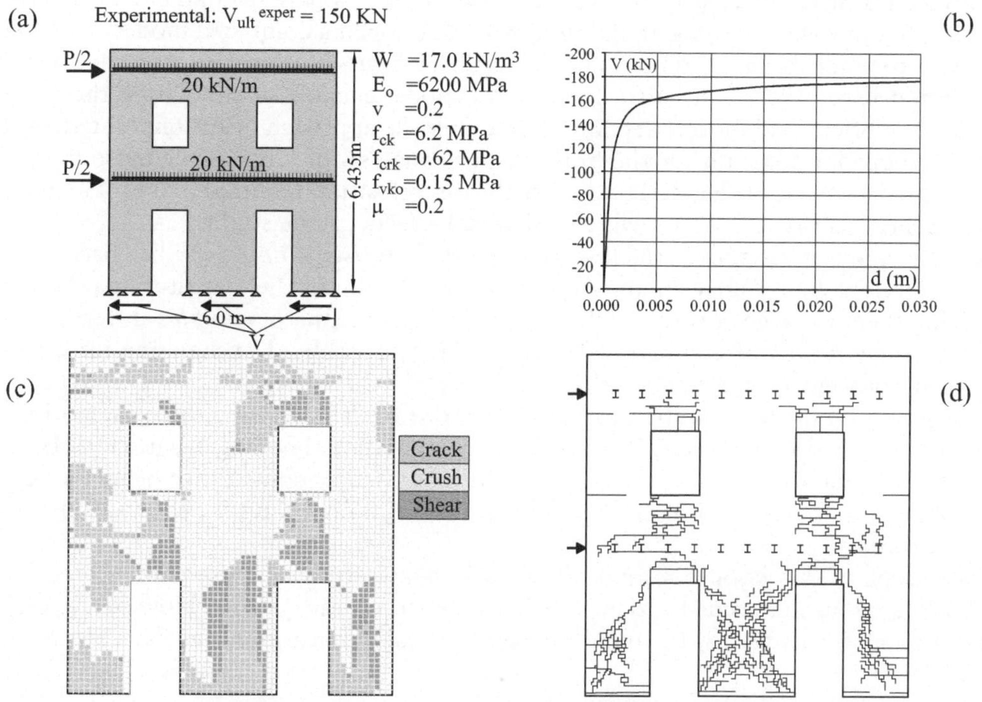

Figure 7.39 Pavia wall. Analytical results versus experimental outcome: (a) geometric and mechanical properties; (b) V–d diagram (analysis); (c) analytical damage pattern for 0.43% drift and (d) experimental damage pattern for 0.43% drift. (Adapted from Penelis, G., et al., 2010.)

The structural model may consist of membrane elements (i.e. for domes), shell elements (for walls) and frame elements (for piers) which are all available in a tailor-made software package (Figure 7.39).

It should be noted that the above procedure is a static inelastic one, and, therefore, it is proper for the analysis for dead loads and for the *push over analysis* for dead loads combined with a step-by-step applied horizontal seismic action up to collapse.

Closing, it should be noted that macromodels have been extensively used for analysing various types of historical constructions (buildings, bridges, mosques and cathedrals) (Pela, et al., 2009; Roca, et al., 2004; Martinez, et al., 2006).

7.6.4 Heterogeneous nonlinear microelement models for 2D structures

Methods based on *heterogeneous nonlinear material assumptions* are the ones that simulate the composite material (masonry) in the most accurate way, by including different microelements of separate mechanical properties for the two components of masonry (bricks, mortar joints) as well as for the joint–brick interface. The most advanced of these models use a smeared cracking approach, which, combined with the refined discretisation required for the approach, help to simulate the crack pattern in an accurate way. Examples of such approaches are as follows: the approach of Ignatakis, Stavrakakis and Penelis (1989) who developed a model for the in-plane analysis of masonry and applied it on a real structure which was already structurally restored, namely Rotunda of Thessaloniki (1992), the approach of Shing et al. (1992) who developed a combined model using linear elements to model bricks and nonlinear interface elements to model mortar, the approach of Ali and Page (1988) who also developed a model capable of simulating the gradual cracking of bricks and mortar separately and later the approach of Antonuci et al. (1989) who developed a program for the finite element analysis of larger-scale masonry walls. The main drawback of the method is that the discretisation required makes it applicable only to small substructures, on which only local effects can be studied.

For the above reason, over the last decades, *the multiscale approach* has been considerably developed aiming at an automatic homogenisation of the constitutional laws and applying them to large regions of the structure. This technique *couples* different scales of interest by means of compatible transition laws, capable of exchanging information between consecutive scales (Giambanco, G., et al., 2018; Nguyen, et al., 2011). The homogenisation method can be classified under one of the following three categories: the analytical (AH), the numerical (NH) and the computational homogenisation (CH).

The last one allows the determination of the macroscopic constitutive behaviour *of the fly* solving a unit cell (UC) boundary value problem (BVP) (Nguyen, et al., 2011; Kenit, et al., 2003).

The scope of this book does not allow an in-depth study of these complicated issues of heterogeneous materials for their use in whole structural systems. However, extensive literature on this issue may be found in recent special publications (Sacco, et al., 2018).

7.7 DESIGN

7.7.1 Distinction between conventional and seismic loading

It should be noted in advance that there is a substantial difference in the design of structures between the case of conventional loading (e.g. gravity loads, earth pressure, wind, hydrostatic pressure etc.) and that of seismic actions.

7.7.1.1 Conventional loading

In the first category, if the *structural resistance* R_k is bigger than the action effects E_k (internal forces) by a specified percentage (safety factor), the structure is considered safe.

This means that if

$$\frac{[R_k]}{[E_k]} \geq \gamma_f \cdot \gamma_m = \gamma_F \tag{7.7a}$$

or

$$\boxed{\frac{[R_k]}{\gamma_m} \geq \gamma_f \cdot [E_k],} \tag{7.7b}$$

then the structure is covered by a safety factor:

$$\gamma_F = \gamma_f \ \gamma_m. \tag{7.8}$$

In the above equations, the meanings of the notations used are as follows:

$[R_k]$ is the characteristic resistance of the structure
$[E_k]$ is the characteristic action effect on the structure
γ_f, γ_m are partial safety factors for loads and materials (both greater than unity).

The above considerations are deemed to ensure the safety of the structure to an acceptable level of probability. The above considerations are valid for *linear elastic* or *inelastic structures.*

7.7.1.2 Seismic loading

In the second category, seismic action is introduced at the base of the structure in the form of a vibration of high frequency and short duration (a few seconds). Consequently, a limited amount of kinetic energy is introduced to the mass of the structure via the elastic or inelastic links of the mass to the foundation ground as long as the seismic motion lasts.

If the structure is in a position to transform the induced kinetic energy into potential strain energy without failure and then to dissipate a considerable percentage of this during its cyclic vibration in the form of hysteretic damping, then the structure can be considered safe.

Therefore, if

$$\frac{\left[W^u_{\text{supply}}\right]}{\left[W_{\text{demand}}\right]} \geq \gamma_w \cdot \gamma_m = \gamma_d \tag{7.9a}$$

or

$$\boxed{\frac{\left[W^u_{\text{supply}}\right]}{\gamma_m} \geq \gamma_w \cdot \left[W_{\text{demand}}\right]} \tag{7.9b}$$

in which

W^u_{supply} is the strain energy that the structure can absorb until failure
W_{demand} is the kinetic energy that is induced to the structure during the vibration
γ_w is the safety factor with respect to energy absorption,

then the structure can undergo the earthquake motion with safety factor

$$\gamma_d = \gamma_w \, \gamma_m. \tag{7.10}$$

It is worth reviewing the preceding remarks in the diagrams of Figures 7.40 and 7.41.

Consider a building (Figure 7.40) which is loaded by permanent gravity loads and a constantly increasing lateral load P_i which in turn causes a *base shear* V_i, which also is increasing continuously until *failure.*

This lateral loading results in a displacement of the top of the building relative to its base and equal to u. If the successive steps of V_i–u, are recorded in a diagram, a curve will be formed, known as the *capacity curve of the structure* (Figure 7.41a).

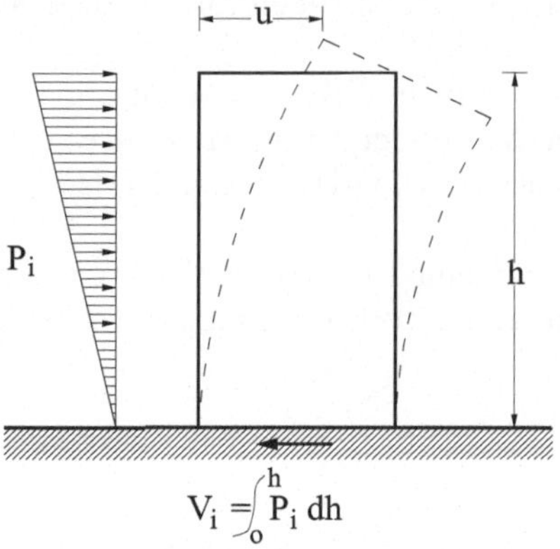

Figure 7.40 Lateral load P_i of a building developing base shear V_i and respective top horizontal displacement u. (Adapted from Penelis, G. and Penelis, Gr. (2019). *Concrete buildings in seismic regions*. CRC Press (A SPON Press Book), London.)

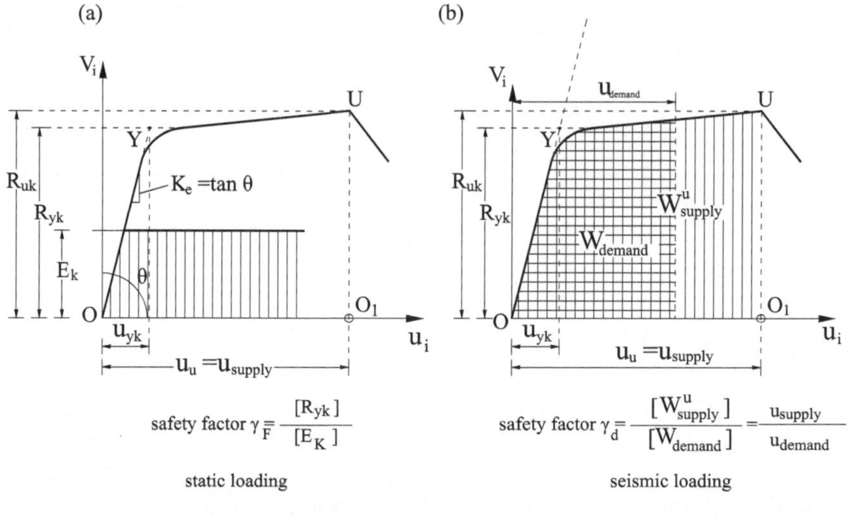

Figure 7.41 The main concept of safety verifications for (a) conventional loads; (b) seismic loads. (Adapted from Penelis, G. and Penelis, Gr. (2019). *Concrete buildings in seismic regions*. CRC Press (A SPON Press Book), London.)

This curve has the following characteristic properties:

- It has first an *almost linear* branch of high slope (stiffness $K_e = \tan\theta$).
- At the end of this branch, there is the *yield point* Y, where the slope of this branch is reduced drastically. The coordinates of this point are known as the *yield strength* ($R_{\gamma\kappa} = V_{\gamma\kappa}$) and the *yield strain* ($u_{\gamma\kappa}$) of the structure.
- The second branch of the diagram starts from the yield point Y with a very small slope (strain hardening) and continues up to point U with coordinates $R_{uk} \cong \left(1.05 \div 1.20 R_{\gamma k}\right)$ known as strength at failure and u_u (strain at failure).

After this point, *collapse takes place,* that is, an abrupt loss of strength capacity for small additional displacement. The area $\left(W_{\text{supply}}\right)$ of the surface under the capacity curve of the structure ($OYUO_1$) represents the potential strain energy that can be accumulated in the structure for monotonous lateral loading up to collapse.

- For almost three centuries, the structural design for almost all types of loading, *except seismic actions*, has been based on the *basic requirement* that action effects $[E_\kappa]$ should not exceed a percentage of the yield strength $[R_{\gamma\kappa}]$ of the structure (Equations 7.7a and 7.7b) (Figure 7.41a). The reasoning behind this basic concept is that, since loading is not instantaneous and acts permanently (static loadings), if it exceeds the yield capacity $[R_{\gamma\kappa}]$, the structure presents big deformations for small changes of loading; in other words, the system stops to be stable, and therefore it fails and collapses. Therefore, the safety margin must be based on strength capacity, and inequalities 7.7a and b must be used.

In case of seismic action (Figure 7.41b), the vibration of the mass of the structure is the result of kinetic energy $\left[W_{\text{demand}}\right]$ that has passed from the foundation ground to the vibrating mass through the elastic or inelastic links between ground and mass. This energy $\left[W_{\text{demand}}\right]$ is *limited*, due to the fact that the seismic duration is limited. Therefore, it is displayed in the form of a number of displacement reversals. In this context, it is accepted that in case that yield strength $[R_{\gamma\kappa}]$ is exhausted, it would be enough if the displacement $\left[u_{\text{demand}}\right]$ demand corresponding to the energy demand $\left[W_{\text{demand}}\right]$ does not exceed a specified percentage of the displacement capacity at failure u_u and therefore a percentage of energy dissipation capacity W^u_{supply}:

$$u_u = u_{\text{supply}},$$

$$W_u = W_{\text{supply}}.$$

The above reasoning is expresses in Equations 7.9a and b. By taking into account that

$$\left[W^u_{\text{supply}}\right] \cong u_{\text{supply}}\left[R_{yk}\right] \tag{7.11a}$$

and

$$\left[W_{\text{demand}}\right] \cong u_{\text{supply}}\left[R_{yk}\right], \tag{7.11b}$$

it is concluded that

$$\frac{\left[W^u_{\text{supply}}\right]}{\left[W_{\text{demand}}\right]} \cong \frac{u_{\text{supply}}}{u_{\text{demand}}} \geq \gamma_w \cdot \gamma_m = \gamma_d. \tag{7.12}$$

Now introducing the equivalent u_{supply} and u_{demand}, i.e.

$$u_{\text{supply}} = \mu_{\text{supply}} u_y, \tag{7.13a}$$

$$u_{\text{demand}} = \mu_{\text{demand}} u_y, \tag{7.13b}$$

where μ is the ductility of the structure $\left(\mu = \frac{u_u}{u_y}\right)$ (Figure 7.41), the following equation is obtained:

$$\frac{\left[W^u_{\text{supply}}\right]}{\left[W_{\text{demand}}\right]} \cong \frac{u_{\text{supply}}}{u_{\text{demand}}} = \frac{\mu_{\text{supply}}}{\mu_{\text{demand}}} \geq \gamma_w \cdot \gamma_m = \gamma_d, \tag{7.14}$$

where γ_w is the safety factor referring to displacements.

From the above analysis, the following conclusions may be drawn:

i. For seismic actions, *and only for these*, it is accepted that, during the earthquake, the yield resistance $[R_{\gamma K}]$ of the structure may be completely exhausted. This means that it is accepted that due to the seismic action, the structure may pass into the inelastic range of the capacity curve.
ii. For seismic actions, *and only for these*, it is accepted that the safety factor *may be satisfied on a displacement basis*, which means that the maximum displacement $\left[u_{\text{demand}}\right]$ must not exceed a specified percentage of $\left[u_{\text{supply}}\right]$ (Equations 7.12 and 7.14)
iii. The above considerations constitute the core of the modern philosophy of seismic design, as it is carried out in the form of *displacement design* (Oliveira and Costa, 2010).

7.7.2 Displacement-based design for seismic loading

7.7.2.1 *Inelastic dynamic analysis and design*

Taking into account the acceptance of inelastic response of the structure during an earthquake in the inelastic range, which means the acceptance of a specified level of damage, the direct approach to the problem of the analysis and design of seismic-resistant structure includes the following steps:

i. Determination of the capacity curve of the structural system
ii. Dynamic inelastic analysis of the structure for a series of normalised accelerograms to a code-specified intensity (SI)
iii. Comparison of u_{supply} of the capacity curve with $u^{\max}_{\text{demand}}$ derived from the dynamic inelastic analysis of the structures (Figure 7.42).

The above procedure should include verifications for all crucial parameters. These are

- Internal forces M, V, N
- Relative displacements
- Rotations θ of the nodes
- Interstorey drift.

For the time being, this procedure is used only in research and in special design cases (e.g. base-isolated systems).

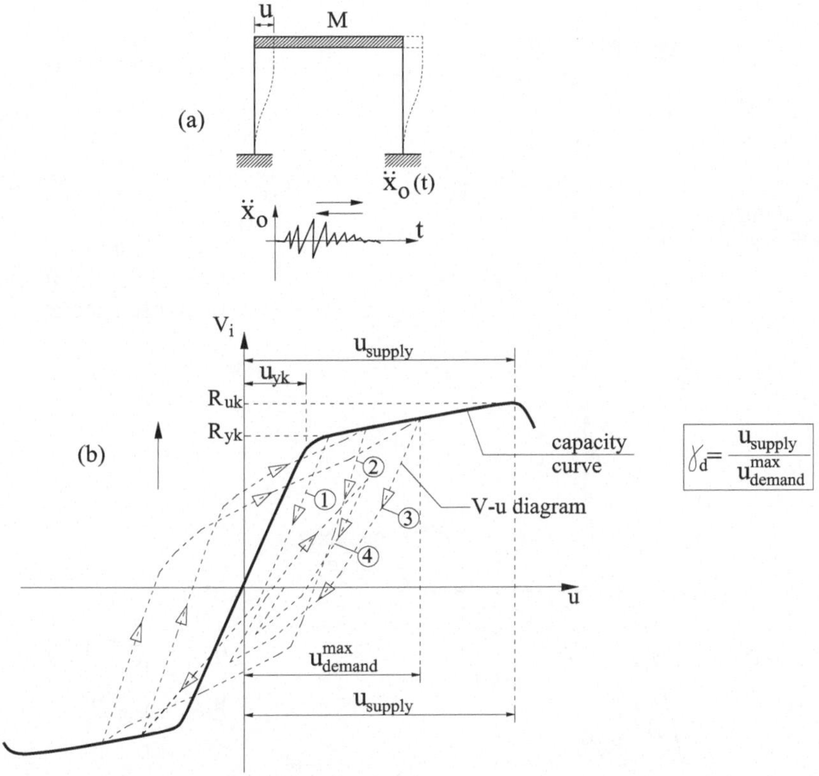

Figure 7.42 Displacement-based design based on inelastic dynamic analysis: (a) the structural system under seismic excitation; (b) V_i–u_i curve for seismic loads and capacity curve of the structure. (Adapted from Penelis, G. and Penelis, Gr. (2019). *Concrete buildings in seismic regions*. CRC Press (A SPON Press Book), London.)

7.7.2.2 Inelastic static analysis and design

This method has been developed in the last 20 years, and it is implemented for the assessment, verification and retrofitting of existing buildings.

A conceptual approach to this method will be given here for the single-degree-of-freedom (SDOF) system depicted in Figure 7.43a.

Step 1: Calculation and plotting of the capacity curve of the system (Figure 7.43b). This procedure assumes the implementation of an inelastic static step-by-step analysis of the system for an increasing lateral force ('pushover analysis').

Step 2: Determination of the fundamental period T_o, assuming linear elastic response $T_o = 2\pi\sqrt{M/K_e}$.

Step 3: Determination of the base shear of the system $V^{\circ}_{\text{demand}}$, assuming linear elastic response, using the elastic acceleration response spectrum (Figure 7.43c), that is,

$$V^{\circ}_{\text{demand}} = \ddot{a}_{\max(T_0)} \cdot M. \tag{7.15}$$

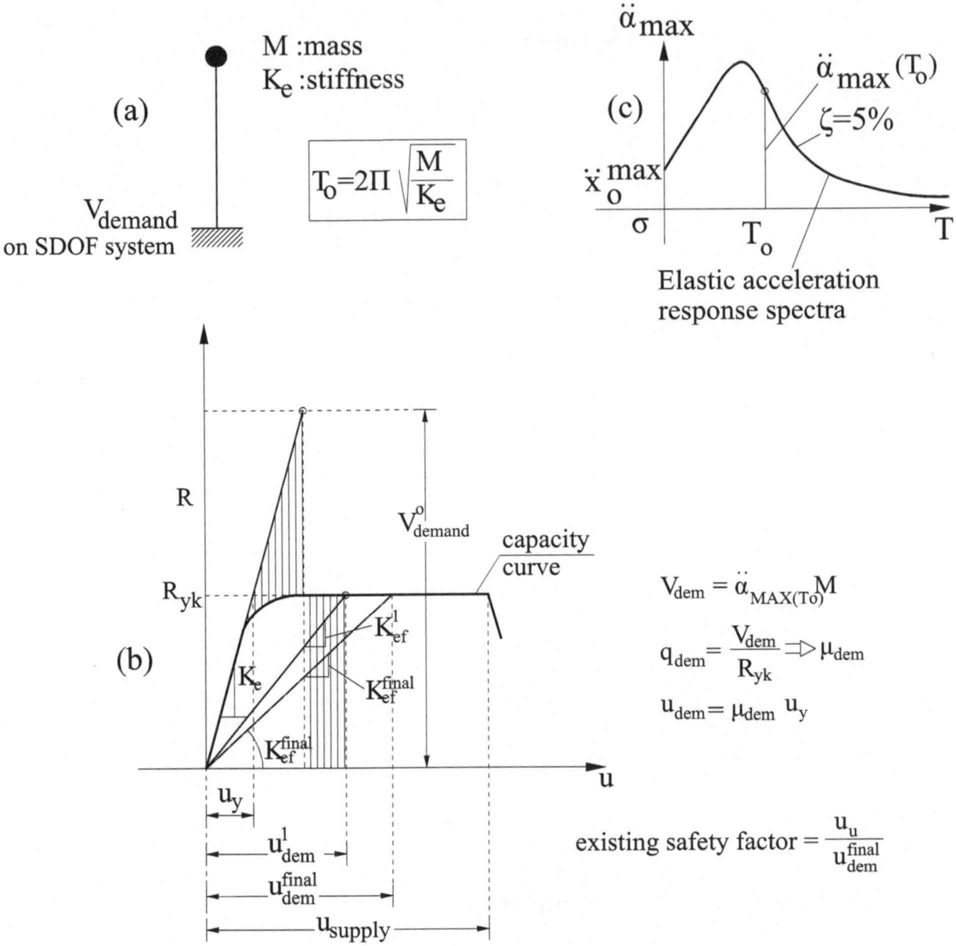

Figure 7.43 Displacement-based design based on inelastic static analysis (pushover analysis): (a) structural system; (b) capacity design curve and successive steps for the determination of u demand and (c) elastic acceleration response spectrum. (Adapted from Penelis, G. and Penelis, Gr. (2019). *Concrete buildings in seismic regions*. CRC Press (A SPON Press Book), London.)

Step 4: Determination of the behaviour factor q_{demand} equal to

$$q_{\text{demand}} = \frac{V^{\circ}_{\text{demand}}}{R_{yk}}. \tag{7.16}$$

From q_{demand}, μ^1_{demand} is determined easily (Penelis, G. and Penelis, Gr., 2019).

Step 5: Using the relation

$$u^1_{\text{demand}} = \mu^1_{\text{demand}} \cdot u_y, \tag{7.17}$$

u^1_{demand} is determined (Figure 7.43b).

Step 6: Determination of the new effective stiffness K_{ef}^1 equal to (Figure 7.43b)

$$\boxed{K_{ef}^1 = \frac{R_{yk}}{u_{\text{demand}}^1}.} \tag{7.18}$$

Step 7: For this new stiffness K_{ef}^1, steps 2 through 6 are repeated until successive iteration steps result in very close values for K_{ef}.

Step 8: For the final $u_{\text{demand}}^{\text{final}}$ (Figure 7.43b), the existing safety factor is determined and compared with a codified safety factor.

$$\boxed{\frac{u_{\text{supply}}}{u_{\text{demand}}^{\text{final}}} = \gamma_{\text{existing}} \geq \gamma_d.} \tag{7.19}$$

The above inequality constitutes the safety verification according to *displacement-based design*. Meanings of the notations used in the above equation are given in Figure 7.43.

This method is presented in detail in various manuals for earthquake engineering design (Penelis, G. and Penelis, Gr., 2019). It should be noted that, in this method, the inelastic static analysis requires the elaboration of the capacity curve of the system following a monotonous increasing of the lateral loading (*pushover analysis*).

It is obvious that, for the elaboration of the capacity curve, it is necessary to know.

- The material properties of the structure
- Its geometry.

This means that the structure has been previously designed, and in this respect, this method for the time being is used mainly for the analysis, safety verification and retrofitting of *existing buildings*.

7.7.2.3 Force-based design

This method is implemented according to codes in effect *for the analysis and design of new buildings*, and it is considered nowadays to be the reference method for the seismic design.

In this method, the following procedure is followed:

i. The acceleration and, therefore, the *design base shear*, which result from the linear acceleration response spectrum, are reduced by the *behaviour factor q* ranging between 1.5 and 5.0. This *reduction factor* is specified by the codes of practice, and its value depends on the structural type (Figure 7.44). For these reduced values of inertial forces, the structural system is analysed *assuming linear behaviour.*
ii. Based on the combinations of load effects that have resulted from these inertial forces reduced by q and from the gravity loads wherein masses have been taken into account for the determination of the inertial forces, using *partial load safety factor* $\gamma_f = 1.0$, the dimensioning and safety verification of the structure are carried out for ultimate limit state (ULS), that is,

$$\boxed{[E_d] \leq \frac{[R_k]}{\gamma_m}.} \tag{7.20}$$

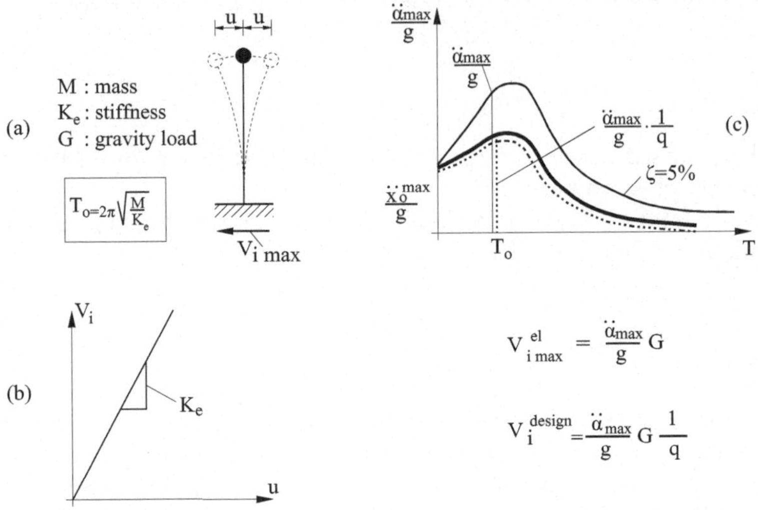

Figure 7.44 Force-based design based on 'model response spectrum analysis' or on 'lateral force method of analysis': (a) structural system under seismic action; (b) V_i–u_i diagram; (c) relation between elastic and design acceleration response spectrum. (Adapted from Penelis, G. and Penelis, Gr. (2019). *Concrete buildings in seismic regions*. CRC Press (A SPON Press Book), London.)

iii. *Modern codes do not provide behaviour factor q for historical buildings.* In this respect, in case this method would be used, *q-factor* must be determined ad hoc (Subsection 7.3.8). In addition, you should be reminded that, due to cracking, the extrapolation of the linear analysis up to failure is unreliable. Therefore, this method is questionable in case of historical masonry structures.

7.7.3 Design considerations

As known, the main task of design is the calculation of the characteristic resistance [R_k] of a structure at all its critical regions, so that the verification of inequity (7.7b) is implemented. It is evident that the determination of the action effects [E_k] (moments, shear, axial forces) has been carried out with the help of the analysis of the structure.

Therefore, either in case of *structural verification* of the original building, or its *redesign*, in case of an intervention, the task is identical, no matter if the structural system is *linear elastic* or *inelastic*.

Particularly, the design for bending with axial force is based on a number of assumptions, no matter what the construction material is. These assumptions are the following:

i. Equilibrium conditions between action effects and internal forces (stress resultants) are considered at the undeformed axis of the structural element (first-order theory).
ii. Stone materials (e.g., concrete masonry) cannot carry tensile stresses.
iii. The principle of *compatibility* is also valid for the deformed cross section. Particularly, in case of 1D structural elements, this principle is expressed by *Bernoulli concept* up to failure (linear distribution of deformations all over the cross section up to failure).
iv. Stress–strain diagrams for design are given by the codes (see Chapter 3), or they are based on experimental evidence.

v. There is perfect bonding between various materials cooperating at a section (steel reinforcement, FRPs etc.) with concrete, masonry or wood.
vi. *Failure criterion* in bending is *the ultimate strain* of the construction material (see Chapter 3).

On the other hand, design for shear is usually carried out using any one of the following conceptual approaches:

i. Implementation of the 'Mohr–Coulomb' failure criterion (see Subsection 3.2.4.1 equation 3.32) (Figure 7.45a):

$$f_v = f_{vo} + \mu_c\ \sigma_d.$$

ii. The failure criterion of exceedance of the critical value of the principal tensile stress (see Subsection 3.2.4.1, equation 3.34) (Figure 7.45b):

$$f_v = \tau_{H\max} = \frac{f_t}{\beta}\sqrt{\frac{\sigma_d}{f_t}} + 1.$$

In case of reinforced texture with steel reinforcement or FRPs, the contribution of reinforcement is added to f_v of the plain texture (masonry or concrete).

It is evident that an attempt for a further exemplification of computational procedures and closed formulas for various materials of the past and contemporary ones in use is meaningless and out of the scope of this book. In fact, the variety of the original materials and the construction techniques of the past, extending to a period of almost 2.5 millennia led to a wide spectrum of case studies. Additionally, the use of modern materials for structural restoration with high strength and the requirement of ensuring their cooperation

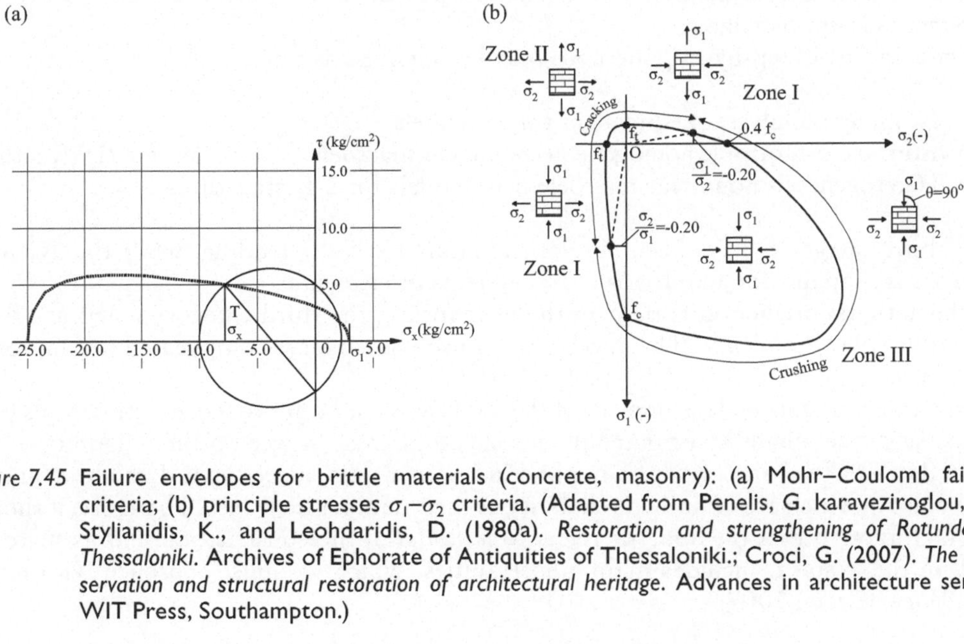

Figure 7.45 Failure envelopes for brittle materials (concrete, masonry): (a) Mohr–Coulomb failure criteria; (b) principle stresses σ_1–σ_2 criteria. (Adapted from Penelis, G. karaveziroglou, M., Stylianidis, K., and Leohdaridis, D. (1980a). *Restoration and strengthening of Rotunda of Thessaloniki*. Archives of Ephorate of Antiquities of Thessaloniki.; Croci, G. (2007). *The conservation and structural restoration of architectural heritage*. Advances in architecture series. WIT Press, Southampton.)

with the original ones calls for the mobilisation of all scientific knowledge and initiative, for a reliable design.

In this respect, an attempt to present calculation procedures for each case would be out of the scope of this book, since the combination of structural elements, intervention techniques and materials would lead to an endless set of case studies with disputable scientific structure. So, we limit our presentation for design within the frame of conceptual issues and only for masonry walls; their design has been discussed in detail in Chapters 3 and 5.

7.8 CONCLUSIONS

The dawn of structural analysis and design coincided with the procedure for the structural restoration of one of the most prominent monuments of Europe, the Saint Peter's dome of Rome (1743 A.D.).

In the subsequent three centuries, *the linear elastic analysis* was developed, basically stimulated by the industrial revolution. In this framework, a sound scientific discipline was born, and this rules the analysis and design of the overwhelming majority of contemporary structures for static or dynamic loads. However, the analysis and design of masonry monumental buildings presents extended inadequacies due to low tensile strength of the material. So, although this method continues to be a useful tool for the analysis and design even of historical buildings, providing a first step of assessment at least for a low level of loading, additional methods have been developed in the last decades for a better approach of the response up to collapse, which are summarised below.

The limit-state method is a genius method which, based on equilibrium principles on one side and on the kinematic theories of the virtual work on the other, allows the easy determination of the collapse load, but without any intermediate information about load–displacement diagrams, shear failure or buckling.

It is the authors' opinion that this method has closed its cycle due to the storming development of computer technology which has allowed the development of inelastic step-by-step methods up to collapse.

These inelastic step-by-step methods may be classified as follows:

- Nonlinear analysis of equivalent frame models
- Isotropic/orthotropic homogeneous nonlinear macroelement models for 2D structures
- Heterogeneous nonlinear microelement models for 2-D structures.

The above categories have been developed mainly for static loading, while the dynamic analysis is available for simple structural systems or elements.

The authors' prediction is that, in the near future, the third category combined with the *multiscale approach* will be developed for use even in large complicated real masonry buildings.

However, any future development of the analysis must be approached with serious concerns, since the reliability of outcome depends to a great degree on the reliability of the input data. As the perplexity of a method in use increases, more input data are required, the values of which are of limited reliability as their number increases. Therefore, it should not be forgotten that the analysis refers to *a model* of the construction and its material and, in this respect, includes many uncertainties, which we must approach with much 'prudence' (Croci, 2007).

Chapter 8

Assessment

8.1 INTRODUCTION

Assessment constitutes the last and the most important step of a laborious procedure towards the formation of the scheme of *structural restoration.*

It is based on two basic pylons, namely (Croci, 2007)

- *Diagnosis*
- *Quantitative evaluation of the structural capacity.*

At this stage, it would be useful to sum up the successive steps of a structural restoration. These steps have been already analysed in detail in the previous chapters but without a consistent recording of the steps of a structural restoration.

8.2 THE SUCCESSIVE STEPS OF STRUCTURAL RESTORATION

The successive steps of structural restoration are given below (Penelis, G., 2002):

i. *Temporary shoring*

   This is a separate step in the restoration procedure, particularly when restoration is decided after an accidental effect like a strong earthquake, a fire or a bombing, which has caused damages endangering the building and eventually leading to partial or total collapse. In this step, enough time is available for a detailed approach to the problem. This step is usually combined with the scaffolding and is necessary for the following steps of investigation and work (Figures 8.1 and 8.2).

ii. *In-situ investigations*

   They include usually the following activities:
   - Geometrical and constructional survey
   - Surveying of the pathology of the structure
   - *In-situ* nondestructive tests (NDT) e.g. hammer testing, ultrasonic measurements, ambient vibration measurements etc.
   - Follow-up of the changes of cracks, inclinations etc. in time
   - *In-situ* soil investigations (boreholes, excavations near to the foundations, cross-hole tests etc.).

Figure 8.1 Rotunda of Thessaloniki: temporary external buttressing with timber struts and lightly prestressed mill steel rings. (From Penelis, G. (1982). *The Masonry Roman and Byzantine Monuments in Greece. UNDP/UNIDO, PROJECT (RER/79/015) National Report of Greece.*)

Figure 8.2 Acheiropoietos church of Thessaloniki: temporary buttressing with timber struts at the north-eastern corner. (Courtesy: Ephorate of Antiquities of Thessaloniki (1978) Archives of Acheiropoietos church, Thessaloniki.)

iii. *Laboratory tests*

They include usually the following activities:

- Determination of chemical and mineralogical properties of the original materials
- Determination of mechanical properties of the original materials
- Composition of mortars and grouts compatible to the original ones
- Soil tests on samples taken from the boreholes
- Model tests (e.g. scaled walls or piers).

iv. *Analysis and design of the original structure*

Analysis and design of the monumental building before intervention is a very important step and includes as prerequisites the following actions:

- Determination of the mechanical properties of masonries and other compound elements, namely, modulus of elasticity, σ–ε curves, σ–τ curves, σ_1–σ_2 curves etc.
- Determination of the dynamic characteristics of the monument, i.e. fundamental period T_o, damping, eigenmodes etc.
- Determination of the ductility factor (q-factor) of the monumental building, in case a force-based design is adopted
- Determination of the scaled design spectrum or the scaled accelerograms for use in the case under consideration.

For the analysis of the structural system, various approaches may be used. The most common ones are the following:

- Static elastic analysis with frame or shell elements.
- Dynamic elastic analysis with frame or shell elements
- Static inelastic analysis using the step-by-step method for the determination of the collapse mechanism and loads (pushover analysis)
- Static inelastic analysis using the upper- and lower-bound theorems for the determination of the collapse loads.

For the design of cross sections, the σ–τ or σ_1–σ_2 curves must be used.

v. *Assessment of the residual resistance of the structure*

Taking into account the pathology of the structure, the time history of cracks and inclinations, the soil investigations and mainly the results of the analysis and design of the original structure, an assessment of the residual resistance of the original structure is made. This estimation, carried out by the structural engineer in charge, is crucial for the decision-making on the type and the degree of the intervention scheme for the structural restoration. This step will be analysed in detail in the subsequent paragraphs.

vi. *Restoration scheme*

The formulation of the restoration scheme constitutes the core of the whole procedure, and it is the outcome of an interrelated approach from the structural, architectural and archaeological points of view. Usually, the structural engineer in charge formulates various alternatives fulfilling the safety requirements based on a global estimate of the *in-situ*, laboratory and analytical investigations. After exhausting discussions with the other cooperating components on the restoration, the final solution is formed and adopted.

vii. *Reanalysis and redesign of the modified structure*

The modified system of the monumental building must be analysed and dimensioned again with the same procedure presented above.

viii. *Drawings, descriptions and specifications*

The whole design is completed with general drawings, detailed drawings, technical descriptions and technical specifications, all necessary for on-site work.

8.3 DIAGNOSIS

Diagnosis, in structural restoration, is the identification of *the nature* and *the cause* of various symptoms, namely various defects in the structural system, like cracks, splits, drifts, settlements etc.

For this task, a qualitative procedure is followed. This procedure comprises four groups of parameters that must be clarified before the *diagnosis* is formed (Croci, 2007; Mann, 1989; Ersoy, 1989; Mainstone, 1989).

i. *Type, extent and progress of defects*
   This group includes the pathology survey of namely
   - Cracks, spalls or splits of the structural system
   - Displacements and drifts
   - Settlements.

   Additionally, it should be noted that the survey of damages evaluated properly, on the basis of scientific knowledge and experience, constitutes a reliable guidance mechanism for focusing on crucial points of interest on the remaining investigations.

ii. *Geometry of the structure and mechanical properties of materials*
   This group includes
   - Geometric characteristics of the structure
   - Mechanical properties of materials at the various regions of the structure
   - Soil properties.

   The information of this group is based on the outcome of
   - The structural survey of the geometry of the building
   - The *in-situ* and laboratory tests.

iii. *Actions*
   This group includes information on
   - Dead loads
   - Live loads
   - Changes in the soil of foundations (e.g. water-table level, underground excavations in proximity of the monumental building)
   - Seismic actions
   - Other recent accidental actions (e.g. fire, bombing etc.).

   The information of this group is based on
   - Outcome of site investigation and laboratory test
   - Code requirements particularly for the seismic action.

iv. *Structural changes in the past*
   This group of information indicates
   - Modifications of the structural systems in the past that have endangered the building and lives
   - Modification of dead loads
   - Registered seismic actions of the past that have caused well-defined damages
   - Registered accidental events of the past like fires, bombing etc. that have caused well-defined damages.

   The information of this group is based on historical documents referring to the building and survey on site.

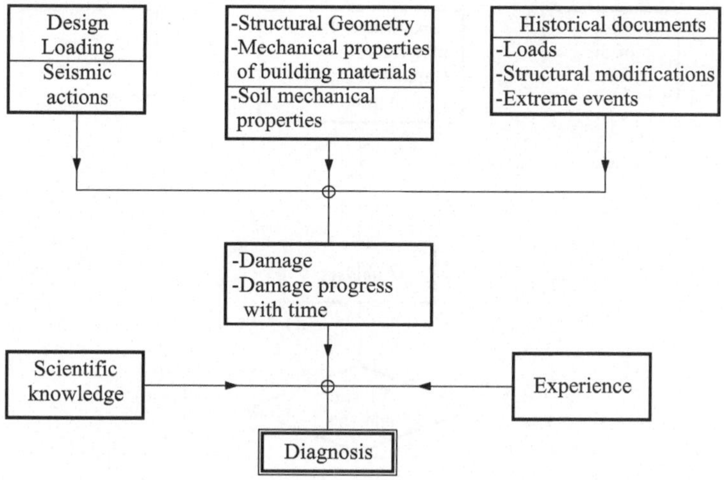

Figure 8.3 Flow chart of 'diagnosis' procedure.

The above four groups of factual parameters, properly interrelated on the basis of scientific *knowledge* and *experience* allow a rather *safe diagnosis* (Figure 8.3). *Diagnosis* is a procedure of paramount importance because it connects *causes* with *effects* based mainly on *facts*. In this respect, it constitutes a *deductive* procedure of reliability higher than that of the quantitative safety evaluation. The only disadvantage of diagnosis is that the outcome of the whole *procedure* is a *qualitative one*. At the same time, diagnosis constitutes an effective tool for the choice of the proper structural model and relevant loading cases for the structural analysis and safety evaluation.

8.4 QUANTITATIVE SAFETY EVALUATION

Basic tool for the quantitative evaluation of the residual structural resistance is *analysis* and *design*.

For gravity loads, the design is based on comparison of required strength to the available one (force-based design).

For the combination of gravity load and seismic actions and particularly for existing buildings, the design is based on the comparison of the required deformations (displacements δ_{req}, rotations ϑ_{req}) to the relevant capacities (displacement-based design), unless a forced-based design procedure is adopted by introducing a q-factor.

It is apparent that safety evaluation is an *abductive procedure* (Croci, 2007) that is based on *a series of assumptions* in relation to the following:

- The structural model (geometry, materials, restrains, cracks)
- The loading (gravity loads, forced displacements at foundations, seismic actions).

It is apparent that the more accurate the above assumptions are, the more reliable are the outcoming results.

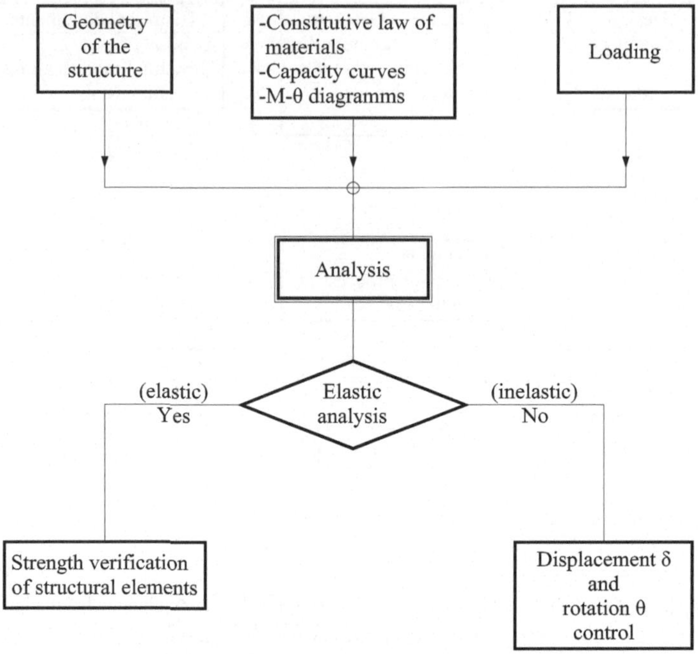

Figure 8.4 Flow chart of quantitative safety evaluation.

The crack pattern that is predicted for the combination of *gravity loads* with *frequent earthquake* (return period of occurrence 95 years, namely 50% of the design seismic action) constitutes a crucial checking of the reliability of the safety evaluation. In fact, if the resulting crack pattern (regions of tensile stresses) coincides more or less with the surveyed crack pattern, there is a high probability that the analysis and design at near-collapse limit state will give reliable results.

A second checkpoint could be the combination of *gravity loads* with the *maximum seismic* action of the past in the region in case there are reliable recorded data. For this combination, the near-collapse *limit state should not be exceeded.* Otherwise the input data for analysis and design should be revised (Figure 8.4) (Penelis, G., 2002).

Finally, it should be noted that a crucial point of the whole procedure for the choice of the input data is the estimate of the *seismic action* that will be taken into account in the analysis. This issue has been discussed in detail in Subsection 7.3.7.

8.5 ASSESSMENT – STRUCTURAL RESTORATION SCHEME

Diagnosis and *quantitative safety evaluation* are the basic pylons for *assessment.* A basic question that arises at this stage is regarding the estimate of the *residual capacity* of the structure, in other words, how close the structure is to local or general collapse.

It is apparent that diagnosis and safety evaluation in the procedure for drawing conclusions are both of the same significance, and their conclusions are cross referenced. In case these outcomes lead to controversial conclusions, the outcome of the diagnosis which is

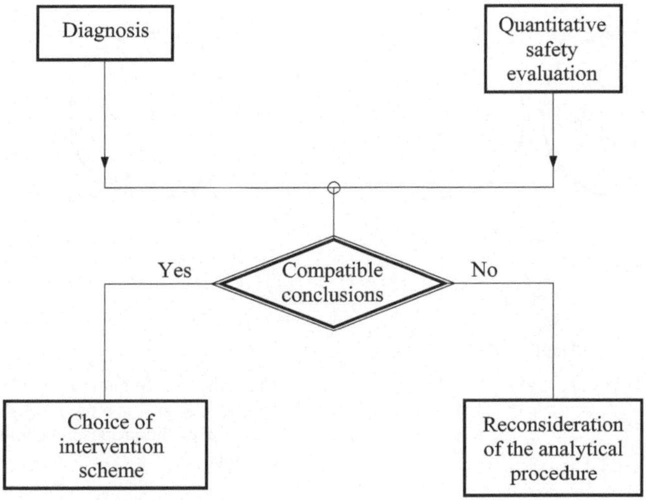

Figure 8.5 Flow chart of assessment procedure.

based on facts (inductive procedure) prevails in the decision-making (Figure 8.4), even if these conclusions are qualitative (Figure 8.5).

Based on the final assessment, a decision is made for the elaboration of two or more *structural restoration schemes.*

For the formulation of these schemes, the following considerations should be taken into account. From a study of a very broad range of monuments and historical buildings, it is concluded that one of the most important causes of damage to their structural system in areas of high seismicity is the earthquake action. It may be said that these buildings, being exposed every now and then for centuries to strong seismic actions, have been subjected to a kind of natural selection, so that only those which were properly designed and constructed have survived (Penelis, G., 2002).

Consequently, in the case of a structural restoration, *the preservation of the original structural system with proper strengthening of its weak elements must be one of the basic concepts of the intervention.* Nobody and no means can guarantee that a strongly modified structural system will have a better chance than the original one to survive in the following centuries.

Based on the preceding remarks, it may be said that the seismic and dynamic analyses of a monument for a probable intervention arises only when the building presents detectable failure symptoms, namely cracks, deviations from the vertical, settlements of the foundations or continuous worsening of the above symptoms.

With all these in mind, an intervention scheme is decided. It usually includes two types of interventions, i.e.

- Local-type interventions
- General-type interventions.

The first type refers to interventions that improve the local strength and integrity of the structural members, e.g. deep rejoints, grouting, stitching of wall with posttensioned (see Chapter 5) rebars etc. The second type refers to interventions that improve but don't

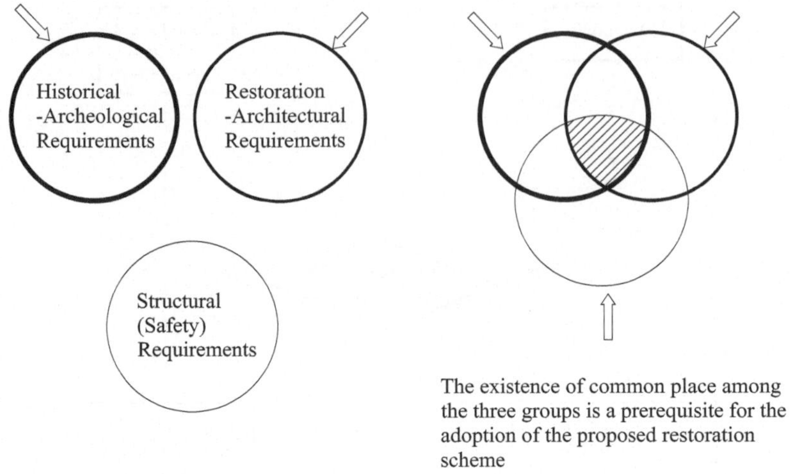

Figure 8.6 Prerequisite for the adoption of the proposed structural restoration scheme.

change the overall structural behaviour of the building, e.g. ties at the springings of arches, rings at the base of domes, external buttresses, improvement of the stiffness of existing diaphragms etc. This second intervention type calls for reanalysis and redesign of the structure (see Chapter 5).

The formulation of alternative intervention schemes *constitutes the core of the restoration design procedure.*

The choice of the final scheme of intervention is the result of not only a multidisciplinary cooperation, but also a compromise between the three specialists involved in the project, namely, the archaeologist, the architectural restorer and the structural engineer in charge (Figure 8.6). All of them must show reasonable flexibility without overlooking their basic requirements to approach each other's considerations so that they can find a common place for the final solution (UNDP/UNIDO, 1984; Penelis, G., 2014).

Finally, for the approved intervention scheme, *reanalysis* and *redesign* follows. Usually, the same model of the original analysis and design is used, which is modified properly at the points where interventions have been decided upon.

Chapter 9

Case studies

9.1 THE ROTUNDA OF THESSALONIKI

9.1.1 Introduction

This case study outlines the seismic behaviour of the Rotunda of Thessaloniki, one of the most significant domed monuments of the Late Roman and Early Byzantine periods that have survived to the present, and discusses the consequences of its behaviour over the centuries, particularly after the earthquake of the 20 June 1978, which pushed the Rotunda to the limits of collapse. It also describes the structural interventions, based on diagnosis and analysis of the monument's seismic behaviour, carried out after 1980 (Penelis, G., 1992, 1982; Penelis, G., et al., 1980a).

The structural restoration design was awarded to the first of the authors who had elaborated the project with his scientific team consisting of (Penelis, G., et al., 1980a):

- Karaveziroglou Maria, PhD, structural engineer
- Stylianidis Kosmas, PhD, structural engineer
- Leondaridis Demitrios, MSc, structural engineer
- Liaris Nicolaos, PhD, chemist.

In the laboratory of R/C and Masonry Structures of AUTH.

The intervention activities were carried out by the 9th Ephorate of Byzantine Antiquities.

9.1.2 Description

The Rotunda was built close to, and as a part of, the Imperial Palace of Galerius in about 300 A.D. It was an imposing circular building resembling the Pantheon in Rome, and it was covered by a huge dome 24.5 m in diameter. The main difference (apart from the smaller size) was that the construction techniques were typical of those in this part of the Eastern Empire rather than those prevalent in early-2nd-century Rome. Figure 9.1.1 illustrates how eight barrel-vaulted niches lightened its 6.25-m-thick walls (Moutsopoulos, 2013, Krautheimer, 1965).

The original structure has survived substantially intact, as shown in Figures 9.1.2 and 9.1.3, although it has undergone a number of adaptations. Built first as a mausoleum or a throne room, it was adapted in about 400 A.D. for use as a church. A vaulted chancel, apse and ambulatory were added. The walls of the niches were pierced to give access to the ambulatory, and the main cylindrical wall drum was cut further at the east to open

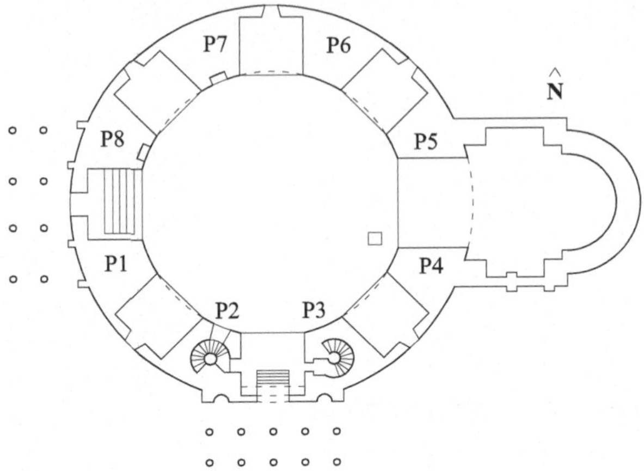

Figure 9.1.1 The Rotunda at Thessaloniki: plan at ground level. (Adapted from Penelis, G. (1992). The Rotunda of Thessaloniki: Seismic behaviour of Roman and Byzantine structures. in *Hagia Sophia from the age of Justinian to the present*, Mark, R. and Çakmak, A. (ed.). Cambridge University Press, Cambridge. pp. 132–157.)

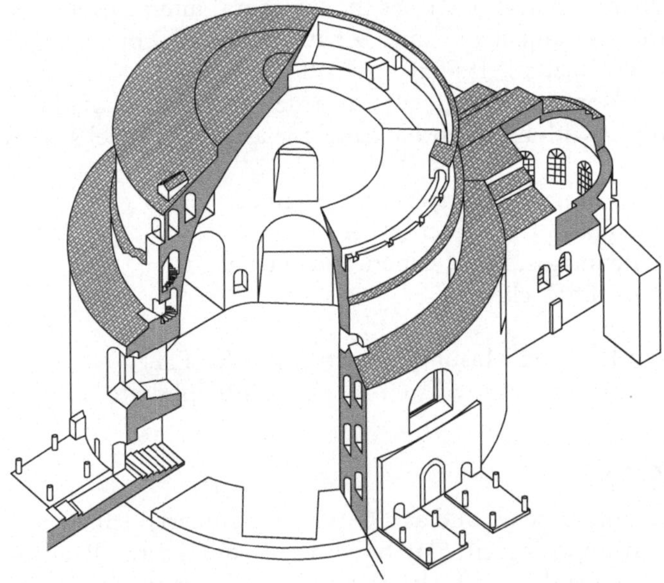

Figure 9.1.2 Cutaway isometric view from the southwest. (Adapted from Penelis, G. (1992). The Rotunda of Thessaloniki: Seismic behaviour of Roman and Byzantine structures. in *Hagia Sophia from the age of Justinian to the present*, Mark, R. and Çakmak, A. (ed.). Cambridge University Press, Cambridge. pp. 132–157.)

the interior to the new chancel (see Figure 9.1.1). This last intervention seems to have led to a partial collapse of the dome and to its subsequent reconstruction. The dome was given a mosaic decoration of saints standing against an architectural background, which has largely survived.

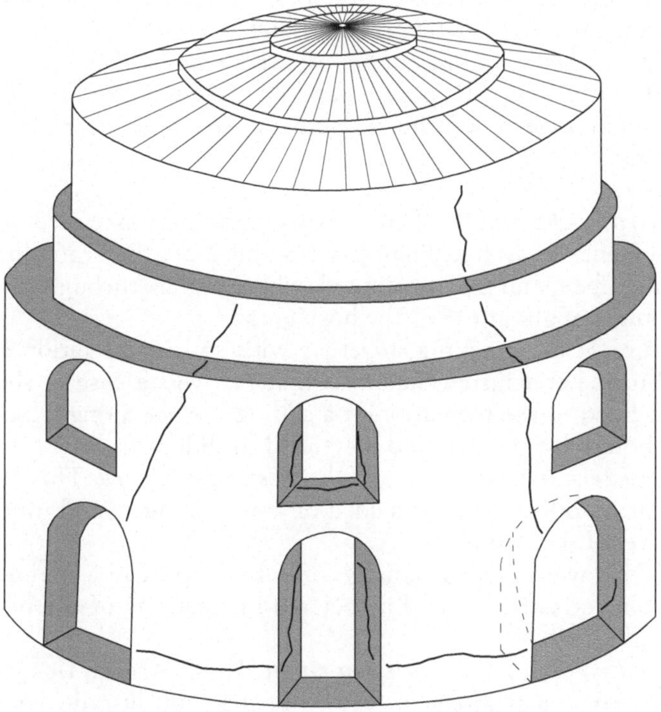

Figure 9.1.3 The main sliding cracks on the southern piers P2 and P3. (Adapted from Penelis, G. (1992). The Rotunda of Thessaloniki: Seismic behaviour of Roman and Byzantine structures. in *Hagia Sophia from the age of Justinian to the present*, Mark, R. and Çakmak, A. (ed.). Cambridge University Press, Cambridge. pp. 132–157.)

Later, the ambulatory was damaged, perhaps partly due to earthquakes. It was removed during the Ottoman occupation, when the church became a mosque. At this time, the pierced walls of the niches were again closed, and a minaret, which still stands nearby, was added.

9.1.3 Pathology

Before the earthquake of 20 June 1978 (epicentre at Lake Volvi, 30 km from Thessaloniki, magnitude 6.2 on the Richter scale, maximum ground acceleration in Thessaloniki 135 cm/s^2), the structure was already extensively cracked, especially in the piers (P2 and P3 in Figures 9.1.1 and 9.1.3) that flank the southern entrance and that were already weakened by the helical staircases within them. The inclined cracks at these piers appear to have had in the beginning of the 20th century A.D. a width of about 120 mm, as indicated by tile and mortar fillings, probably, dating from the beginning of this century (Figure 2.29). These cracks widened an additional 25 mm over the eight decades before 1978.

Concern over the poor structural condition of the Rotunda was expressed in several reports from 1976 onwards, but no action was taken when the earthquake of 20 June 1978 caused further damage, including further opening of the cracks. Unfortunately, the absence of measurements before this earthquake precludes a quantitative estimation of the damage caused by this earthquake. But what is certain is that, since then, the monument had shown a tendency to collapse. *The main crack at piers P2, P3 widened 1 mm per month, indicating* that the part of the dome bound by piers P2 and P3 had begun to slide slowly.

Anyhow, the main defects of the monument surveyed after the earthquake are summarised below:

a. *The southern piers P2 and P3*. The cracks of these piers extended over the full height of the monument. As seen in Figures 9.1.3 and 2.29, they ran almost vertically up to the outer surface, and they continued, where visible, through the thickness of the piers at an angle of about 61° to the horizontal.

   This picture of the cracking, together with soil investigation and static analysis, has led to the conclusion-diagnosis that this was a case of shear–compression failure, which led to the formation of a sliding surface along which the part of the monument bound by piers P2 and P3 tended to slide.
b. *The north-western pier at the left of the western entrance*. This pier had a vertical crack near the middle, which extended on both the outer and inner faces up to the springing level of the dome.
c. *The dome*. This was cracked radially in its lower part in a manner that would be expected from the cracking and associated inclinations of the piers (Figures 2.38 and 4.31).
d. *The walls and vaults of the Christian chancel and apse*. In this part of the monument, there were cracks along the crown of the vault intrados associated with the outward inclinations of the walls. There were also radial cracks above the windows of the apse, extending up to the springing level of the semidome of the apse, associated also with an outward inclination of the wall of the apse (Figure 4.48).

9.1.4 Emergency measures

Because of the apparently serious structural condition after the earthquake, and in order that time be given for further systematic investigation and full consideration of a scheme for final measures and strengthening, the following emergency measures were undertaken (Figure 8.1):

- Erection of heavy-timber shoring between piers P1 and P2, P2 and P3 and P3 and P4, to support the load from the dome directed towards piers P2 and P3
- Erection of shoring on all sides of piers P2 and P3 to prevent their collapse
- Internal cross-bracing, with heavy timber, of the helical staircases in piers P2 and P3
- Installation of temporary post-tensioned external rings of 26-mm-diameter mild steel bars to further assist in preventing continued outward displacements of piers P2 and P3
- Support of the chancel vault by metal frames
- Shoring of the south porch
- Erection of steel scaffolds, both inside and out, to provide access for detailed measurements and, later, for final repair and strengthening.

It should be noted that after the installation of the temporary posttensioned rings, the sliding of piers P2 and P3 stopped (Figures 6.2 and 6.3).

9.1.5 Structural restoration

9.1.5.1 General

The structural restoration of the monument after the earthquake of 20 June 1978 followed the procedure described in Chapter 8, and it is presented in short below.

9.1.5.2 Documentation

The documentation of the historical phases and the descriptive information of damages was carried out by the restorer architect in charge, Prof. Moutsopoulos (2013), and was evaluated structurally by the first of the authors of this book. The information presented in Subsection 9.1.2 above was based on this documentation.

9.1.5.3 In-situ investigations

i. *Architectural and constructional survey*
   The architectural and structural survey of the monument was carried out using standard survey techniques including photogrammetric procedures. Additionally, cross sections were facilitated at the level of the foundations (Figure 2.41), at the dome and at the vault of the Christian part.
ii. *Survey and recording of the damage*
   All cracks were mapped on architectural drawings with all details referring to their width and depth.
   To give a more comprehensible overall picture of damage, the main cracks were marked on a transparent model of the structure to a scale of 1:100. In this way, the kinematic of the sliding part of the monument bound by piers P2 and P3 became clearly apparent.
   The changes of the cracks width in time were recorded by means of mechanical extensometers installed at the critical cracks (Figures 6.2 and 6.3).
   Finally, it should be mentioned that the deviations from the vertical of all piers at various levels were recorded and evaluated for damage diagnosis (Figure 6.4).
iii. *Sample taking and in-situ tests*
   For the direct destructive test (DT) of structural materials, nineteen cores of 10 cm diameter were taken from various locations (Roman and Christian parts) (Figure 6.5).
   The core test results were combined with nondestructive (NDT) hammer tests (600 tests (Figure 6.9)), as well as ultrasonic measurements (300 measurements) (Figure 6.8). The correlation of hammer testing with corresponding compressive strengths determined in the laboratory was found to be very satisfactory ($r = 0.867$) (Table 6.1). The correlation of ultrasonic measurements to corresponding results of compressive tests found in the lab did not provide satisfactory results ($r = 0.500$). Nevertheless, ultrasonic measurements have given very good correlation for the dynamic modulus of elasticity.
   The statistical evaluation of the mortar and brick strengths and dynamic moduli of elasticity is given in Table 9.1.1.
   The following can be seen from the evaluation:
   - The mean strength of the mortar in the Christian part is considerably higher than that in the Roman part, despite its higher variability.

Table 9.1.1 Strength and modulus of elasticity of construction materials

	Compressive strength		
Material	*Mean value (MPa)*	*SD (MPa)*	*Dynamic modulus of elasticity (MPa)*
Roman mortar	2.876	0.65	1.890
Christian mortar	3.814	1.25	2.840
Roman bricks	13.120	2.33	10.700
Christian bricks	10.209	2.97	6.000

Source: Adapted from Penelis, G. (1992). The Rotunda of Thessaloniki: Seismic behaviour of Roman and Byzantine structures, in *Hagia Sophia from the age of Justinian to the present*, Mark, R. and Çakmak, A. (ed.). Cambridge University Press, Cambridge. pp. 132–157.

- By contrast, the mean strength of the bricks in the Roman part is considerably higher than that in the Christian part.
- The standard deviations (SDs) of the strengths of materials of the Christian period are higher than those of the Roman period.

iv. *Natural period of the monument*

To determine the natural period of the monument, ambient vibration measurements were carried out in cooperation with the laboratory of engineering seismology and earthquake engineering, National Technical University of Athens (Figures 6.16 and 6.17). These measurements, properly elaborated by a 'fast Fourier transform' procedure, indicated a natural period of $T_o = 0.42$ s for the radial direction. These measurements were repeated after the monument's restoration. The outcome of the new measurement had a reduced value of $T_o = 0.36$ s, which means that the monument's stiffness after the structural restoration was improved radically (36% improvement).

v. *Soil borings*

Four borings were facilitated around the monument, and soil samples were taken for laboratory tests. These soil investigations, together with the soil sections in the neighbourhood of the foundations, ascertained that the monument was founded on the very stiff clay subgrade of Thessaloniki, and, therefore, damage could not be attributed to soil settlements.

9.1.5.4 Laboratory tests

i. *Chemical analyses*

The chemical analyses of the construction materials had the following tasks:

- Determination of the proportions between the binder and the aggregates in the mortar (Table 2.3)
- Determination of hydraulic elements in the mortar, their origin, namely, whether they were due to the addition of pozzolanic ash or to brick powder
- Determination of the composition of the bricks (Table 6.2) and, more specifically, whether they were of the same composition as the mortar's brick powder and whether they had incorporated hydraulic material (Table 6.3)
- Chemical analysis of the materials that could eventually be used for the composition of grout, for example, hydrated lime, pozzolanic ash, cement and brick powder.

ii. *Mechanical properties of structural materials*

Using the cores taken from two parts of the monument, the Roman and the Christian one, mortar specimens were prepared in the Laboratory of Reinforced Concrete and Masonry Structures of Aristotle University of Thessaloniki (AUTH), with dimensions as prescribed by codes (Figure 2.7). For these specimens, flexural strength, compressive strength, modulus of elasticity and bonding to bricks were determined (Figure 6.20, Tables 6.5, 6.6 and 6.7).

The final results of strength and moduli of elasticity of original materials are given in Table 9.1.1

iii. *Mix design of the grouts and testing of their properties*

Based on the chemical analyses and tests of the mechanical properties of the monument's mortars, the composition of adequate grouts was attempted at the Laboratory of Reinforced Concrete and Masonry Structures of Aristotle University of Thessaloniki (AUTH), using traditional materials with mechanical properties compatible with those of the original mortar of the monument. The following materials were used: hydrated lime, pozzolanic ash, powder cement, fine sand and additives.

iv. *Soil tests*

Soil samples were tested in the laboratories of the Ministry of Public Works. Main outcome from *in-situ* and laboratory tests are given below:

- Soil description: very stiff inorganic sandy clay (CL)
- S.P.T: >50
- W_p–W_c–W_4: 12%–17%–34%
- Internal friction $\varphi = 15°$
- Unconfined compressive strength: $q_u = 0.4\,\text{MPa}$
- Compression index: $C_c = 0.07$.

9.1.5.5 Evaluation of the mechanical characteristics of the structure

i. *General*

As known, a series of mechanical characteristics of the structure must be determined in advance so that a structural analysis and design may follow. So, the following characteristics were determined.

ii. *Strength and stress–strain relations of masonry*

Having in mind the mechanical properties of bricks and mortars that were determined as mentioned above, the load-bearing capacity and modulus of elasticity of the masonry were assessed, using various semiempirical formulas (see Chapter 3). The results are given in Table 9.1.2.

Table 9.1.2 Strength and modulus of elasticity of original masonry

	Comprehensive strength nominal value (MPa)	*Tensile strength nominal value (MPa)*	*Bond strength nominal value (MPa)*	*Modulus of elasticity mean value (MPa)*
Roman part	2.5	0.3	0.3	2.450
Christian part	2.5	0.3	0.3	3.100

Source: Adapted from Penelis, G. (1992). The Rotunda of Thessaloniki: Seismic behaviour of Roman and Byzantine structures, in *Hagia Sophia from the age of Justinian to the present*, Mark, R. and Çakmak, A. (ed.). Cambridge University Press, Cambridge. pp. 132–157.

For Poisson's ratio, an estimated value from the bibliography was used, namely $\nu = 0.19$.

At the same time, the σ_1–σ_2(±) and σ_y–τ_{xy} envelopes were determined, as the failure mode of piers P2–P3 was attributed to shear–compression failure.

The formulation of these envelopes was based on the following:

- Compressive strength of masonry
- Tensile strength of mortar determined in the lab
- Bonding between the mortar and the brick determined in the lab
- Existing curves of σ_1–σ_2 (±) of low-strength concrete available at that time were used for the analogue formulation of σ_1–σ_2 (±) curves of masonry. In Figure 9.1.4 (Kupfer, 1973), these curves are referred to as *conventional.*

Ten years later, extensive research has been carried out to find a better approach to the problem of evaluating the mechanical properties of masonry, by means of finite element micromodelling of mortar, bricks and bonding links. The results of this analysis are also given in Figure 9.1.4 for comparison.

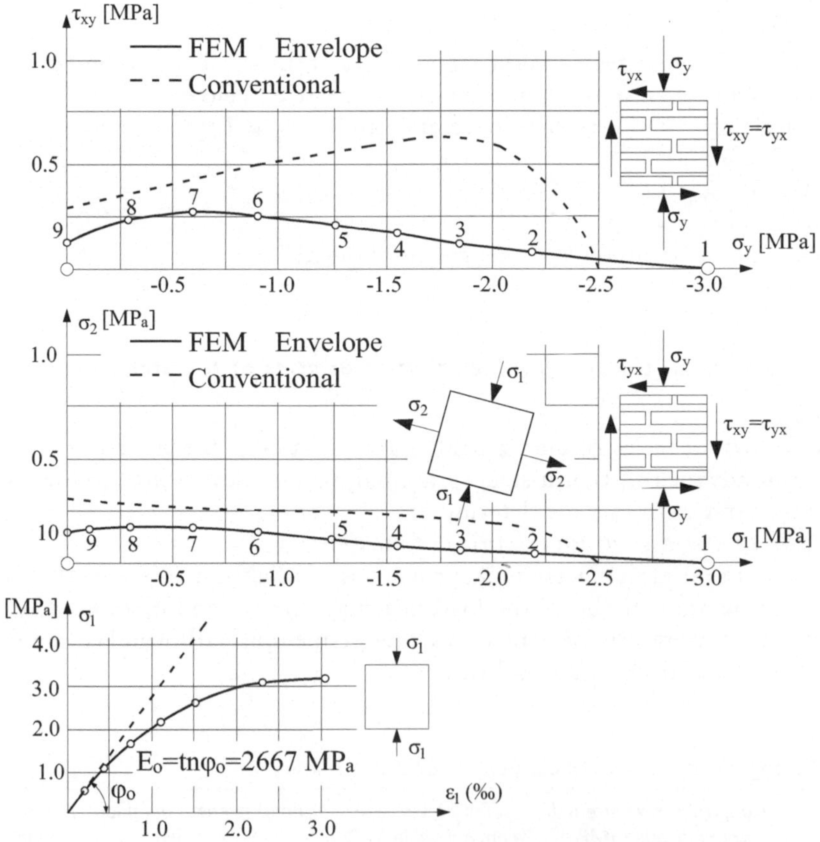

Figure 9.1.4 The τ_{xy}–σ_y and σ_1–σ_2 envelope of the original masonry and the respective σ_1–ε diagrams (finite element model). (Adapted from Penelis, G. (1992). The Rotunda of Thessaloniki: Seismic behaviour of Roman and Byzantine structures. in *Hagia Sophia from the age of Justinian to the present*, Mark, R. and Çakmak, A. (ed.). Cambridge University Press, Cambridge. pp. 132–157.)

iii. *Ductility*

The ductility factor, for estimating the design spectrum (force-based design procedure), was the one associated with outward rotations of the piers. This factor was determined both for a free cantilever model of a pier and for analytical models where rotation of the piers at their base was restrained by posttensioned steel rings of various types of steel at the levels of the masonry rings where the steel rings would eventually be installed in the restoration stage. The resulting ductility values for various cases of steel rings were as follows:

- Free cantilever: $\mu_{avail} = \frac{\delta_u}{\delta_y} = \frac{4.46}{1.50} = 2.97$
- Cantilever with steel rings ST220/340: $\mu_{avail} = \frac{\delta_u}{\delta_y} = \frac{4.46}{1.70} = 2.62$
- Posttensioned steel rings ST220/340: $\mu_{avail} = \frac{\delta_u}{\delta_y} = \frac{4.46}{1.25} = 3.57$
- Posttensioned steel rings ST420/700: $\mu_{avail} = \frac{\delta_u}{\delta_y} = \frac{4.46}{1.25} = 3.57.$

iv. *Natural period and damping*

The natural period of the monument, necessary for the evaluation of the seismic coefficient, was determined analytically by means of the Rayleigh method. Two different mechanical models were used. The first was a series of free cantilevers with a masonry ring at the mid-height that vibrated radially simultaneously. This model simulates the behaviour of the piers, given that the upper ring, supporting the dome, is extensively and radially cracked. The second was a free cantilever pier corresponding to piers P2 and P3. The analytical results of these two models were found to be in a very good agreement with the results of *in-situ* ambient vibration measurements: $T_{\text{free cantil.}} = 0.53\,\text{s}$, $T_{\text{ambient vibr.}} = 0.42\,\text{s}$ and $T_{\text{ring + cantil.}} = 0.33\,\text{s}$. The degree of damping for the monument was estimated from other similar structures to be $\xi = 0.10$.

v. *Seismic action*

Taking into account the seismic risk analysis of the region and using the aforementioned data of soil, natural period, damping, ductility and importance of the monument, the seismic loading was estimated according to the Greek code (in preparation at that time), the SEAOC code and the response spectrum derived from the accelerogram of the 1978 Thessaloniki earthquake. This approach resulted in a value of the base shear due to earthquake ranging from $V = (0.11–0.14)W$, where W is the dead load of the monument. Finally, a value $V = 0.12\ W$ was chosen. *For this value of base shear V, the massive piers that have not failed for seventeen centuries presented a safety factor of 1.5 (e.g.* $V_{avail} \approx 1.50 \times 0.12\ W = 0.18\ W$*).*

9.1.5.6 Analysis and design of the Roman part of the structure

Initially, several elastic analyses were facilitated for the following:

- Vertical gravity loading
- Independent ring-type loadings at three levels where there was an intention of installing permanent peripheral posttensioned rings
- Horizontal seismic loading.

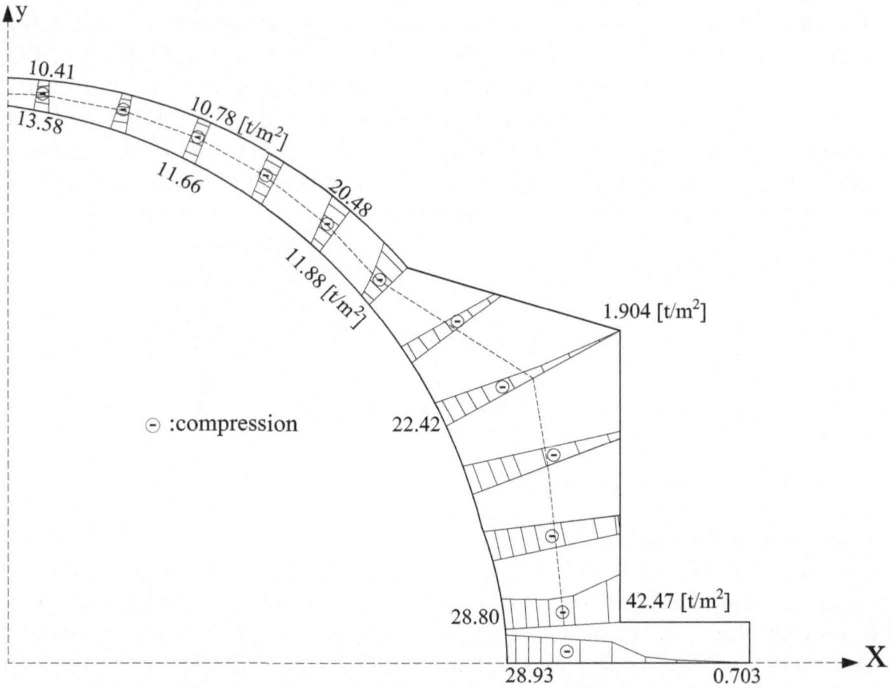

Figure 9.1.5 Calculated distribution of the meridional stresses in the dome without taking into account the meridional cracks. (Adapted from Penelis, G. (1992). The Rotunda of Thessaloniki: Seismic behaviour of Roman and Byzantine structures. in *Hagia Sophia from the age of Justinian to the present*, Mark, R. and Çakmak, A. (ed.). Cambridge University Press, Cambridge. pp. 132–157.)

The structure as a whole was analysed first, using an axisymmetric finite element model. The presence of the niches in the cylindrical lower part of the structure was simulated by introducing a reduced modulus of elasticity there. Stresses in the dome, as calculated in one of these analyses, are plotted in Figure 9.1.5. Ten years after the original restoration design and its execution, an alternative elastic analysis of the Roman part was facilitated, using finite element modelling and taking into account the meridional cracks and the openings (Ignatakis, et al., 1993). The results of this analysis are plotted in Figure 9.1.6.

The support system of the dome was then analysed as a space frame (Figure 9.1.7) with 1D elements.

This analysis was performed both for the initial state of the structure with uninterrupted supporting ring at the intermediate level in the east and for the latter state with the ring cut here to give access over its full height to the Christian chancel.

Elastic analyses, however, provide only an approximate estimate of collapse loads. Therefore, the application of *the upper- and lower-bound theorems* led to the conclusion that the collapse mechanism would involve interruption of the continuity of the rings of masonry which link the piers and the formation of hinges at the top and bottom of the piers. Ultimate loads were therefore calculated for the piers acting as a free cantilever (Figure 9.1.8).

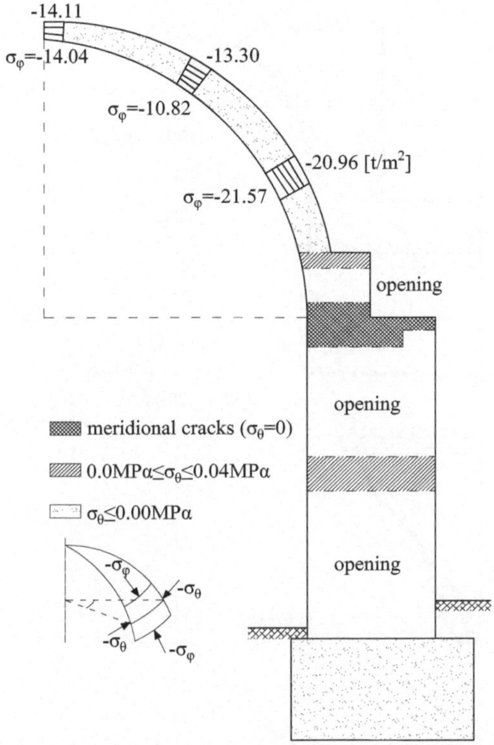

Figure 9.1.6 A latter elastic analysis of the Roman part using finite element modeling and taking into account the meridional cracks and the openings (Ignatakis, et al., 1993, 1989). (Adapted from Penelis, G. (1992). The Rotunda of Thessaloniki: Seismic behaviour of Roman and Byzantine structures. in *Hagia Sophia from the age of Justinian to the present*, Mark, R. and Çakmak, A. (ed.). Cambridge University Press, Cambridge. pp. 132–157.)

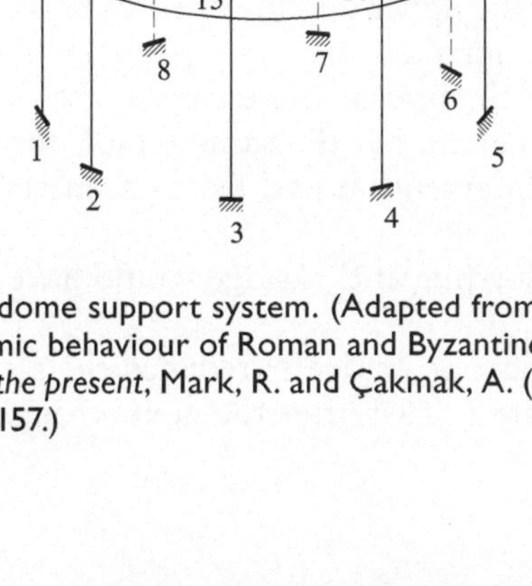

Figure 9.1.7 Frame model of the dome support system. (Adapted from Penelis, G. (1992). The Rotunda of Thessaloniki: Seismic behaviour of Roman and Byzantine structures. in *Hagia Sophia from the age of Justinian to the present*, Mark, R. and Çakmak, A. (ed.). Cambridge University Press, Cambridge. pp. 132–157.)

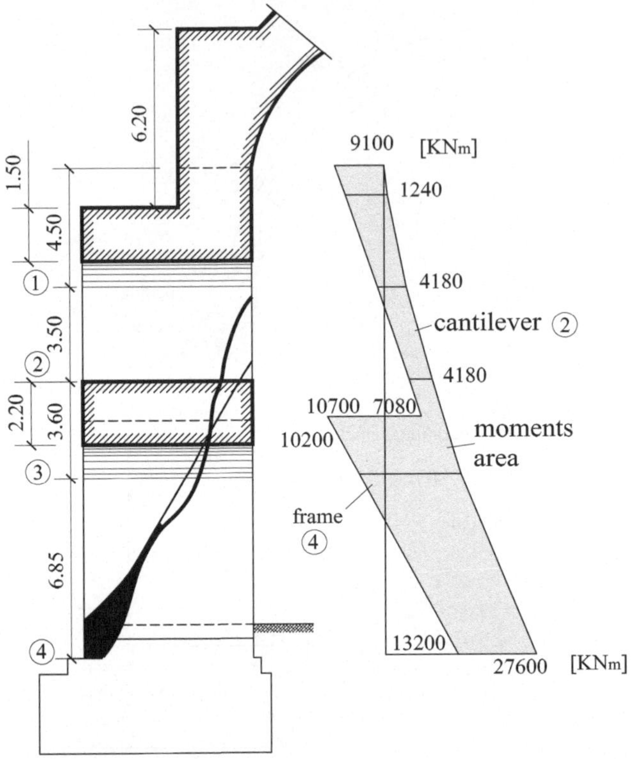

Figure 9.1.8 Cracked pier with staircase and corresponding moment diagram. (Adapted from Penelis, G. (1992). The Rotunda of Thessaloniki: Seismic behaviour of Roman and Byzantine structures. in *Hagia Sophia from the age of Justinian to the present*, Mark, R. and Çakmak, A. (ed.). Cambridge University Press, Cambridge. pp. 132–157.)

9.1.5.7 Assessment for the Roman part of the monument

i. *Superstructure*

From these analyses, the following conclusions were drawn for the structural behaviour and safety of the structure at different periods of the monument's life:

- As the building was first constructed, all structural members had a safety factor under gravity load within the range of about 2.70–3.10. Thus, they had fairly a uniform utilisation factor at all critical points, similar to that which would be provided today for masonry construction under seismic plus gravity load. However, cracks of the upper and lower rings between the piers might appear in case of a strong earthquake.
- After the cutting of the intermediate ring in the east for the Christian chancel, a rupture of the upper ring at the same region should be expected, even under gravity load. This intervention had led to a partial collapse of the dome at its eastern part.
- Additional seismic loading at this stage would have caused undoubtedly rupture of the upper ring.

The Christian modification critically reduced the structure's load-bearing capacity, and it is certain that, at least after the next strong earthquake, the piers would have had to act as free cantilevers. Thus, the subsequent behaviour should be considered only on this basis. The following further conclusions were drawn for the piers acting as cantilevers:

- Gravity load could be carried with a safety factor of about 2.15.
- Under gravity plus seismic action, piers P2 and P3 should fail in shear from a crack inclined at 61° degrees to the horizontal for a base shear due to earthquake with a lower bound of about 8.5% of the gravity load plus dome thrust (Figure 9.1.8). As can be seen in Figure 9.1.8, this calculated crack inclination agreed closely with the measured inclinations of the observed cracks in these piers. The remaining piers (which are not weakened by internal stairs) should safely withstand a roughly 70% greater base shear, due to an earthquake (13.5% of the gravity load). This is consistent with the observed lack of damage and the estimate of a base shear of $V_R = 0.135\,W + V_{\text{dome thrust}}$.
- Failure of piers P2 and P3 activated the sections of infill wall of the flanking niches, and these prevented complete collapse of the piers by improving the combined seismic shear resistance to about 14% of the gravity load plus dome thrust.
- The addition of the temporary posttensioned rings significantly improved the safety margins for the transient period between earthquake event and final restoration.

For comparison, we shall present the results of a finite element model inelastic analysis of the two types of pier (Figure 9.1.9a and b) carried out by means of a computer program developed in laboratory of AUTH. This analysis was performed with the failure envelope σ_1–σ_2 and the curve illustrated in Figure 9.1.4.

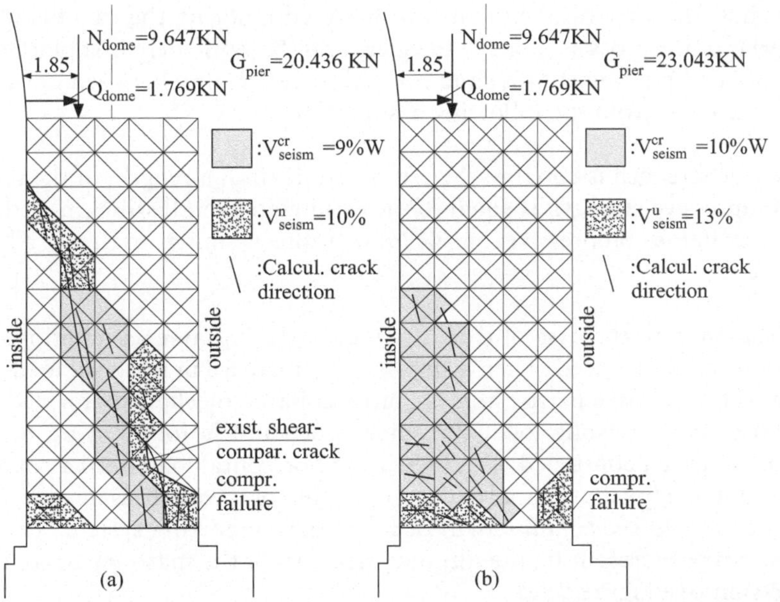

Figure 9.1.9 Predicted failure pattern of the piers of the Roman part using an inelastic finite element model analysis: (a) cracked pier with staircase; (b) uncracked massive pier. (Adapted from Penelis, G. (1992). The Rotunda of Thessaloniki: Seismic behaviour of Roman and Byzantine structures. in *Hagia Sophia from the age of Justinian to the present,* Mark, R. and Çakmak, A. (ed.). Cambridge University Press, Cambridge. pp. 132–157.)

The failure pattern and the crack inclination of the weakened pier with the internal staircase, as determined analytically by this procedure, are in very good agreement with the real picture of damage. Furthermore, the base shear resistance of both types of pier was found to be very close to that found through the shear compression model 12 years ago: $V_R = 0.10\,W + V_{\text{dome thrust}}$ for the weakened piers and $V_R = 0.13\,W + V_{\text{dome thrust}}$ for the regular piers.

ii. *Foundations*

On the basis of the soil investigations, the *in-situ* survey of the foundation and the computed stresses and deformations, the following conclusions were drawn in reference to soil stresses:

- For dead loads, no eccentric action has been determined at the foundation level. The uniform required compressive strength was calculated to be 0.3 MPa, a value well below $\sigma_{avail} = 0.6$ MPa, which was based on the soil tests.
- For a seismic action corresponding to $V = 0.12W$, a small inactive area existed at the soil level with maximum stress at the outer boundary of the foundation equal to 0.79 MPa, which is within permissible limits.
- Finally, the ultimate seismic load that could be carried by the foundation, before the strengthening of the monument, was found to be $V = 0.185\,W$, which is greater than the ultimate seismic strength of the superstructure before strengthening.

9.1.5.8 Analysis, design and assessment of the Christian part of the structure

The analysis of the Christian part was performed using a step-by-step ultimate load method. At first, the structural system was analysed ignoring the two flying buttresses. Failure criteria for the cross sections of the piers and the vault were first plotted as interaction curves of bending moment (M) and normal force (N) and between normal stress (σ) and shear (τ), derived from the following assumptions:

- The tensile strength is equal to the estimated flexural strength; of the mortar.
- The compressive strength is equal to the estimated overall strength of the masonry.
- Behaviour in the compressive–tensile (σ_1–σ_2) zone is similar to that of a low-strength concrete.

It is quite clear that, in the case of the Christian part, the same assumptions as those of the Roman part could be used. Then a step-by-step increase in the dead load was applied to determine the successive plastic hinges until a collapse mechanism formed.

For response under seismic load, the same procedure is followed up to full service (gravity) load and then continued with increasing horizontal load until a collapse mechanism was formulated (pushover analysis). At each step of the analysis, however, it is necessary to ensure that *no shear failure* will occur at the supporting system. This is done by means of the curve representing the ultimate strength of the masonry in combined shear and compression (see Figure 9.1.4).

An initial analysis of the chancel according to the procedure just described indicated a factor of safety of about $\nu = 0.67$ for gravity load alone. It seemed unlikely that the apse would contribute sufficiently to the total strength, helping to prevent collapse. This suggested that either the flying buttresses were erected at the same time with the chancel or that the present heavy covering of the vault was a latter addition and the buttresses were added when this was decided. Hence, a further analysis was made including the flying buttresses in the model. This still indicated a low factor of safety of about 1.20 for gravity load alone.

9.1.5.9 Proposals for repair and strengthening

i. *Roman part*

- Local repair and strengthening of piers P2 and P3

  The following interventions were proposed:

  - Application of an internal reinforced concrete (gunite) skin to each helical stairway, using as reinforcement high-strength deformed bars of an austenitic stainless steel alloyed with chromium, nickel and molybdenum. This proposal was later modified upon request of the Central Archaeological Council, and the reinforced concrete skin was replaced by a steel cage (reversible intervention), as shown in Figure 9.1.10.
  - Stitching of the piers throughout their height as shown in Figure 9.1.10 using posttensioned bars 2Φ26/1.5 m, $V = 335$ kN/1.5 m of the same steel (high-strength deformed bars ST 420/700 of an austenitic stainless steel) protected by means of a cement grout of high alkali content.

- General type of strengthening

  It included the following interventions:

  - Grouting of all cracks in the piers and the dome (Tables 5.8, 5.9, and 5.10)
  - Installation of posttensioned rings of stainless austenitic steel bars of type Cr–Ni–Mo (ST 420/700). The positions, the diameters of the bars and their numbers are depicted in Figure 9.1.11. These bars were posttensioned with an original stress of $\sigma_v = 165.0$ MPa, which corresponds to 40% of the yield strength of the steel rebars. The posttensioning forces are also depicted in Figure 9.1.11.

ii. *Christian part*

The major weakness of the Christian part has been found to lie in the resistance of the piers to horizontal thrusts exerted by the vault. The following proposals aimed chiefly at remedying this weakness:

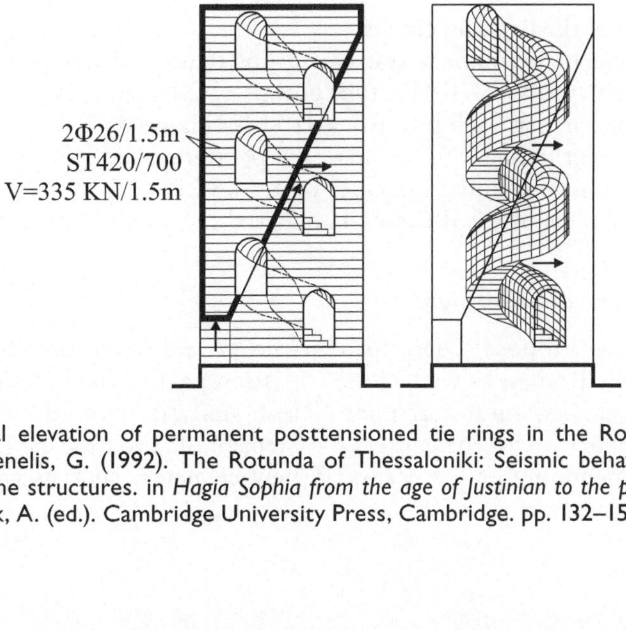

Figure 9.1.10 External elevation of permanent posttensioned tie rings in the Roman part. (Adapted from Penelis, G. (1992). The Rotunda of Thessaloniki: Seismic behaviour of Roman and Byzantine structures. in *Hagia Sophia from the age of Justinian to the present*, Mark, R. and Çakmak, A. (ed.). Cambridge University Press, Cambridge. pp. 132–157.)

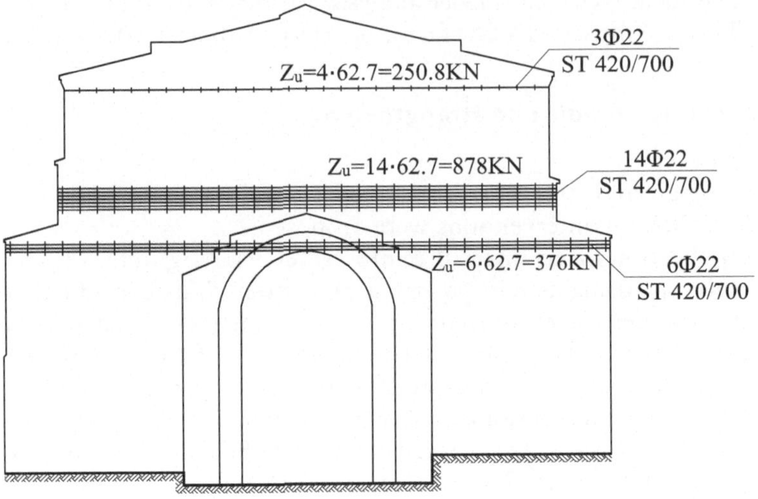

Figure 9.1.11 Reinforcement diagram of piers P2 and P3 of the Roman part. (Adapted from Penelis, G. (1992). The Rotunda of Thessaloniki: Seismic behaviour of Roman and Byzantine structures. in *Hagia Sophia from the age of Justinian to the present*, Mark, R. and Çakmak, A. (ed.). Cambridge University Press, Cambridge. pp. 132–157.)

Alternative A

- Grouting of existing cracks
- Installation of temporary transverse ties at the springing level of the vault
- Strengthening of the existing flying buttresses
- Construction of an additional pair of flying buttresses
- Removal of the temporary ties
- Grouting of any fresh cracks
- Local strengthening of the apse by posttensioned peripheral ties 3Φ26/ST 420/700, initial posttensioning force per tie $S = 88$ kN.

Alternative B

- Grouting of the existing cracks
- Installation of permanent transverse posttensioned ties of stainless austenitic steel of the type Cr–Ni–Mo, two groups of 2Φ30 mm each, ST 420/700, posttensioning force at each group $S = 320$ kN (Figure 5.38)
- Local strengthening of the apse by posttensioned peripheral ties 3Φ26/ST 420/700, posttensioning force per tie $S = 88$ kN.

The Central Archaeological Council approved the second alternative.

9.1.5.10 Reanalysis and redesign

On the basis of the proposals described in the preceding section for the repair and strengthening of the Roman as well as the Christian part of the monument, a reanalysis and redesign were carried out for each part. The reanalysis involved the overall structural behaviour of the structure for the consecutive steps of interventions, whereas the redesign involved the proportioning of the necessary new elements, such as rings, ties, posttensioned stiches and steel cages.

i. *Roman part*

It was concluded that the whole structure, after the installation of the posttensioned rings, presented a safety factor of $\nu = 2.58$ for dead loads instead of $\nu = 2.15$ which was the safety factor for dead loads before the failure of piers P2 and P3.

It should be mentioned that the local interventions to the piers were designed for a safety factor $\nu = 2.04$ for the steel stitches (shear–compression).

For the seismic action, the ultimate seismic resistance of the base of the piers (bending with axial force) was assessed as follows:

$$V_{\text{ult}} = (0.205 - 0.185)W \qquad \text{for the massive piers,}$$

$$V_{\text{ult}} = (0.218 - 0.200)W \qquad \text{for the piers P2 and P3.}$$

It should be noted that the bending resistance for piers P2 and P3 resulted in values higher than those of the other ones because of their bigger cross sections.

ii. *Christian part*

Reanalysis and redesign of the Christian part were performed using the same step-by-step ultimate load procedure followed for the analysis of the original structure. The reanalysis indicated a safety factor for gravity loads of $\nu = 2.10$ instead of $\nu = 1.20$ of the original structure.

9.1.6 Execution

When a detailed design had been completed for all proposals, full detailed drawings and specifications were prepared: These included a technical description of the existing structure and details of the procedures to be followed. The work of repair and strengthening was authorised on this basis in December 1981 and completed in 1992 (Figure 9.1.12). The execution was carried out by the 9th Ephorate of Byzantine Antiquities.

Figure 9.1.12 The Rotunda of Thessaloniki after structural interventions. (From Penelis S.A., Archives, 2019d.)

9.2 STRUCTURAL RESTORATION OF THE ACHEIROPOIETOS BASILICA IN THESSALONIKI

9.2.1 Introduction

The church of Panagia Acheiropoietos is one of the biggest Basilicas of the eastern Mediterranean Basin, and it dates back to the 5th century B.C. During the strong Thessaloniki earthquake in 1978, the monument suffered extensive damage, namely excessive cracking and drifts of masonry walls and colonnades. A first attempt at analysing and designing the remedies, within the framework of the structural restoration, led to an extensive and heavy intervention, which was rejected by the authorities responsible for the monument. As a result, Prof. Penelis G. and Prof. Stylianidis K. were assigned to re-examine the problem. The methodology and the results of the new attempt are briefly presented in this chapter. The approach that was followed proves that *the best criterion for accepting or rejecting the results of an analysis is the degree to which they justify the existing crack pattern, the deformations and the natural period of the building.* Finally, a much lighter intervention scheme was proposed and was approved by the authorities. The restoration works carried out by the 9th Ephorate of Byzantine Antiquities had been completed by 2002 (Figure 9.2.1 and Figure 2.39).

9.2.2 Historical background and structural form

Acheiropoietos is the best preserved basilica in Greece having an almost intact 5th century interior. In its current state, five major periods of construction, including modifications, repairs due to earthquakes and modern reconstruction, can be distinguished. The first period corresponds to the initial erection (third quarter of the 5th century) and the last one to the drastic repair in 1912–1914, during which significant reconstruction was carried out (Raptis and Zombou, 2010; Theocharidou, 1985).

Figure 9.2.1 General view of the Acheiropoietos church of Thessaloniki after the recent restoration. (From Penelis, S.A., 2019a.)

From the architectural point of view, the Acheiropoietos is a three-aisled Christian basilica with a narthex (Figure 9.2.2) covered by timber roof. The three aisles are formed by two colonnades in the east–west direction. The overall dimensions of the church are 51.90 m length, 30.8 m width and 22.0 m height at the top of the roof and 14.0 m at the top of the external walls. The thickness of the brick and stone masonry walls is about 1.0 m (Figures 2.39 and 2.40).

From the structural point of view, the vertical load-bearing system consists of the external walls and the two colonnades (Figures 9.2.3 and 9.2.4). At the mid-height of the two side aisles, there is a wooden floor carrying the loads from the women's galleries. The structure is covered by three timber nonthrusting truss systems.

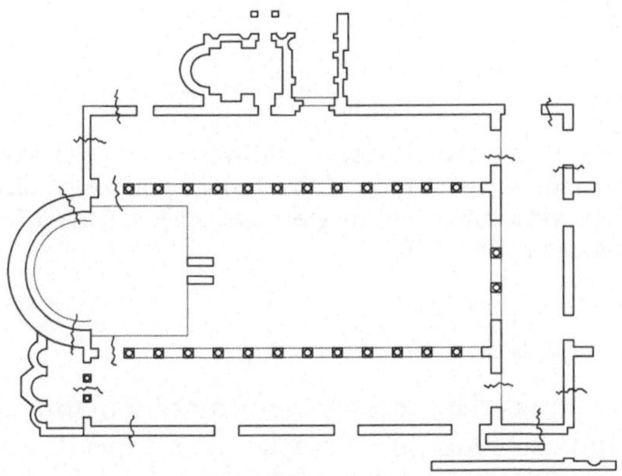

Figure 9.2.2 Plan of the church showing the regions of the main vertical cracks. (Adapted from Penelis, G. and Stylianidis, K. (2001). Structural restoration of the Acheiropoietos basilica in Thessaloniki – Greece. *Proceedings of the 2nd International Congress on Studies in Ancient Structures*. (ed. Arun, G. and Seçkin, N), Istanbul. pp. 847–856.)

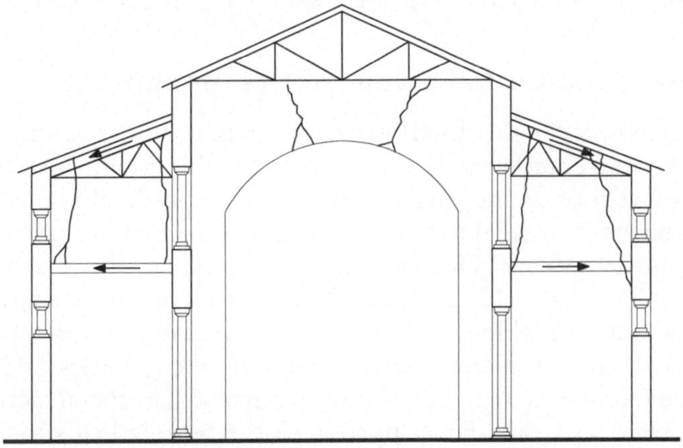

Figure 9.2.3 North–south cross section of the church. (Adapted from Penelis, G. and Stylianidis, K. (2001). Structural restoration of the Acheiropoietos basilica in Thessaloniki – Greece. *Proceedings of the 2nd International Congress on Studies in Ancient Structures*. (ed. Arun, G. and Seçkin, N), Istanbul. pp. 847–856.)

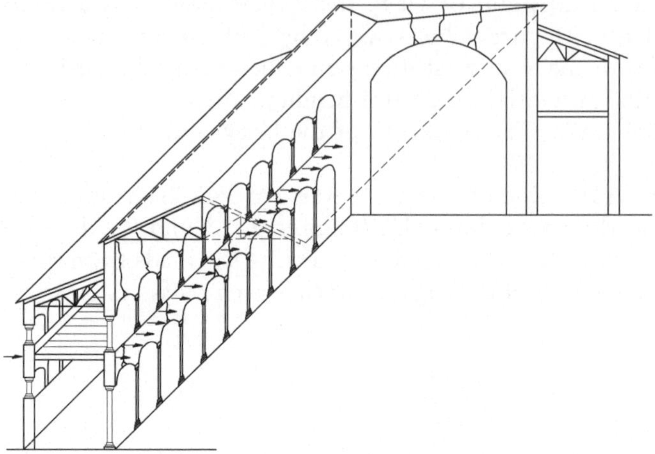

Figure 9.2.4 Cutaway isometric view of the church from southwest. (Adapted from Penelis, G. and Stylianidis, K. (2001). Structural restoration of the Acheiropoietos basilica in Thessaloniki – Greece. *Proceedings of the 2nd International Congress on Studies in Ancient Structures.* (ed. Arun, G. and Seçkin, N), Istanbul. pp. 847–856.)

9.2.3 Pathology

The major problem of the structure is the excessive deformation mainly appearing in the weak north–south direction and the consequent cracks. More specifically, as a result of their rocking, the colonnades incline outwards at their ends and inwards at their mid-height, the maximum inclination being 3.5%. The rotation of masonry arches connecting the columns is also apparent. Drifts of the north and south walls (especially of the latter) in the order of 3.0% are observed at many regions. As a consequence, vertical cracking of the walls especially near the corners of the structure have been induced. There is also an extensive vertical cracking of the masonry arches connecting the columns and at the keystone of the quarter-spherical dome of the sanctuary (Figures 4.41, 4.42 and 4.43).

9.2.4 Static and dynamic behaviour of the monument

It is reasonable to expect for this kind of structures that, under seismic loads, the walls behave as free-standing cantilevers in their transverse direction, since the floors at mid-height and the timber roof are not likely to behave as rigid diaphragms, transferring the seismic loads to the walls parallel to the seismic direction at the narthex and the chancel and so minimising the damage. However in the case of cantilever behaviour, excessive damage (or even collapse) should be expected, the main cracking being horizontal at the bases of the walls, due to rocking, but this kind of cracking was not actually observed. This remark leads to the suspicion that the floor of the galleries and the roof of the building presented an intermediate behaviour, in any way beneficial, acting partly as diaphragms and transferring a significant portion of the seismic loads to the walls parallel to the seismic action. In the light of this hypothesis of partial diaphragmatic action of the floors and the roof, the expected cracking pattern matches very well with the existing one (Penelis, G. and Stylianidis, K., 1998, 2001).

To verify this scenario, the predominant natural period of the structure was calculated based on the hypothesis of cantilever action and was found to be 2.13–3.30 s, depending on the degree of fixity on the ground. However, the natural period was measured *in situ* and was found to be 0.60–0.65 s in north–south directions and 0.24 s in east–west, respectively, meaning that the actual structure is much more rigid than that of a free-standing cantilever.

9.2.5 Proposal of intervention scheme

In the light of the aforementioned remarks, a new scheme of general and local interventions was proposed, estimating that it can minimise the seismic sensitivity of the monument.

General nature interventions aim at providing controlled strength and stiffness to the already existing galleries' floors and roofs. Strength and stiffness can be ensured by several alternative structural techniques. The resulting diaphragms transfer a significant portion of the transverse seismic load to the walls of the chancel and the narthex, which are parallel to the seismic action. In order to withstand this action, these walls have to be lightly strengthened by external tie rods. Previously mentioned interventions improve the static and dynamic behaviour of the monument drastically. The resulting system reaches the serviceability limit state under a seismic coefficient $\varepsilon = 0.12$ and the ultimate limit state under $\varepsilon = 0.16–0.18$, which corresponds to a return period of 500 years (Figures 5.25, 5.26 and 5.32). The magnitude of the improvement is obvious, considering that, despite the interventions of the first restoration attempt, the ultimate limit state corresponded to a capacity of $\varepsilon = 0.05$.

Local nature interventions aim at restoring the structural integrity of extensively cracked walls and repairing those regions of the structure which have presented damage due to aging. These interventions include injections using traditional materials, repointing, local reconstruction, stitching etc.

9.2.6 The type of chosen diaphragms

Several alternative types of diaphragms were considered and evaluated. The Central Archaeological Council approved the proposal of a wooden diaphragm strengthened by a metal tube beam at the perimeter. Initially it was anticipated that the diaphragms of the wooden floors and roofs of the galleries would be constructed by successive rows of planks, water-resistant nine-plywood panels and planks, 2.50 + 2.10 + 2.50 cm thick, connected with woodscrews and with the existing joists of the floors and the semitrusses of the roof with steel shear joints. The steel tube beam (RHS) at the perimeter of each diaphragm with a cross section of 140 × 80 × 7 mm was connected to the masonry with anchored metal bars (see Figures 9.2.5, 9.2.6 and 9.2.7).

It should be noted that during the construction of mockup difficulties in the application, misses in the allocated wooden layers of the timberwork were observed especially at their conjunction with the steel shear joints. So the modification of the diaphragm was decided by the design team by using two successive layers –instead of one– of compressed impregnated waterproofing nine-plywood panels 2.10 cm each (Raptis and Zombou, 2010).

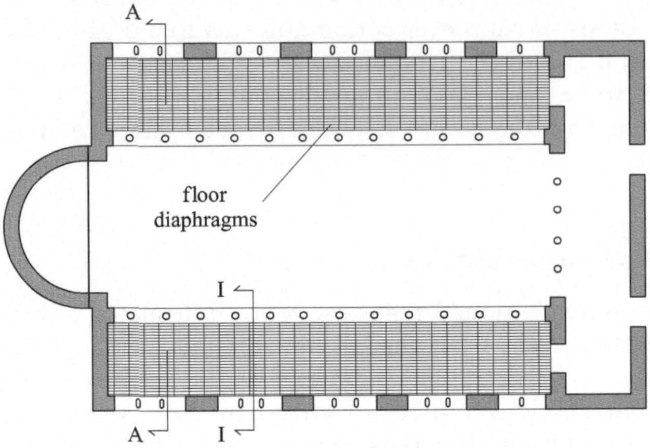

Figure 9.2.5 The layout of the floor diaphragms in plan. (Adapted from Penelis, G. and Stylianidis, K. (2001). Structural restoration of the Acheiropoietos basilica in Thessaloniki – Greece. *Proceedings of the 2nd International Congress on Studies in Ancient Structures.* (ed. Arun, G. and Seçkin, N), Istanbul. pp. 847–856.)

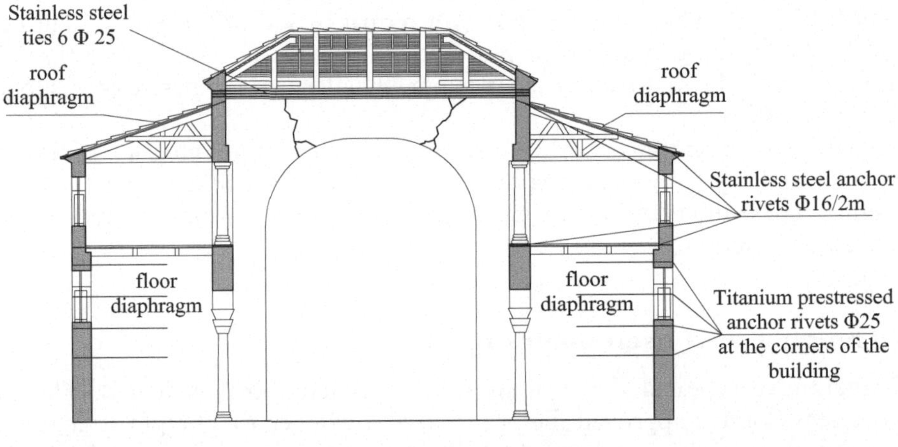

Figure 9.2.6 The layout of the diaphragms in cross section. (Adapted from Penelis, G. and Stylianidis, K. (2001). Structural restoration of the Acheiropoietos basilica in Thessaloniki – Greece. *Proceedings of the 2nd International Congress on Studies in Ancient Structures.* (ed. Arun, G. and Seçkin, N), Istanbul. pp. 847–856; Raptis, K. and Zombou, A. (2010). The consolidation and restoration project of Acheiropoietos basilica in Thessaloniki. *Proceedings of the 8th International Symposium on the Conservation of Monuments in the Mediterranean Basin Monubasin 8,* Patras, Greece, pp. 411–428.)

9.2.7 Structural analysis and design

9.2.7.1 *Basic assumptions*

For the analysis of the monument, the SAP90 program was used on the assumption of elastic behaviour. The monument was modelled as a 3D structure using shell elements for the masonry walls and frames for the columns.

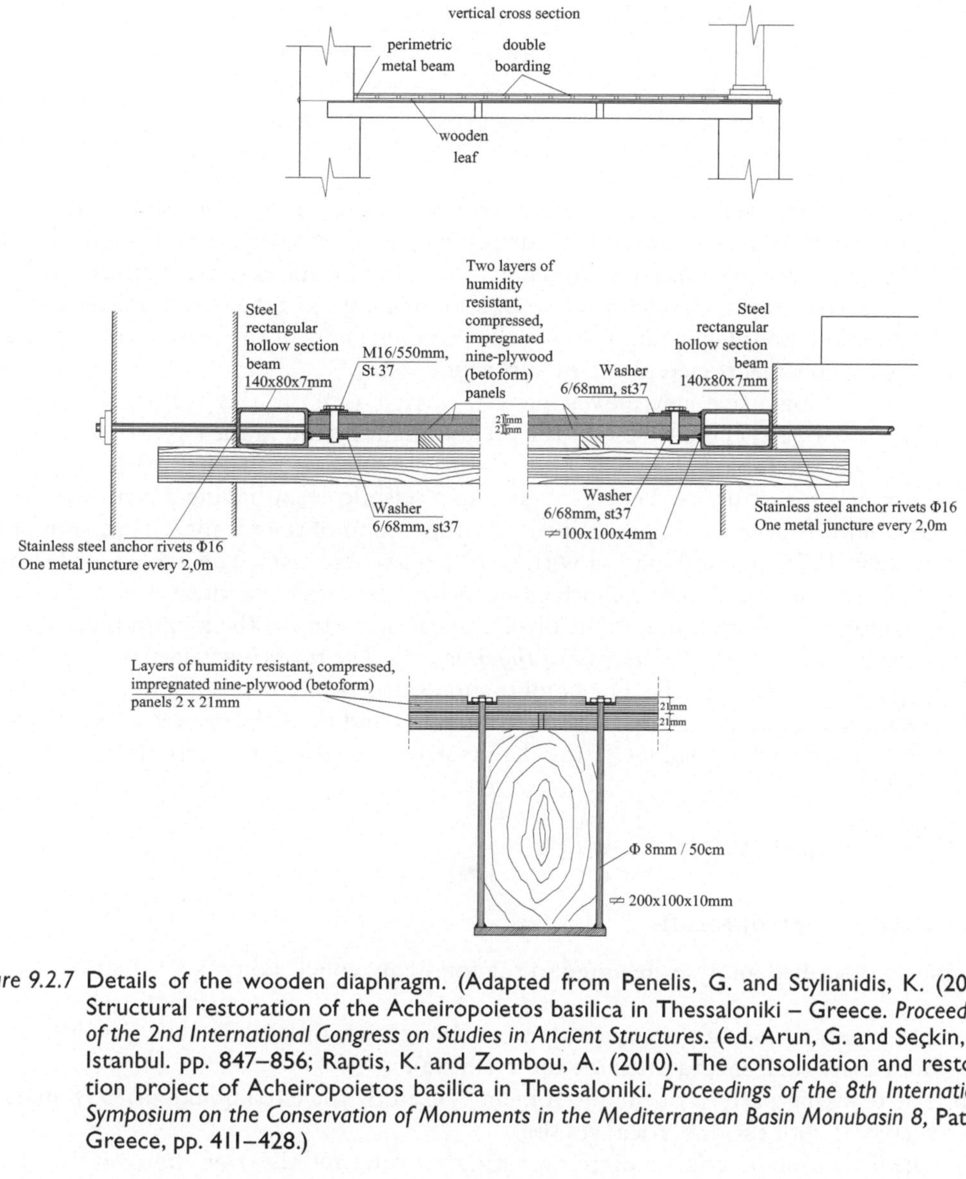

Figure 9.2.7 Details of the wooden diaphragm. (Adapted from Penelis, G. and Stylianidis, K. (2001). Structural restoration of the Acheiropoietos basilica in Thessaloniki – Greece. *Proceedings of the 2nd International Congress on Studies in Ancient Structures.* (ed. Arun, G. and Seçkin, N), Istanbul. pp. 847–856; Raptis, K. and Zombou, A. (2010). The consolidation and restoration project of Acheiropoietos basilica in Thessaloniki. *Proceedings of the 8th International Symposium on the Conservation of Monuments in the Mediterranean Basin Monubasin 8,* Patras, Greece, pp. 411–428.)

The roof of the central aisle was modelled with frame elements, whereas the roof and gallery floors of the side aisles were modelled in two different ways. More specifically, in the case of the existing structure, the roof and the gallery floors were modelled with frame elements, whereas in the case of their transformation in wooden diaphragms of controlled strength and stiffness, they were modelled by shell elements. It is obvious that two different models were formed, one corresponding to the current state and another one to the state after the intervention. In the second case, the determination of the strength and stiffness of the finite elements of the diaphragms was critical. These values were determined using extensive analysis and design partially based on DIN 1052 – sheets 1 and 2 (DIN 1052, 1976) and on the work of K. Möhler (1976).

Table 9.2.1 Predominant period after the restorations

Direction	*Measured (s)*	*Calculated (s)*
North–south	0.46	0.48
East–west	0.18	0.185

The analysis under seismic actions was performed using equivalent static analysis. For the determination of the seismic action, the fundamental natural period was calculated to $T_i = 0.63$ s for the north–south direction and $T_2 = 0.24$ s for the east–west direction. These values were derived from a model response spectrum analysis of the structural system in its original situation, and they comply very well with the results of the *in-situ* ambient vibration measurements mentioned above in Subsection 9.2.4.

Recently, an extensive new investigation carried out on the restored monument (Salonikios, et al., 2017) has given for the predominant period in the two main directions the results listed in Table 9.2.1.

The response spectrum used corresponded to a seismic event having a return period of 500 years and was derived via probabilistic extrapolation of the response spectrum of the earthquake in 1978, in combination with *in-situ* cross-hole tests. The behaviour factor of the monument was calculated for both cases (before and after the intervention) by means of P–δ diagrams, corresponding to the displacement of the top of the longitudinal walls *on the assumption of inelastic behaviour of the masonry.* The resulting values were $q_{\text{init.}} = 1.5$ and $q_{\text{strength.}} = 2.0$, respectively. The combination of the fundamental periods, the elastic response spectrum with a return period of 500 years and the behaviour factors led to the adoption of the following values of the design seismic coefficient (ratio of the behaviour factors 1.5/2.0):

$$V_{\text{base}}^{\text{original}} = 0.16\,W, \quad V_{\text{base}}^{\text{strengthened}} = 0.12\,W.$$

9.2.7.2 *Evaluation of results*

Previously described analyses resulted in the following conclusions:

a. The combination of gravity loads to the north–south seismic action is critical for both cases (before and after the intervention).
b. The cracking pattern of the monument caused by the earthquake in 1978 matches very well with the analytical results.
c. The formation of wooden diaphragms at the floors and the roof improves drastically the stresses and the deformations of the bearing system to such an extent that the monument can withstand the design earthquake without even cracking.

9.2.8 Intervention materials

The intervention scheme included two categories of work:

a. Reversible works, which include the wooden diaphragms with their metal perimetric beam
b. Irreversible works, which include mass injections, rejointing and stitching of the cracked corners of the church.

For reversible works, common materials were used, without having to fulfill specific requirements, such as plywood for external use, galvanised steel beams etc. For irreversible works, materials fulfilling the requirements of compatibility and durability were used. Therefore, as far as the injections and mortars are concerned, specific compositions studied at the laboratories of Aristotle University of Thessaloniki were used. These compositions were based on research of the chemical, mineralogical and mechanical properties of the existing mortars of the monument. Specifically, for injections, a thin colloid composition was selected for the improvement of the penetration of the material to the fine pores of the original mortar (Miltiadou, 1990). Such compositions were, at the time of restoration, the most modern approach to the problem of penetration to fine pores, yet they demand specific high-speed mixers for the preparation of the colloid mixture. Finally, titanium alloy was used for all metal elements (connectors, stiches) embedded into the masonry.

9.2.9 Conclusions

On the occasion of the present structural restoration of an important monument, the following general validity conclusions were once more confirmed:

a. At regions of high seismic risk, earthquakes are the main reason for damages or collapse of monuments and historical buildings. These buildings have suffered the consequences of seismic events for centuries, and some of them have survived under a process of natural selection. Only well-designed and well-constructed buildings have survived to our times.
b. As a result, in the case of intervention, the structural system should not be altered radically. Only local interventions or improvements of the virgin structural system (as in the present case of strengthening of wooden diaphragms) can be accepted, since there is no guarantee that the modified structural system will have greater probabilities to survive than the older one, which has already been physically tested.

   In the case of the Acheiropoietos church, by increasing rigidity and strength of the diaphragms, the stiffness of the structural system as a whole, as it comes out from recent measurements (see Table 9.2.1), was improved by 70%.
c. To estimate the bearing capacity of a monument, in the perspective of interventions to the structural system, *in-situ* measurements and research, laboratory tests and complex and detailed static and dynamic analyses are necessary. Only deep knowledge of the virgin structural materials and the structural behaviour of the monument can justify an intervention.
d. In case an intervention is decided, it must not be forgotten that the main aims of structural anastylosis are the conservation and the preservation of the historical and aesthetic values of the monument. As a result, any intervention should be based on respecting the virgin materials and the authentic elements. This choice imposes strict limitations on the selection of materials and techniques of interventions.
e. The key to the selection of the materials and the techniques to be used in a structural anastylosis is the distinction of those techniques into reversible and irreversible ones. The materials used in reversible techniques are subjected to very few limitations. On the contrary, the materials used in irreversible techniques have to be durable and compatible to the existing materials.
f. Finally, structural anastylosis constitutes an intervention of high skill which demands cooperation of specialists coming from several scientific fields and strong support of laboratories and computational facilities.

9.3 THE STRUCTURAL RESTORATION OF THE NATIONAL LIBRARY OF GREECE IN ATHENS

9.3.1 Introduction

The National Library of Greece is located in a magnificent neoclassic complex of URM buildings built at the beginning of the 20th century, which have been damaged by the Athens quake of 7 September 1999. The authors of this book were awarded with the structural restoration design of the complex. In order to achieve a light and reversible intervention, the design consisted of extensive *in-situ* and laboratory tests concerning the mechanical, chemical and mineralogical characteristics of masonry; special considerations for the composition of the enemas to be used in order to prevent any impact on the frescos and sophisticated analytical models and procedures including a pushover analysis and evaluation using the ADRS methodology.

9.3.2 Description

During the 7 September 1999 Athens quake on mount Parnitha, a lot of buildings had been damaged within the city limits. The National Library building is a neoclassic complex of buildings located in the centre of the city of Athens on Panepistimiou Street and was designed and built in the beginning of the 20th century by the famous architects Hansen and Schiller (Figure 9.3.1). The complex is composed of three independent rectangular parallel wings and three walkways connecting the two side wings to the main one (Figure 9.3.2). The main buildings are built with URM dressed externally with smoothened marble megastones combined with marble columns, while the floors are composed

Figure 9.3.1 View of the National Library building. (Penelis, G. and Penelis, Gr. (2001). The structural restoration of the National Library of Greece in Athens. *Proceedings of the 3rd International Congress of Studies in Ancient Structures*. Istanbul. pp. 857–865.)

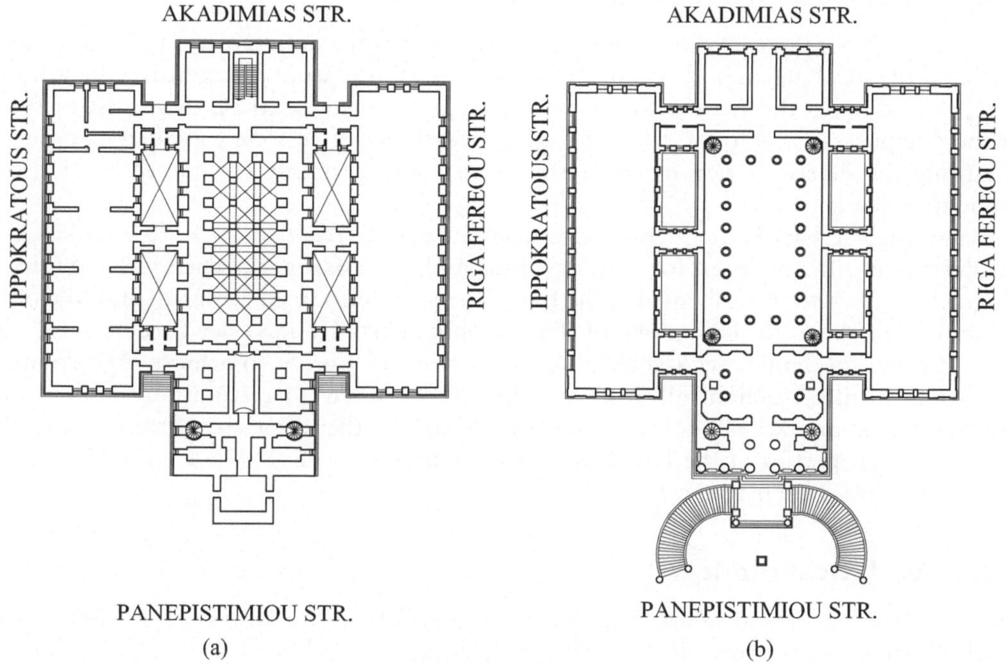

Figure 9.3.2 Plan view: (a) ground Floor; (b) first floor. (Adapted from Penelis, G. and Penelis, Gr. (2001). The structural restoration of the National Library of Greece in Athens. *Proceedings of the 3rd International Congress of Studies in Ancient Structures*. Istanbul. pp. 857–865.)

of steel beams filled with brick vaults in between. The main wing is decorated with beautiful frescos, which should obviously be preserved (Penelis, Gr. and Penelis, G., 2001).

The building is the actual National Library of Greece, and therefore besides being a monument, it is being used every day by hundreds of people who, after the earthquake, were at great risk as it could be concluded from the pathology.

9.3.3 Building pathology

The buildings had three different types of damage: (I) time related, (II) damage from the 1981 quake which had not been repaired and (III) damage from the 1999 quake.

In all three wings, the type (I) damage consisted mainly of water leakage and oxidisation of the steel beams, while the damages of types (II) and (III) are different for each wing.

More specifically, the two side wings presented inclined cracks at the edges of the transverse walls typical of independent out-of-plane behaviour of these walls due to the lack of diaphragm constraint at the floor and roof levels. Furthermore, horizontal cracks were presented at the foot of the piers of the longitudinal walls due to in-plane flexure. The spandrels of the longitudinal walls suffered small vertical cracks indicative of tensile stresses.

The main wing had significantly less damage which can mainly be attributed to its higher rigidity and robustness (longitudinal walls) and the intermediate transverse walls which prevented independent out-of-plane behaviour of walls.

9.3.4 Site investigations and laboratory tests

The site and laboratory investigations and tests aimed at the determination of the ground-type and the mechanical and chemical characteristics of masonry. Regarding the ground type, according to the Athens Metro boreholes performed by MECASOL just in front of the complex, it was found to be a slightly weathered limestone classified as type 'A' regarding the determination of the seismic forces and certainly capable of bearing the foundation loads.

Regarding the URM, the investigation consisted of destructive (six core samples) and nondestructive (hammer test for masonry) tests and analysis. Special chemical and mineralogical analyses were performed at the Laboratory of Reinforced Concrete and Masonry Structures of Aristotle University of Thessaloniki (AUTH) in order to determine the chemical and mechanical characteristics of the mortar in order to achieve high compatibility (chemically, mechanically and aesthetically) between the restoration enemas to be used, having in mind the importance of the building and the sensitivity of the frescos. The mechanical properties of the URM as derived according to EN1996 1-1 (2005), on the basis of the tests, are listed in Table 9.3.1.

9.3.5 Analytical models

Two different analytical approaches were selected for the buildings. In the first stage, all three wings were analysed using elastic dynamic spectral analysis of a finite element mesh of plane (shell) elements in order to identify the areas of stress concentration and the overall behaviour of each wing. In the second stage, nonlinear static (pushover) analysis of the two side wings, which presented extensive cracking during the earthquake, was performed, using equivalent frame models, since their cracking suggested nonlinear behaviour. For both the analyses, the seismic loads were determined by the Greek seismic code spectra for seismicity zone II (Figure 9.3.3).

9.3.5.1 Dynamic finite element analysis

a. *Side wings*

Regarding the two side wings, the results were similar and are therefore presented together. So, the building had a low period of 0.21 s and presented only transitional modes due to its symmetry. Tensile stresses were developed locally at the spandrels explaining the cracks presented there, and higher tensile stresses were developed at

Table 9.3.1 Mechanical properties of URM, characteristic values

Compressive strength	$f_{wk} = 4.50$ [MPa]
Shear strength	$f_{vk} = 0.22 + 0.40\sigma_d$ [MPa]
Tensile strength	$f_{wt} = 0$
Tensile strength (out of plane)	$f_{wx1} = 0.18$ [MPa] (horizontal cracks) $f_{wx2} = 0.31$ [MPa] (vertical cracks)
Elastic modulus	$E = 4500$ [MPa]
Shear modulus	$G = 1800$ [MPa]
Poisson's ratio	$\nu = 0.25$

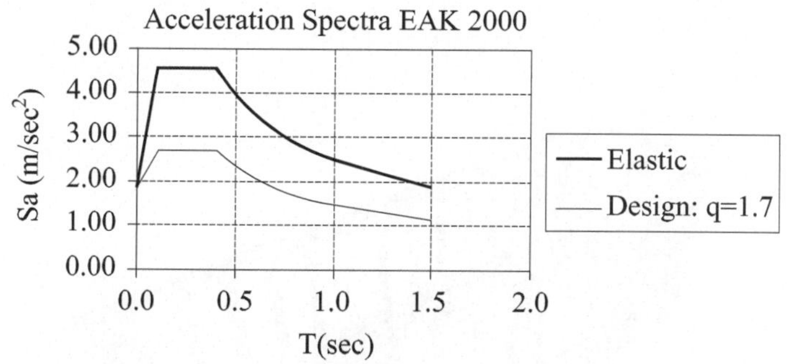

Figure 9.3.3 Elastic and design spectra as defined in Greek seismic codes. (Adapted from Penelis, G. and Penelis, Gr. (2001). The structural restoration of the National Library of Greece in Athens. *Proceedings of the 3rd International Congress of Studies in Ancient Structures*. Istanbul. pp. 857–865.)

the piers justifying the horizontal cracks presented there (Figure 9.3.4). Of course, the analysis of the model did not produce the pathology attributed to the out-of-plane behaviour of the two transverse walls after their separation from the longitudinal ones, since in the analytical models they were connected with the longitudinal ones, which suggested that the main deficiency of the structural system is this connection which, obviously, had to be restored.

b. *Main wing*

The main wing building was found to be stiffer than the two side wings with a natural period of 0.165 s. It presented only transitional modes due to its symmetry and high torsional restraint attributed to the perimetric masonry walls. The stresses were significantly lower than the ones developed at the side wings (Figure 9.3.5), since there are no openings at the first floor, smaller openings at the ground floor and the ratio of URM wall over square metres (m^2) of plan is higher in both directions. These results were in complete agreement with the minimal damage observed at this wing and led to the conclusion that a more sophisticated analysis was not required for this wing and only local restorations would be needed.

9.3.5.2 Nonlinear static analysis

The pushover analysis of the two side wings was performed using plane equivalent frames. The model used was a frame-element model with areas of concentrated inelasticity. The nonlinearity was concentrated in rotational springs located at the ends of the piers and the spandrels. The moment–rotation (M–θ) diagrams of the URM elements had been derived based on a procedure elaborated by the second of the authors (Penelis, Gr., 2006a), and these had been used as constitutive laws for the nonlinear springs (see Subsection 7.6.2).

The use of two independent plane equivalent frames (one per direction) is obviously a conservative assumption since the participation of the transverse walls is neglected and, as shown in the dynamic analysis, torsional modes do not exist.

The evaluation of the pushover analysis and the determination of the target displacement were based on the acceleration–displacement capacity–demand (ADRS) spectrum procedure prescribed by ATC-40 (ATC-40, 1996). The inelastic spectra have been derived

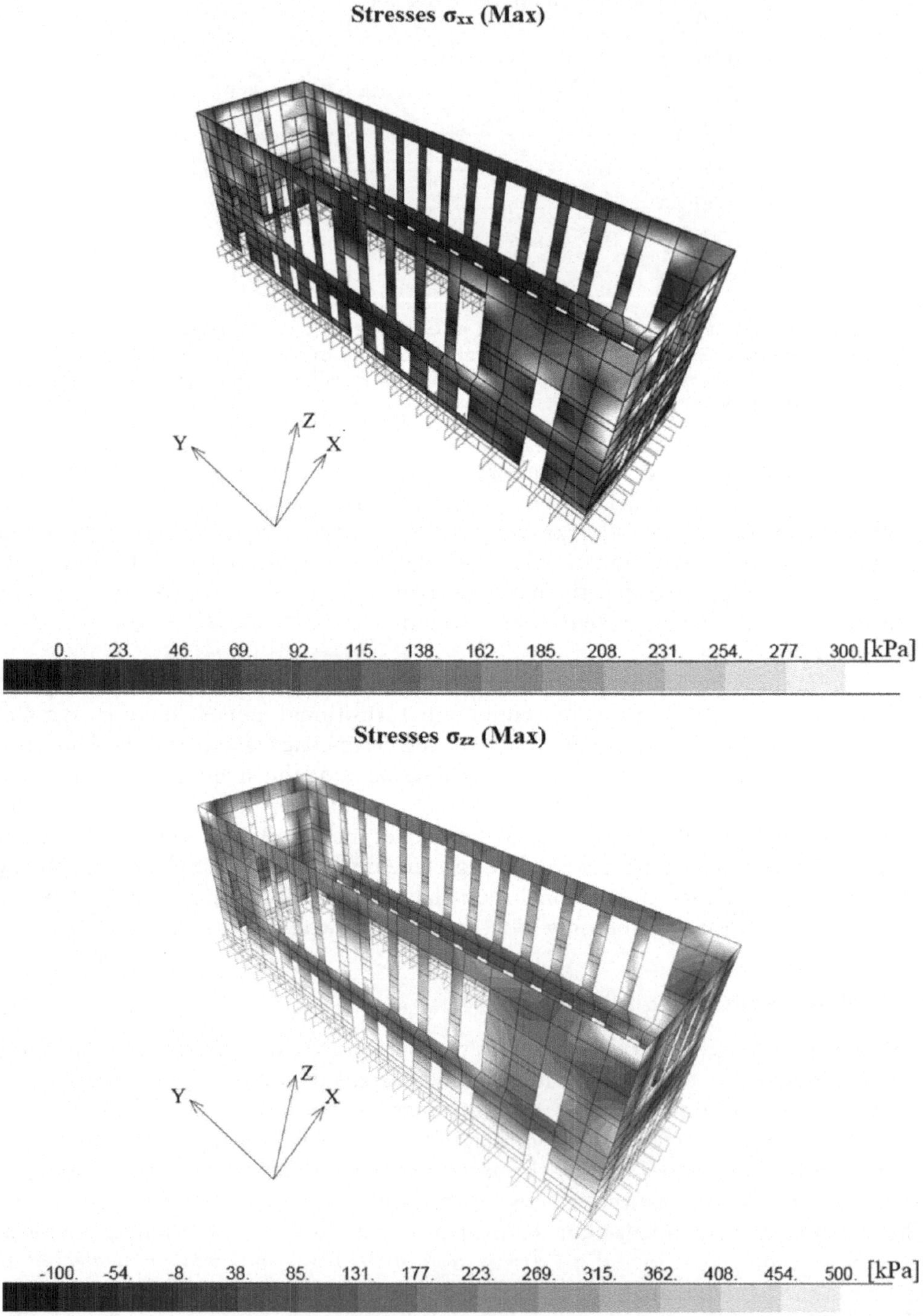

Figure 9.3.4 Elastic dynamic analysis of side wings. (Adapted from Penelis, G. and Penelis, Gr. (2001). The structural restoration of the National Library of Greece in Athens. *Proceedings of the 3rd International Congress of Studies in Ancient Structures*. Istanbul. pp. 857–865.)

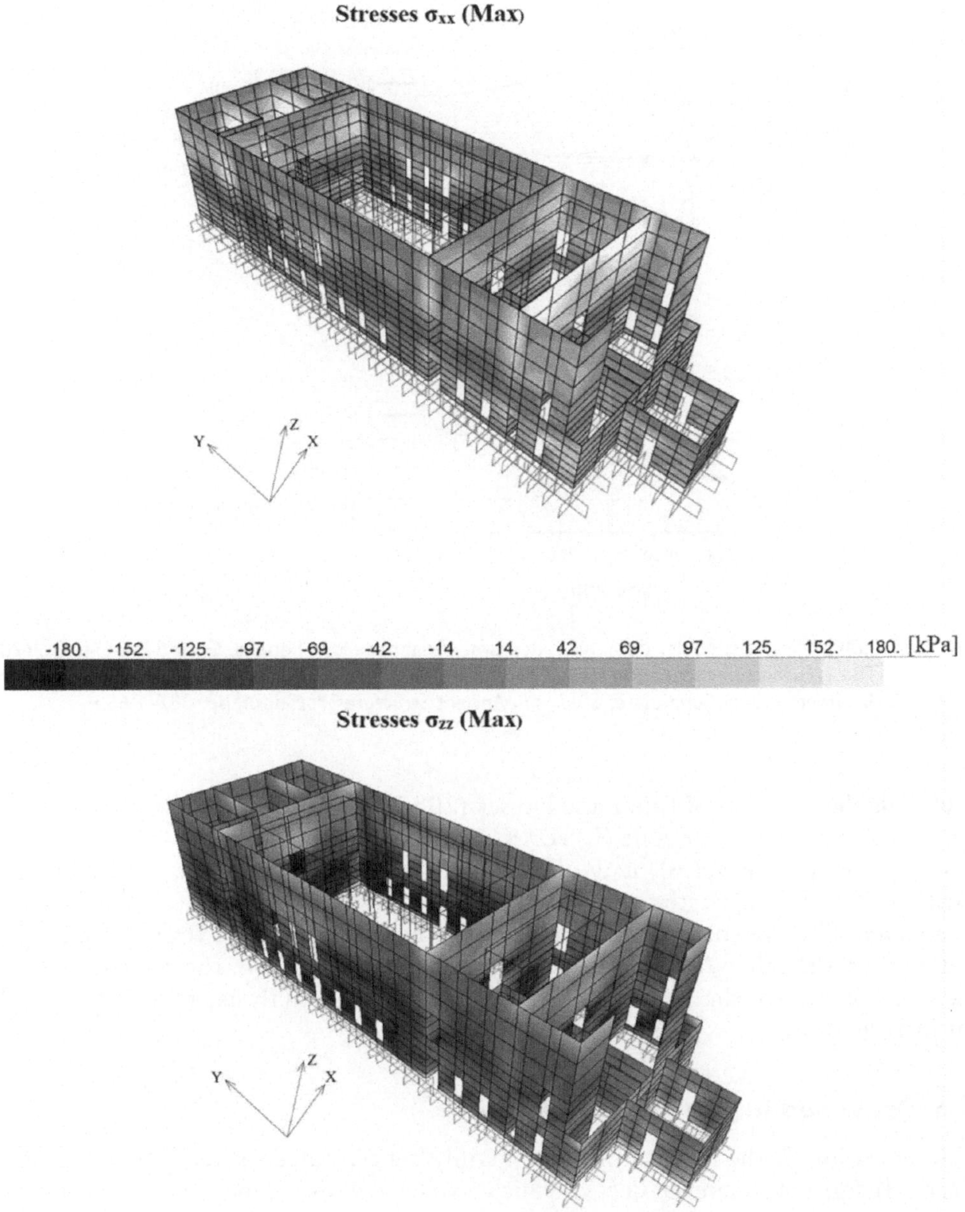

Figure 9.3.5 Elastic dynamic analysis of main wing. (Adapted from Penelis, G. and Penelis, Gr. (2001). The structural restoration of the National Library of Greece in Athens. *Proceedings of the 3rd International Congress of Studies in Ancient Structures*. Istanbul. pp. 857–865.)

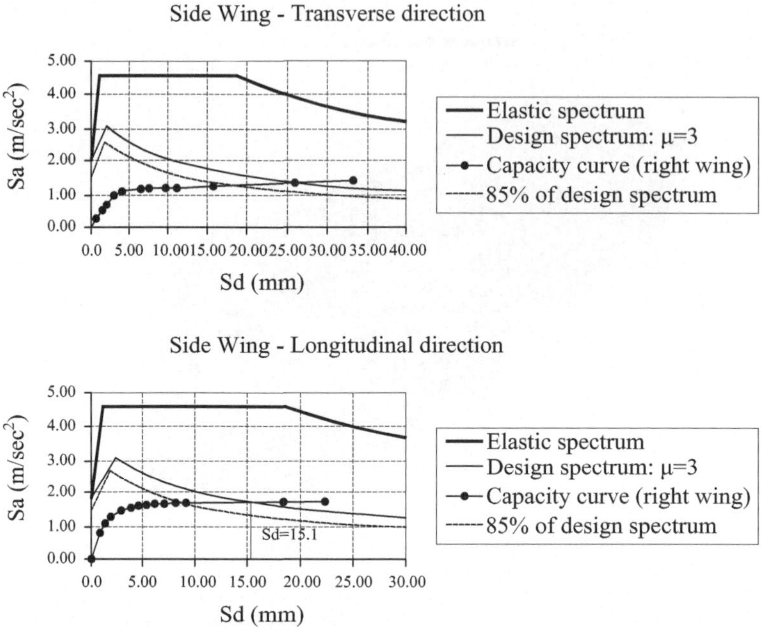

Figure 9.3.6 ADRS format capacity demand spectrum. (Adapted from Penelis, G. and Penelis, Gr. (2001). The structural restoration of the National Library of Greece in Athens. *Proceedings of the 3rd International Congress of Studies in Ancient Structures*. Istanbul. pp. 857–865.)

by applying the formulas of Fajfar and Dolsek (2000) on the elastic acceleration spectra of the Greek seismic codes for zone (II) (EAK, 2000).

From these (Figure 9.3.6), it is obvious for both directions that the building can withstand the design earthquake but it does so by consuming most of its inelastic capacity. This means that, in both directions, the building would sustain under the design earthquake with serious but not critical damage. So, the pushover analysis validated completely the results of the elastic dynamic analysis, pointing towards a light intervention.

9.3.6 Proposed intervention scheme

As aforementioned, the intervention was mainly concentrated on the side wings, except for type (I) damages (material decay), which also existed in the main wing. Furthermore, all the intervention has been directed towards being light and reversible.

So, in that context, for the two side wings, the structural restoration consisted of the implementation of a perimetrical, internal zone consisting of beams on the roof level, in order to restore the diaphragm constraint, either using concrete with stainless steel as reinforcement or stainless steel profiles in order to avoid future corrosion. Furthermore, four titanium stitches per corner per wall were suggested in order to secure their cooperation and the uniform behaviour of the longitudinal and transverse walls (Figure 9.3.7). The cracks on URM were filled with enema with specially selected composition in order

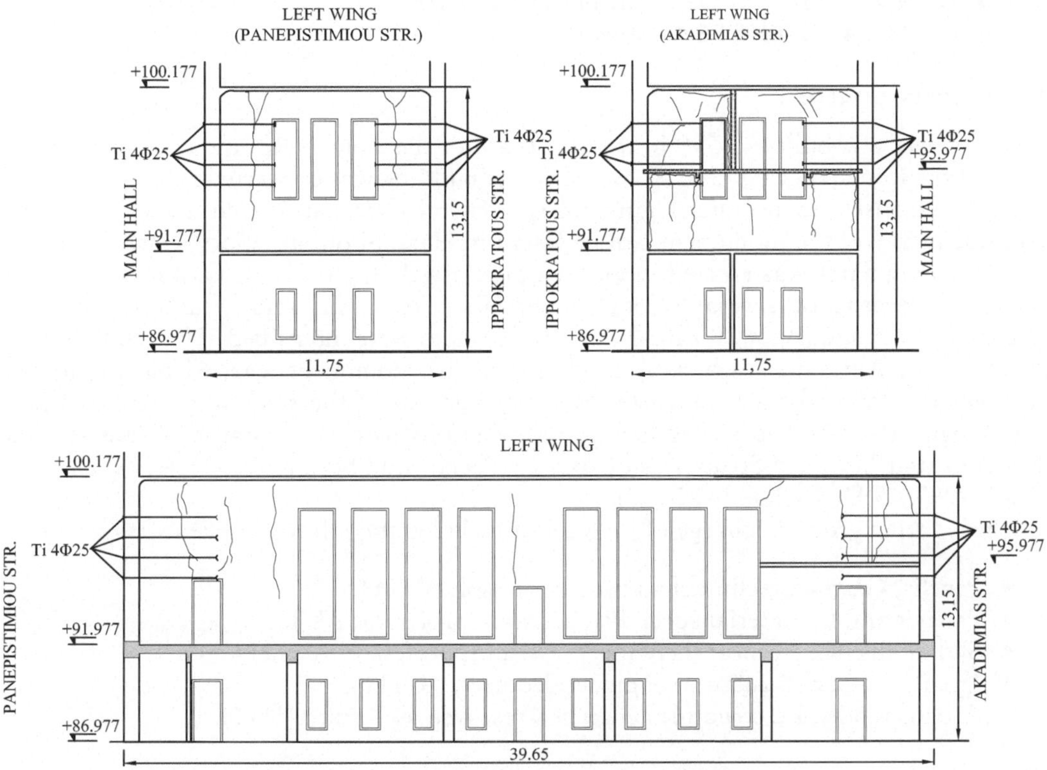

Figure 9.3.7 Titanium stitches layout. (Adapted from Penelis, G. and Penelis, Gr. (2001). The structural restoration of the National Library of Greece in Athens. *Proceedings of the 3rd International Congress of Studies in Ancient Structures*. Istanbul. pp. 857–865.)

to conform chemically, mechanically and aesthetically to the existing mortar using sophisticated laboratory tests that had been performed and the existing database at the Laboratory of Reinforced Concrete and Masonry Structures of Aristotle University of Thessaloniki (AUTH).

9.3.7 Conclusions

All the restoration techniques used in the case of the National Library building in Athens may not be innovative, but the fact that the state of the art has been used in laboratory and *in-situ* tests and analyses successfully and efficiently (time wise, within 2 months) in order to securely reach a very light intervention scheme for the restoration of a monument, which is not only being admired as a monument and landmark of Athens but also being visited every day by hundreds of people, is certainly interesting.

Regarding analysis, it has been shown that the most recent trends in earthquake engineering can indeed be used in the analysis of URM monuments and increase the level of confidence and safety of the practicing engineer who is confined (by the authorities), with the limitations of a light intervention.

9.4 STRENGTHENING OF THE MINARET OF ROTUNDA IN THESSALONIKI

9.4.1 Introduction

This minaret was built in 1590 in the courtyard of the Roman Rotunda, a great monument of the Roman period (300 A.D.), when the building was converted from a Christian church to an Islamic mosque (Figure 9.4.1). General information and a detailed architectural survey of the monument can be found in Hatziantoniou report (Hatziantoniou, 1988). The minaret was erected in an area of relatively high seismicity, and it has been damaged because of several strong earthquake excitations and weathering effects. During the earthquake of 20 June 1978, the damages were intensified. The 9th Ephorate of Byzantine Antiquities, officially in charge of the monument, carried out the necessary emergency works and entrusted the investigation of the structural condition and the design of repair and strengthening interventions to the Laboratory of Reinforced Concrete and Masonry Structures of AUTH (Penelis and Stylianidis, 1994; Penelis, G., et al., 1997, 1993; Stylianidis, 2000).

The working group in charge of the project included the following scientists:

- Penelis, G., chief technical advisor, professor, AUTH
- Papayianni, I., material specialist, assistant professor, AUTH
- Stylianidis, K., structural engineer, assistant professor, AUTH
- Ignatakis, Chr., structural engineer, lecturer, AUTH
- Athanassiadis, L., structural engineer, research assistant, AUTH.

9.4.2 In-situ and laboratory research

The 9th Ephorate provided to the research team a series of architectural drawings, based on detailed surveying of the monument, the results of 64 hammer tests on bricks and mortar and a preliminary report concerning the structural condition of the monument. In order to enrich the available data, a research project including constructional pathology surveys, *in-situ* and laboratory measurements and tests, *in-situ* measurements of the fundamental period of the structure and geotechnical and seismological studies of the territory were carried out.

9.4.2.1 Geometry of the structure

The minaret is located at a distance of 3.50 m western of Rotunda (Figure 9.4.1a). It consists of a 4.50 × 4.50 m square cross-sectioned base, 7.50 m in height and founded on a thick, stiff, red clay layer approximately 6.00 m below the ground level. The base supports the trunk of the minaret which consists of two successive parts. The lowest one is a twenty-sided truncated cone of 22.0 m total height, crowned by the muezzin's balcony. The highest one above the balcony is a fifteen-sided cylinder of 6.00 m height (Figure 9.4.2a). The external diameter of the trunk decreases gradually from 3.25 m to 2.10 m and finally to 1.75 m at the highest cylindrical part. The narrow entrance (0.85 × 1.95 m) is located at the northern face of the base. The internal helical staircase with gradually decreasing diameter (D = 1.80–1.45 m) leads up to the balcony at a height of 29.50 m. The small Muezzin's doorway (0.50 × 1.50 m) is located at the southern face of the cylindrical trunk. The narrow staircase (D = 1.65 m) goes on till the top of the minaret. Five narrow loopholes are located along the height of the southern side of the trunk.

(a)

(b) (c)

Figure 9.4.1 (a) Rotunda and minaret: southern view; (b) the balcony and the upper cylindrical part; (c) internal staircase. (Adapted from Stylianidis, K. (2000). Strengthening of the minaret of Rotunda in Thessaloniki. G. Penelis. *International Symposium of Concrete and Masonry Structures*, Stylianidis, K. and Kappos, A. (eds.). Thessaloniki, Greece pp. 335–348; Penelis, G., et al., 1994.)

The base of the minaret consists of successive stone and brick masonry layers, while the masonry of the trunk is pure brickwork built by flattened bricks (30 × 40 × 4 cm) with thick mortar joints (4 cm). The balcony consists also of brickwork supported by successive horizontal wooden beams tangentially projecting from the cylindrical trunk. The staircase is built together with the surrounding trunk. At the edge of every step, a wooden beam is located diametrically and is embedded in the masonry of the trunk, while along the axis of the staircase, a masonry column ($D = 0.25$ m) is constructed (Figure 9.4.1c).

9.4.2.2 Pathology – emergency interventions

As it can be easily seen in Figure 9.4.2, the base of the minaret is not plastered. The plaster covering has been detached at the major part of the trunk, while over the rest of the surface, as well as at the upper cylindrical trunk, it is heavily cracked.

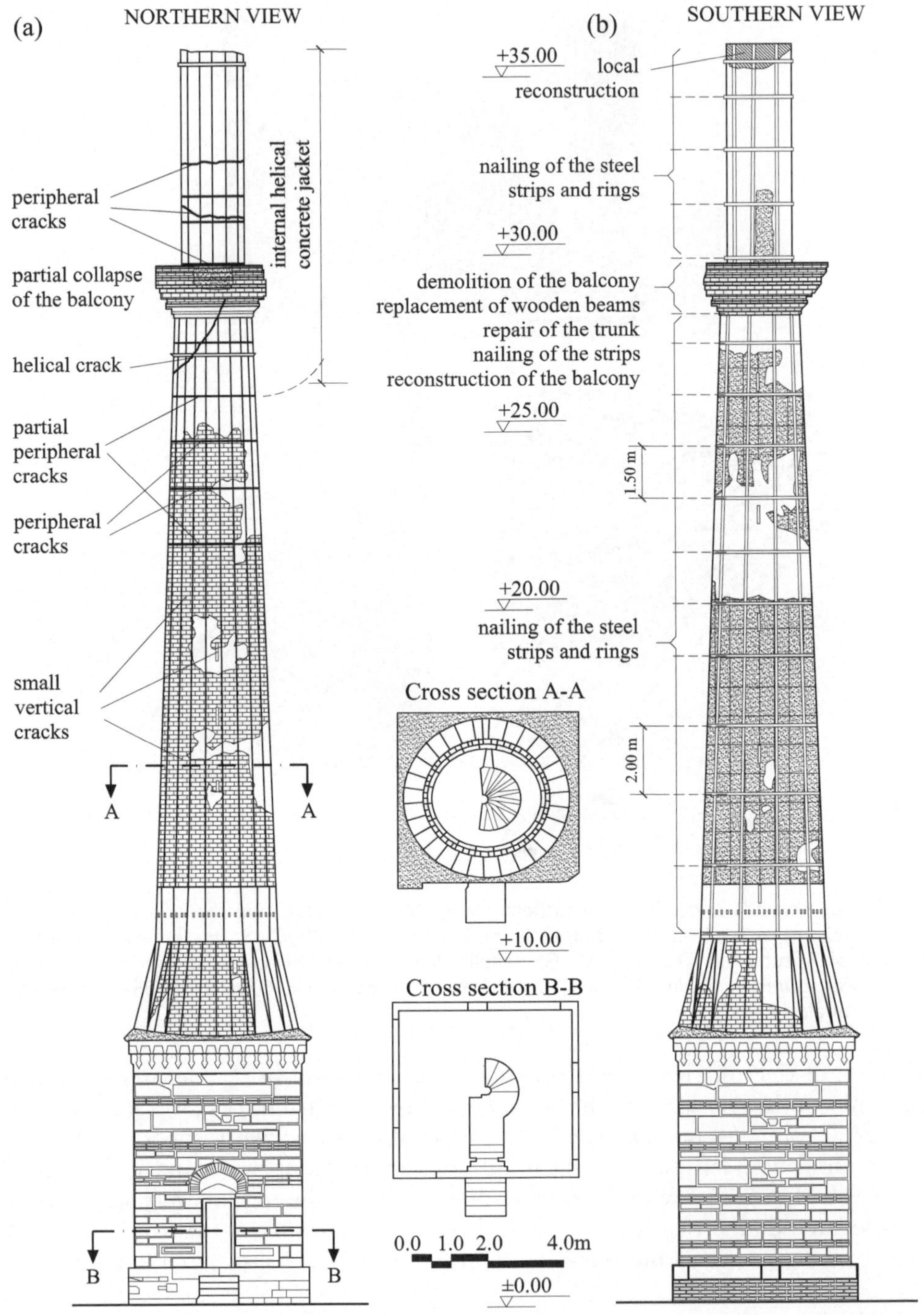

Figure 9.4.2 (a) Northern view of the minaret (geometry, pathology); (b) southern view of the minaret. General layout of external metal strips. (Adapted from Stylianidis, K. (2000). Strengthening of the minaret of Rotunda in Thessaloniki. G. Penelis. *International Symposium of Concrete and Masonry Structures*, Stylianidis, K. and Kappos, A. (eds.). Thessaloniki, Greece. pp. 335–348; Penelis, G., et al., 1994.)

Limited superficial deterioration of masonry, especially at the northern face of the trunk has been detected and confirmed by hammer tests. The wooden beams at the highest steps of the staircase have heavily rotted due to rain water entering through the open top, as the wooden acute conical roof of the minaret has fallen down, probably during an earthquake of the past, at the beginning of the 20th century. Successive horizontal peripheral cracks have been detected at the upper two-fifths, of the minaret. There is also one helical crack starting from the highest loophole, following the ascending steps and cutting the balcony. All the cracks under vertical compression are just visible. During the earthquake of 20 June 1978 (epicentre 30 km from Thessaloniki, magnitude 6.5 grades in Richter scale), a part of the balcony collapsed, and the upper cylindrical trunk was slightly translocated (Figure 9.4.2a). It must be pointed out that, despite the damages, no permanent horizontal drift of the structure has been ascertained and no foundation problem has been detected.

Emergency works were undertaken just after the 1978 earthquake which included general external scaffolding and temporary external confinement by steel rings, wooden beams and inclined tendons, as well as internal helical concrete jacket at the upper 10 m of the minaret.

9.4.2.3 Mechanical characteristics of bricks, mortar and masonry

Nondestructive tests in combination with a limited number of laboratory destructive tests were applied to make an estimation of the mechanical characteristics of bricks, mortar and masonry. In this direction, a new series of hammer and ultrasonic tests on bricks and mortar, beyond those performed by the 9th Ephorate, were carried out. Three masonry triplets were extracted from the base and the trunk of the minaret and were used for the laboratory tests. The results are listed in Table 9.4.1.

From chemical analyses and granulometric graduation tests on mortar sample, a significant difference between the composition of mortars in the highest cylindrical part and that in the rest of the minaret below the balcony has been detected, although the compressive strengths of the mortars were almost equal. Consequently, either the minaret had been erected in two phases or the part over the balcony had probably collapsed and rebuilt in the past.

In Figure 9.4.3, the characteristic masonry stress–strain curve adopted for the analyses and the stress–strain curves of concrete and steel for the interventions are shown.

Table 9.4.1 Mechanical characteristics of bricks, mortar and masonry

Material	*Compressive strength (MPa)*	*Flexural tensile strength (MPa)*	*Dynamic modulus of elasticity (MPa)*
Bricks	$f_{bc} = 17.56$	$f_{bt} = 3.52^{a(1)}$	$E_{bd} = 12{,}300^{c}$
Mortar	$f_{mc} = 1.28$	$f_{mt} = 0.36^{a}$	$E_{md} = 1{,}150^{c}$
Masonry	$f_{wc} = 2.61$	$f_{wt} = 0.11^{b}, 0.19^{e}$	$E_{wd} = 2{,}600^{d}$

[a] Wide scattering
[b] Direct unsticking of bed joints
[c] Estimated from ultrasonic tests
[d] Calculated by using a specific formula
[e] Pure shear strength of joints $f_{wso} = 0.19$ MPa.

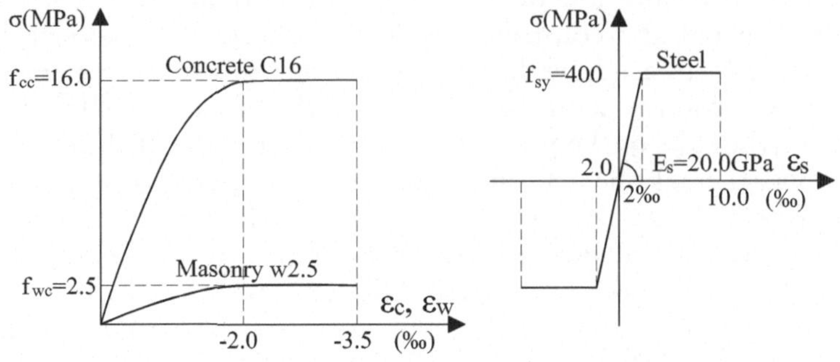

Figure 9.4.3 Characteristic stress–strain curves for masonry and for intervention materials (concrete, steel). (Adapted from Stylianidis, K. (2000). Strengthening of the minaret of Rotunda in Thessaloniki. G. Penelis. *International Symposium of Concrete and Masonry Structures*, Stylianidis, K. and Kappos, A. (eds.). Thessaloniki, Greece. pp. 335–348; Penelis, G., et al., 1994.)

9.4.2.4 Fundamental period of the minaret

Using the ambient vibration method, based on the local traffic effect, with the aid of a sensitive portable system, the fundamental period of the minaret was found equal to 0.80 s before the emergency interventions and 0.97 s 12 years later (1991). Under the reasonable assumption that the metal scaffolds, loosened by the time, have no influence on the response of the structure, the increase of the fundamental period must be attributed to the added mass of the internal concrete jacket on the top of the minaret.

9.4.2.5 Geotechnical and seismological study of the territory

To determine the soil's shear modulus G throughout the foundation depth, two boreholes were drilled very close to the minaret, and the cross-hole technique was applied. Resonant Column Tests on foundation red clay samples were also performed. Based on the obtained values of G, the deformational characteristics of appropriate soil springs, necessary for a reliable analytical model, were estimated.

The analytical procedure was performed using four elastic acceleration response spectra. Three of them were based on the seismic excitation of 1978 at the region of Thessaloniki. The motion at bedrock of this earthquake was multiplied by the factors 1.000, 1.267, 1.756, and then, using the SHAKE program (Schnabel, et al., 1972), the ground motions at the site of the minaret were determined. The first motion (THES 1978: $a_{max} = 0.172$ g) corresponds to the earthquake of 1978, while the other two correspond to similar earthquakes for reference periods of 75 (THES 75: $a_{max} = 0.218$ g) and 500 years (THES 500: $a_{max} = 0.302$ g) with 10% probability of occurrence. For the estimation of the latter maximum ground accelerations, a statistical evaluation of the strong ground motions at the territory from 1500 till 1990 was performed. The adoption of such a long reference period (500 years) was necessary because of the monumental character of the structure and the need for a long-life intervention. The fourth response spectrum corresponds to the 1989 draft of the new Greek seismic codes (1992) (EAK, 2000). These response spectra, after being appropriately smoothed and normalised, were multiplied by the foundation factor $\theta = 0.80$ and were divided by the behaviour factor q (estimated value 1.5) to get design spectra (Figure 9.4.4).

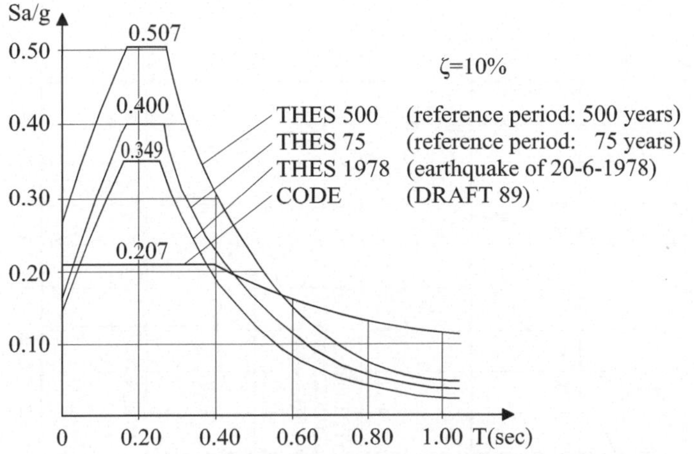

Figure 9.4.4 Design seismic acceleration spectra. (Adapted from Stylianidis, K. (2000). Strengthening of the minaret of Rotunda in Thessaloniki. G. Penelis. *International Symposium of Concrete and Masonry Structures*, Stylianidis, K. and Kappos, A. (eds.). Thessaloniki, Greece. pp. 335–348; Penelis, G., et al., 1994.)

9.4.3 Bearing capacity at virgin state

9.4.3.1 Analytical model and verification

The analytical model at virgin state was a vertical cantilever, consisting of 33 overground and 6 underground beam elements. The mass was assumed to be concentrated at the 6-DOF nodes of the model. The stiffness of the foundation was represented by horizontal, vertical and rotational linear elastic springs. To simulate the mass and stiffness imposed on the structure due to concrete jacket intervention, additional mass at appropriate nodes and new beam elements connected to the same nodes were taken into consideration.

The type of analysis adopted was the linear dynamic spectral analysis, as it is recommended for tall and slender structures (UNDP/UNIDO, 1984). The platform SAP80/86 was used for the analysis which, at the time of design, was a modern one. From pilot analyses, it was proved that, for the adapted values of modulus of elasticity of the masonry and stiffness of the soil springs corresponding to $0.8G_{max}$, the calculated fundamental periods were in a very good agreement with those measured *in situ*.

9.4.3.2 Bearing capacity at virgin state

Since the shape of the structure practically does not permit the existence of significant live loads, the only considerable vertical load is the self-weight of the structure. On the other hand, the minaret is well protected all around against winds. Consequently, the only combination to be considered was that of the seismic action and the dead load.

The structure at its virgin state was analysed under the four design spectra of Figure 9.4.4. The corresponding results are presented in Figure 9.4.5 by means of

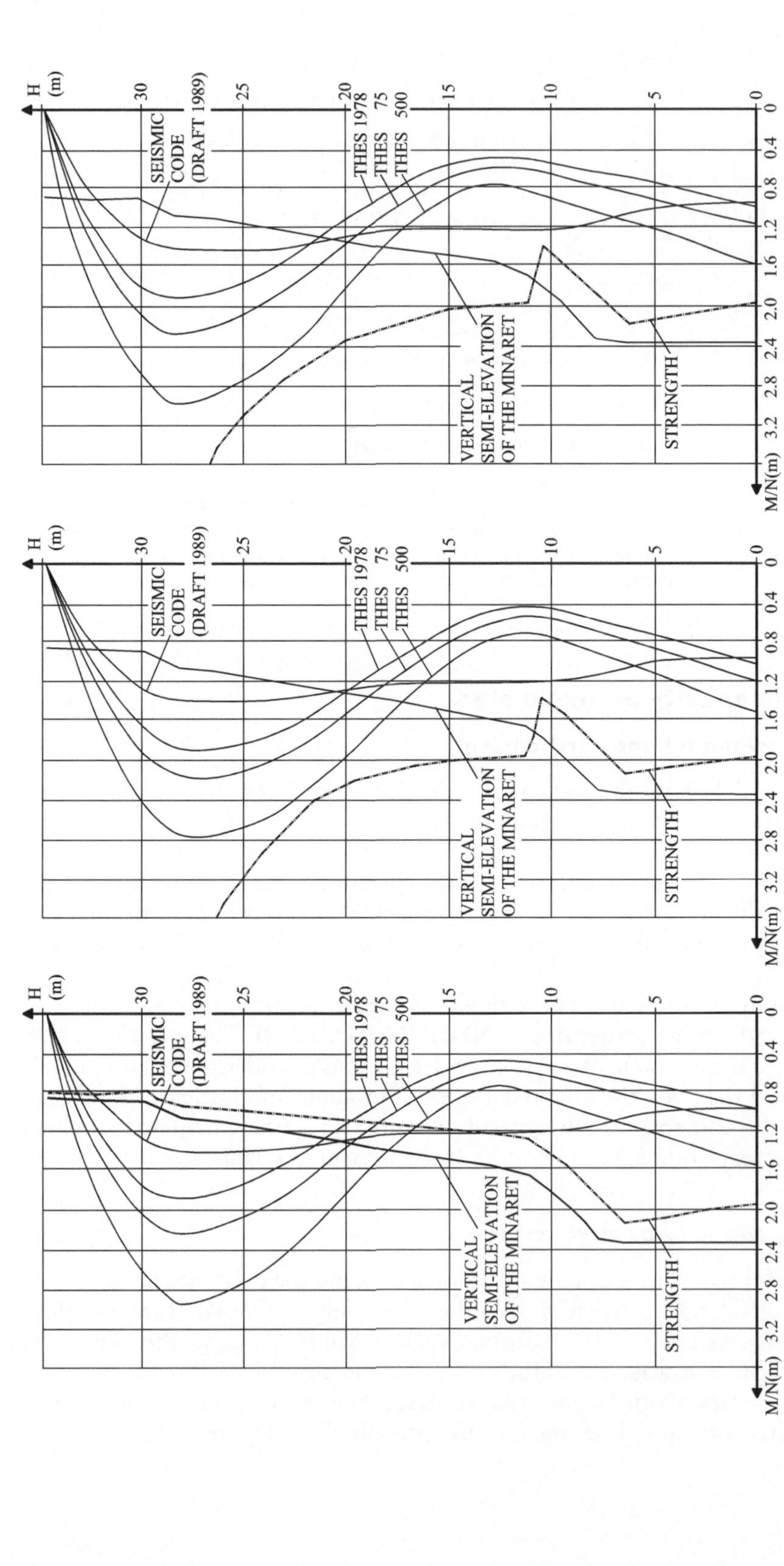

Figure 9.4.5 Eccentricity curves under seismic spectra and strength eccentricity curves: (a) virgin state; (b) concrete jacket; (c) external steel strips. (Adapted from Stylianidis, K. (2000). Strengthening of the minaret of Rotunda in Thessaloniki. G. Penelis. *International Symposium of Concrete and Masonry Structures*, Stylianidis, K. and Kappos, A. (eds.). Thessaloniki, Greece. pp. 335–348; Penelis G., et al., 1994.)

eccentricities of the dead load axial forces due to seismic bending moments ($e = M/N$). It must be pointed out that these diagrams are more informative than the conventional diagrams of bending moments and axial forces since they include their combined action on the structure. From Figure 9.4.5a, it is obvious that over the height of 17 m, where the eccentricity curves surround the semielevation of the structure, the masonry, having an almost zero tensile strength, is not adequate to carry the pair of (N) and (M) since the axial forces act outside the cross section. In order to determine more precisely the region of the structure which needs to be strengthened, the strength curve shown in Figure 9.4.5a was determined with the aid of a specific computer program which can calculate the ultimate bending moment under a given axial force, of a polygonal section composed of various materials (masonry, concrete, steel), and given stress–strain relations of the materials (Figure 9.4.3). It can be seen that, under the height of 14 m, the structure is adequate to carry out the combination of dead and seismic loads. In order to secure a reasonable safety factor slightly greater than unity, it was decided that the region to be strengthened extends from 12 m height to the top of the structure.

It must be pointed out that the conclusions derived from the calculated curves, concerning the region of inadequacy of the minaret, match very well with the pathology of the structure.

9.4.4 Alternative strengthening proposals

The following alternative strengthening intervention proposals have been considered and evaluated:

- Internal vertical posttensioned tendons
- Internal 3D steel truss
- External inclined lightly posttensioned cables
- Internal concrete jacket
- External vertical steel strips nailed to the masonry.

The presence of the internal staircase gives rise to difficulties in the application of any internal intervention. The vertical posttensioning was proved to be unacceptable since the posttensioning force needed to adequately decrease the seismic eccentricity was too high for the low-strength masonry, especially at the top of the structure. Concerning the application of external, inclined, lightly posttensioned cables, it must be pointed out that there is not enough space around the structure to give to the cables the appropriate inclination. On the other hand, the aesthetics of the monument would be rather unacceptable.

The existence of the internal concrete jacket at the top of the minaret, which is very difficult to be removed, was a positive argument for the proposal of concrete jacketing to be chosen. Anyhow, it must be pointed out that the vertical reinforcing bars of the jacket had to penetrate the staircase helix to secure the performance of the jacket as a tube. Moreover, to secure the unified action of the composite section of masonry and concrete, the bond of their interface should be increased by deepening the mortar joints. From what is mentioned above, it is clear that concrete jacketing is an almost irreversible intervention.

The external reinforcing by means of steel strips nailed to the masonry, although not a conventional strengthening method, especially for relatively low-strength materials, is a method of rather low inconvenience and almost fully reversible.

After these preliminary considerations and the rejection of the first three strengthening proposals, a detailed analytical investigation of the latter two proposals was performed and is presented in the following. As mentioned before, the region to be strengthened extends from 12 m to the top of the minaret.

9.4.4.1 Internal concrete jacket

The helical concrete jacket, constructed after the 1978 earthquake at the upper 10 m of the structure, has a thickness of 80 mm and a longitudinal reinforcement Φ14/150 mm, S400 (f_{yk} = 400 MPa). It was decided that the already constructed jacket should be extended, having the same geometrical, mechanical and constructional characteristics (Figure 9.4.6a), by welding the existing longitudinal bars to the new ones. Using the modified analytical model mentioned in Subsection 9.4.3.1, it was proved that the jacketing had a limited influence on the dynamic characteristics of the structure. In Figure 9.4.5b, the eccentricities of the axial forces due to seismic bending moments under the design spectra of (Figure 9.4.4) are shown. It is obvious that, in comparison to the relative curve of (Figure 9.4.5a), the differences are small. The determination of the ultimate bending moment of the cross sections under the acting force was based on the computer program mentioned in Subsection 9.4.3.2. From the strength curve shown in Figure 9.4.5b, *it can be concluded that the strengthening proposal was adequate.*

The shear stress acting on the interface between masonry and concrete has a maximum value of 0.06 MPa, which can be easily sustained by the bond strength of the masonry to concrete interface (see also footnote (5) of Table 9.4.1), especially after deepening of the

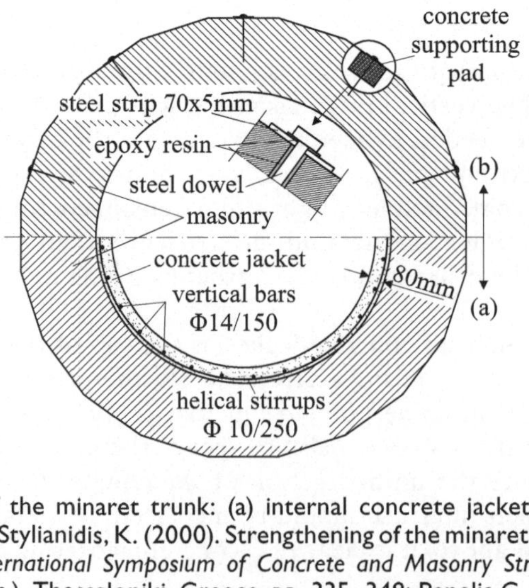

Figure 9.4.6 Semisection of the minaret trunk: (a) internal concrete jacket; (b) external steel strips. (Adapted from Stylianidis, K. (2000). Strengthening of the minaret of Rotunda in Thessaloniki. G. Penelis. *International Symposium of Concrete and Masonry Structures*, Stylianidis, K. and Kappos, A. (eds.). Thessaloniki, Greece. pp. 335–348; Penelis G., et al., 1994.)

mortar joints before concreting. To secure the durability of the intervention, the use of austenitic stainless steel for the reinforcement bars was recommended.

9.4.4.2 External steel strips nailed to the masonry

After some pilot calculations, what was adapted was the use of ten longitudinal austenitic stainless steel strips, grade S400, having a total/effective cross section 70 × 5/50 × 5 mm (Figure 9.4.6b), nailed along the trunk of the minaret, together with seventeen horizontal confining rings of the same cross section. The strips and the rings were placed along the upper 25.0 m of the height of the minaret (Figure 9.4.2b). It must be pointed out that the strips have to penetrate the external balcony in order to come up to the upper cylindrical trunk, which means that, considering also the bad condition of the balcony, the demolition and reconstruction of the balcony was rather inevitable. Because of the small increase in stiffness and the practically zero increase in mass due to the steel strips, the response of the structure under seismic loads was assumed to be equal to that of a virgin state. Using the same program, the strength curve shown in Figure 9.4.5c was calculated, from which it can be seen that it is almost identical with that of Figure 9.4.5b, which means that the selected strengthening proposals were equivalent in strength.

The maximum shear force per unit length, which has to be transferred from each strip to the masonry, was found to be equal to 29 kN/m. It is obvious that the greater the number of the dowels, the more uniform and easier the transfer of the shear force is, on the one hand, but on the other hand, the number of intervention points becomes greater. Finally, it was decided to nail standard dowels M16 every 0.50 m along each strip, after the local replacement of the existing mortar around the nailing points by high-strength, nonshrinkable, ready-mixed mortar, without replacing the bricks (Figure 9.4.7). To prevent buckling, the strips were glued to the masonry by means of epoxy resin.

After thorough consideration, the Supreme Archaeological Council of the Ministry of Culture approved this proposal of external steel strips nailed to the masonry because it was considered to be a method of rather low inconvenience and almost fully reversible. After this approval, further investigation on the details was carried out which covered the following issues:

- Analytical evaluation of the shear capacity of nailed connection (Ignatakis, et al., 1989a)
- *In-situ* verification of the shear capacity of nailed connection
- Design of the balcony in its residual form for the combination of dead load, posttensioning of external tendons and temperature changes.

For the first stage, a finite element micromodel was used (Ignatakis, et al., 1989a) (Figures 9.4.8 and 9.4.9).

For the second stage, a special instrumentation was designed for the *in-situ* measurements (Figures 9.4.10 and 9.4.11).

Finally, for the third stage a solid-state analysis was carried out for the critical part of the minaret and the balcony (Figure 7.9). The minaret after the restoration is depicted in Figure 9.4.12.

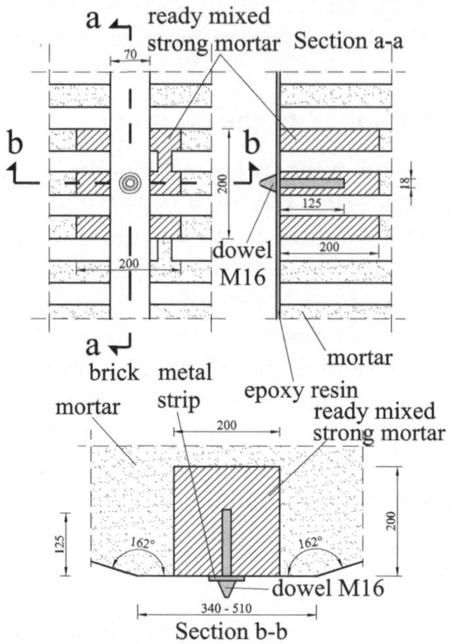

Figure 9.4.7 Details of the nailing. (Adapted from Stylianidis, K. (2000). Strengthening of the minaret of Rotunda in Thessaloniki. G. Penelis. *International Symposium of Concrete and Masonry Structures*, Stylianidis, K. and Kappos, A. (eds.). Thessaloniki, Greece. pp. 335–348; Penelis G., et al., 1994.)

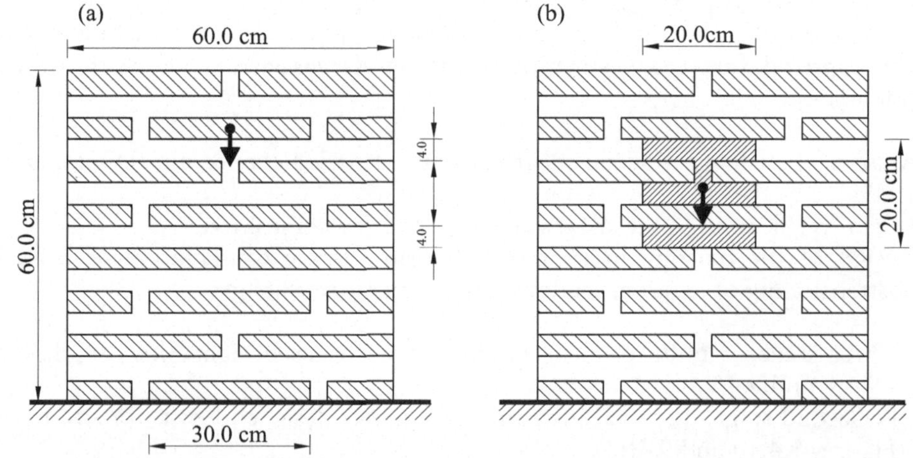

Figure 9.4.8 Masonry finite element model: (a) virgin state, (b) strengthened state. (Adapted from Stylianidis, K. (2000). Strengthening of the minaret of Rotunda in Thessaloniki. G. Penelis. *International Symposium of Concrete and Masonry Structures*, Stylianidis, K. and Kappos, A. (eds.). Thessaloniki, Greece. pp. 335–348; Penelis G., et al., 1994.)

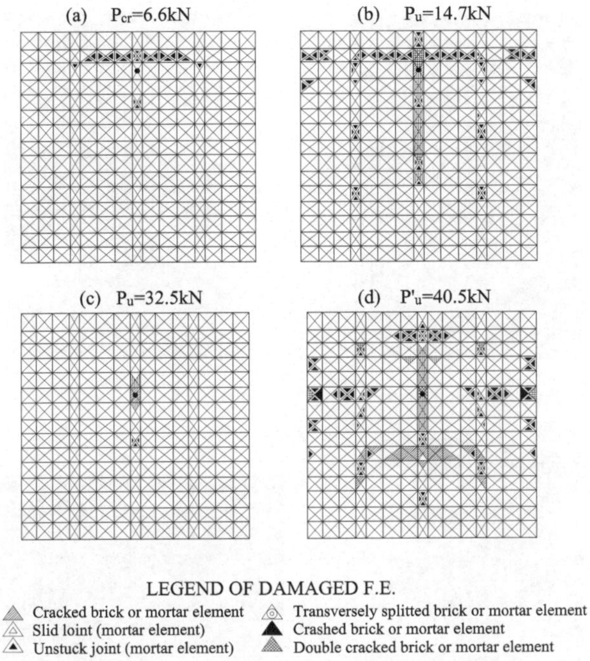

Figure 9.4.9 Damage patterns of the masonry finite element models under dowel force. Virgin state: (a) first crack; (b) failure pattern. Strengthened state: (c) localised failure pattern; (d) general failure pattern. (Adapted from Stylianidis, K. (2000). Strengthening of the minaret of Rotunda in Thessaloniki. G. Penelis. *International Symposium of Concrete and Masonry Structures*, Stylianidis, K. and Kappos, A. (eds.). Thessaloniki, Greece. pp. 335–348; Penelis, G., et al., 1994.)

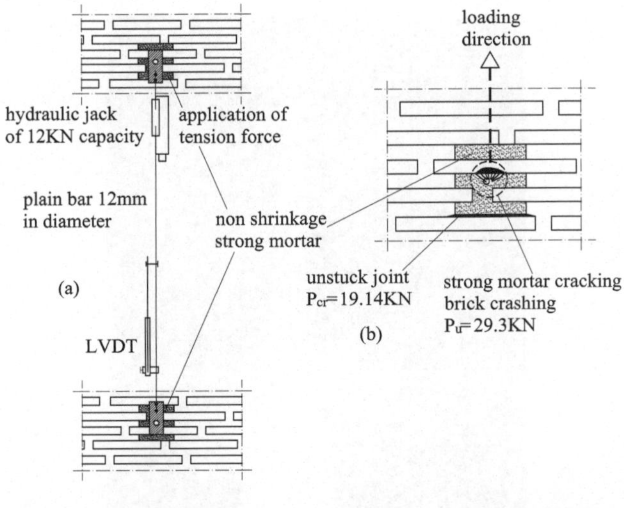

Figure 9.4.10 (a) Schematic instrumentation of the *in-situ* measurements; (b) schematic failure pattern around the dowel. (Adapted from Stylianidis, K. (2000). Strengthening of the minaret of Rotunda in Thessaloniki. G. Penelis. *International Symposium of Concrete and Masonry Structures*, Stylianidis, K. and Kappos, A. (eds.). Thessaloniki, Greece. pp. 335–348; Penelis, G., et al., 1994.)

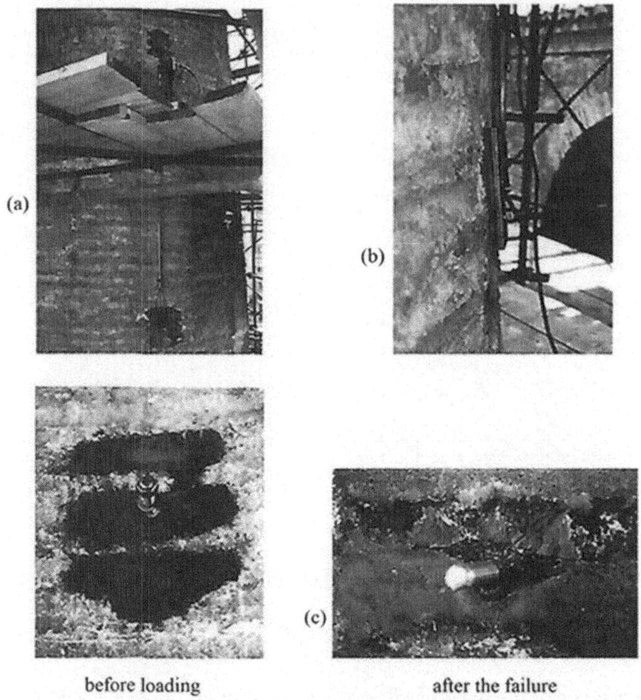

Figure 9.4.11 (a) Instrumentation of the *in-situ* measurements; (b) bending of the dowel failure; (c) view of the failure around the dowel. (Adapted from Stylianidis, K. (2000). Strengthening of the minaret of Rotunda in Thessaloniki. G. Penelis. *International Symposium of Concrete and Masonry Structures*, Stylianidis, K. and Kappos, A. (eds.). Thessaloniki, Greece. pp. 335–348; Penelis, G., et al., 1994.)

Figure 9.4.12 General view of the minaret of Rotunda after its restoration. (From Penelis S.A., Archives, 2019d.)

9.5 HAGIOS ANDREAS CHURCH IN PERISTERA, GREECE

9.5.1 Introduction

The church of Hagios Andreas is one of the smaller Byzantine monuments in and around Thessaloniki that were damaged at the same time as the Rotunda by the earthquake of June 1978. Damage was less serious, but a full investigation was undertaken as part of the general programme of research that has already been referred to in the Rotunda study. Aspects of the research that were fully described there are mentioned only briefly here. But more detail is given here on ultimate load analysis, particularly in reference to the time of implementation, when a computer automatic inelastic analysis of masonry structures had not been developed yet.

The project of the structural restoration was awarded by the authorities to the Laboratory of Reinforced Concrete and Masonry Structures of Aristotle University of Thessaloniki (AUTH). The working group in charge of the project included the following scientists (Penelis, G., et al., 1980b, 1984; Penelis, G., 1982):

- Penelis, G., chief technical advisor, Ordinary, professor, AUTH
- Karaveziroglou, M., PhD, structural engineer
- Stylianidis, K., Diploma, structural engineer
- Leontaridis, D., Diploma, structural engineer.

The design project was finalised by the end of 1980.

9.5.2 Description

The main body of the church was built in about 870 A.D. with a central dome and four slightly smaller ones around it over the arms of a cross, as seen in Figures 9.5.1 and 9.5.2. In 1915, a narthex was added at the west, and a buttress was built against the south arm of the cross at about the same time. The interior is frescoed.

Figure 9.5.1 General view from the north of Hagios Andreas, Peristera, Thessaloniki. (Courtesy of Ephorate of Antiquities of Thessaloniki Region.)

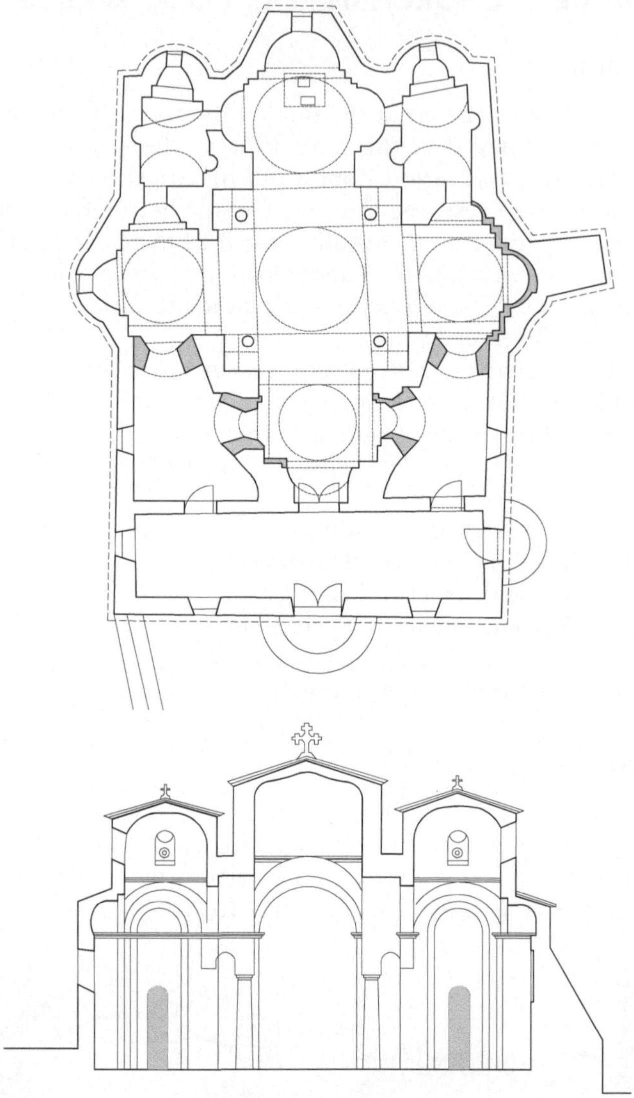

Figure 9.5.2 Plan and cross section looking east. (Adapted from Penelis, G., Karaveziroglou, M., Stylianides, K., and Leondaridis, D. (1984). The church of Hagios Andreas, Peristera case study in the *UNDP/UNIDO manual for repair and strengthening of historical monuments and building in urban nuclear.* UNDP/UNIDO, project RER 79/015 Vol. 6, Vienna; UNDP/UNIDO (1984). *Repair and Strengthening of Historical Monuments and Buildings in Urban Nuclei*, 6, Vienna.)

9.5.3 Structural condition

Before the earthquake, cracks already existed in the domes and their drums and in the supporting arches. The southern part of the church was also separated from the main body, as indicated both by the construction of the new buttress against it and by fillings of tile and mortar in old cracks.

The earthquake led to a slight worsening of these conditions. There was, for instance, a 2.6% deviation from the vertical at the SE corner. But there was no apparent danger of collapse, and no emergency shoring was considered necessary apart from placing of some posttensioned steel rings around the drums of the domes.

9.5.4 Detailed investigations

The investigations were similar to those undertaken for the Rotunda, though somewhat less extensive.

9.5.4.1 Architectural and constructional survey

A precise architectural survey was carried out using standard survey techniques. To determine foundation conditions, several sections were cut alongside the external walls. Two sections were cut into the walls to determine the type of construction and to obtain samples of bricks and mortar for chemical analysis. The quality of construction in the interior of the walls was inferred from the type of the visible surfaces.

9.5.4.2 Survey of the damage

The survey of the damage included a detailed survey of the visible cracks and measurements of the drifts of the walls and piers.

9.5.4.3 Test on materials

Core-taking for destructive tests on the mechanical properties was not attempted because the mortar joints were too thin for the extraction of adequate test samples. Instead, hammer and sonic tests were facilitated over the external surface. (They were not practicable inside of the building because of the risk of damage to the frescoes.) The correlation curves derived from the tests on the mortars of the Rotunda were then used to determine the strengths here. The sonic tests were used for a statistical evaluation of the dynamic modulus of elasticity to assist in the choice of suitable grouting mortars.

Chemical analyses were performed to determine the characteristics of the bricks and mortar as in the Rotunda investigation.

9.5.4.4 Static analyses

The domes and their drums were first analysed as linearly elastic systems of thin shells. The central dome was modelled as a conical shell at the top, merging with a lower doubly curved frustum shell standing on a cylindrical drum. The other domes were modelled as semielliptical shells of revolution, likewise on cylindrical drums.

These analyses (taking into account the internal forces at the interfaces between shells) indicated high tensile stresses at some sections of the drums to cause meridional cracking – especially at the central drum.

Reanalysis, with the drums modelled as rings of independent cantilevers, led to the conclusion that no tensile stresses would exist on horizontal sections at the base for gravity load alone and that stresses sufficient to cause cracking here would increase for a seismic base shear of 8%.

For the investigation of overall strength and stability, it was necessary to perform ultimate-load analyses of the combined systems of principal (dome-bearing) arches and their supporting piers. These were modelled as shown in Figure 4.13.

Failure criteria for the cross section of the piers and arches (CS I and CS II) were first plotted as interaction curves between moment (M) and axial force (N) (Figure 4.13c) and between normal stress (σ) and shear stress (τ) (Figure 9.5.3), based on the following assumptions:

- The tensile strength is equal to the estimated flexural strength of the masonry.
- The compressive strength is equal to the estimated compound strength of the masonry.
- Constitutional law (σ–ε curve) in the compressive zone is similar to that of low-strength concrete.

For these analyses, a step-by-step procedure was followed – initially for service (gravity) load and subsequently for seismic load superimposed on this. Referring to Figure 4.13c and d, a small multiple of the service load, say 0.2, is applied first. For this loading, the internal forces and moments are determined at the critical sections and plotted on the M–N interaction curves (points F_1, D_1, B_1 and A_1). Extension of the line OF_1 to $\bar{F}_1$ on the M–N curve then gives the position of the first hinge and the corresponding load factor (i.e. point F and $O\bar{F}_1/OF_1$). Proportionate extension of the lines $O\bar{D}_1$, $O\bar{B}_1$ and $O\bar{A}_1$ gives the corresponding force and moments at the other critical sections. Repetition of the procedure for the new structural system with a hinge at F gives the location of the successive hinges at B and B', the second partial load factor $\bar{B}_1 \bar{B}_2 / \bar{B}_1 B_2$ and the corresponding forces and moments at the remaining critical sections D and A. A further repetition for the system with hinges at both F, B and B' gives the location of the final hinges necessary to allow collapse at A and A' and the final load factor.

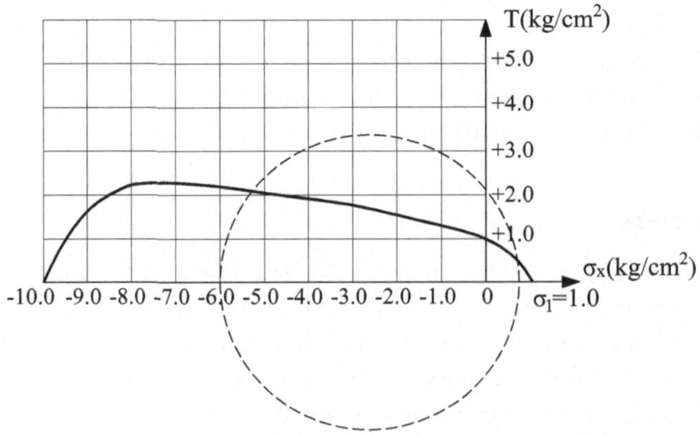

Figure 9.5.3 Strength of the masonry in combined shear and compression. (Adapted from Penelis, G., Karaveziroglou, M., Stylianides, K., and Leondaridis, D. (1984). The church of Hagios Andreas, Peristera case study in the *UNDP/UNIDO manual for repair and strengthening of historical monuments and building in urban nuclear.* UNDP/UNIDO, project RER 79/015 Vol. 6, Vienna; UNDP/UNIDO (1984). *Repair and Strengthening of Historical Monuments and Buildings in Urban Nuclei*, 6, Vienna.)

For response under seismic load, the same procedure is followed up to full service (gravity) load, and then it is continued with increasing horizontal load until a collapse mechanism forms. At each step of the analysis, it is now necessary, however, to check that no shear failure will occur in the supporting system. This is done by means of the curve representing the ultimate strength of the masonry in combined shear and compression reproduced in Figure 9.5.3.

These analyses indicated a load factor of 1.73 for service (gravity) load and showed that, at full unfactored gravity load, collapse would be caused by a seismic load of 12%.

9.5.5 Repair and strengthening

The level of safety indicated by the analyses, together with the evidence of past performance of the structure, showed that only limited intervention was desirable. The proposals were as follows:

- Grouting of cracks
- Installation of permanent posttensioned steel rings around the drum of the domes (Figure 9.5.4)
- Retention of the buttress on the south side.

On the basis of the tests on the existing mortars and the tests on grouts, carried out for the Rotunda, the proper ones were selected for filling small and wide cracks, respectively.

The posttensioned tie rings could easily be removed and replaced, if necessary, at some future date. Their installation (like that of similar rings around the dome of the Rotunda) was thus a reversible intervention, for which mild steel ST240/340 MPa coated with lead was considered appropriate or, alternatively, stainless steel as in the Rotunda dome. The second was preferred by the authorities. The posttensioning was specified to 50 MPa, while the posttensioning tendons were 6Φ16 mm, ST 400/700 MPa. The arrangement of the rings around the drum of the central dome is shown in Figure 9.5.4.

Since no change in the overall structural system was proposed, the only further analysis that was called for was that needed to determine the cross sections and desirable posttensioning of these rings.

9.5.6 Specifications

When detailed design had been completed, drawings and specifications were prepared in the same way as done for the Rotunda.

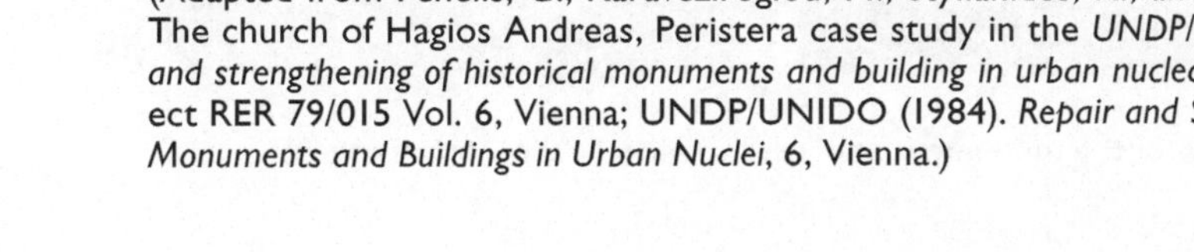

Figure 9.5.4 Development of the permanent posttensioned tie rings of the drum of the central dome. (Adapted from Penelis, G., Karaveziroglou, M., Stylianides, K., and Leondaridis, D. (1984). The church of Hagios Andreas, Peristera case study in the *UNDP/UNIDO manual for repair and strengthening of historical monuments and building in urban nuclear.* UNDP/UNIDO, project RER 79/015 Vol. 6, Vienna; UNDP/UNIDO (1984). *Repair and Strengthening of Historical Monuments and Buildings in Urban Nuclei*, 6, Vienna.)

9.6 RESTORATION OF THE MAR GIRGIS CHURCH AND THE ROMAN TOWER UNDERNEATH IN CAIRO

9.6.1 Introduction

Task of this case study is the presentation of the pathology and the restoration of the Mar Girgis church and the Roman tower underneath in Cairo.

The pathology is focused on three main issues:

- Settlements and cracks of the monument due to dewatering activities for lowering of the water table of the region of old Cairo
- Cracks of the church due to structural deficiencies of the Christian part of the monument
- Decay of materials due to the age of the monument.

The interventions referred to

- The structural system
- The architectural parts
- The paintings
- The wooden parts and furniture
- The mechanical, electrical and plumbing (MEP) of the monument.

The authors of this book were responsible for the project management of the design and execution of works and for the structural restoration of the monument.

The project was initiated in 2009 and was finalised by the end of 2015. The project received recently (2017) the ENR Global Best Project Award of Merit for cultural heritage projects of the year.

9.6.2 General information

The church of St. George (Mar Girgis) is located in old Cairo and more specifically at the Mar Girgis area (Mar Girgis is a paraphrase of the Greek Aigiorgis meaning St. George) and belongs to the Greek Orthodox Patriarchate of Alexandria and Whole Africa. It is an unreinforced masonry (URM) Rotunda 27 m diameter at the base, 13 m dome diameter, 11 m dome and drum height and total height 22 m, built at the beginning of the 20th century upon the ruins of a previous church. The wall paintings have been created by the famous Greek painter *Parthenis* (Figure 9.6.1).

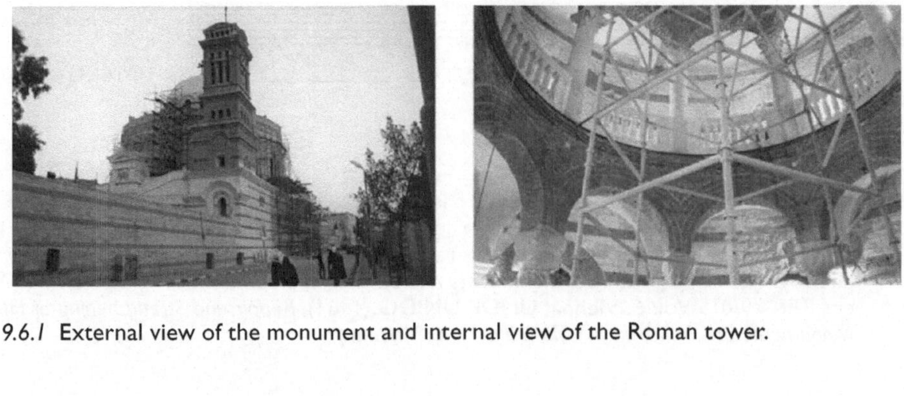

Figure 9.6.1 External view of the monument and internal view of the Roman tower.

The Rotunda is based upon *a three-storey Roman tower circular in plan,* built around 1st century A.D. The tower with a 27 m external diameter consists of two concentric rings of thick stone walls interconnected by radial walls.

At the end of the 19th century A.D., a small church of St. George that existed upon the tower was ruined by fire as well as the wooden floors of the tower. At the beginning of the 20th century A.D., the Roman tower was restored from the fire with the addition of R/C floors combined with standard profile steel beams, and then, the imposing church in the form of Rotunda was erected above the restored tower (Figure 9.6.2).

9.6.3 Historic context

The whole complex, today partially being the St. George Convent, is located at the place of the Babylon Fortress, on the Nile bank, as fully described initially by Strabo in 26 B.C. The Babylon fortress had been built upon a previous installation of the Ptolemies, which was used as a harbour entrance for the canal linking the Nile to the Red Sea (246–46 B.C.), a canal first attempted, unsuccessfully, by the Pharaohs of the twenty-sixth dynasty (610–595 B.C.) and first achieved by Darius I (521–486 B.C.). This canal was in use in the era of Trajan (107 A.D.) who was probably the one that commissioned the two towers designating the entrance to the port, while Diocletian significantly expanded the fortress installation (298–302 A.D.). The church of St. George lies on one of the two towers of the harbour entrance (Sheehan, 2010) (Figure 9.6.3).

9.6.4 Project I.D.

This project has been realised due to the personal interest and efforts of His Holiness Pope and Patriarch of Alexandria, Theodoros II.

Design Team	
Preliminary design	E. Lekkas, N. Chiotinis
Architectural design	Mohamed Awad
Structural design	Penelis Consulting Engineers S.A. & Hafez Akhrass (Hamdi El Swat)
MEP design	EESCO & MAKTE AE
Project management	Penelis Consulting Engineers S.A.
Design funding	Hellenic Ministry of Foreign Affairs – HELLENIC AID, Patriarchate of Alexandria and Whole Africa while Penelis Consulting Engineers S.A. donated their services
Construction	
Contractor	The Arab Contractors (Mahmoud Saudi, site manager)
Project management	Penelis Consulting Engineers S.A.
Supervision	Penelis Consulting Engineers S.A.
	Notis Lysseos (site engineer)
	Konstantinos Maganiotis (site engineer)
Consultants	Hellenic Ministry of Culture, Department of Preservation lead by Mr. N. Minos (wall paintings)
	Fillip Abbott, Julius Abbott (MEP)
	Konstantinos Maganiotis & Gallis Lighting (lighting design)
	Elias Papageorgiou & Assoc. (museum design)
Construction funding	The project with 6.500.000€ budget was completed in 2015 and funded solely by the benefactors, *Marina and Athanasios Martinos*

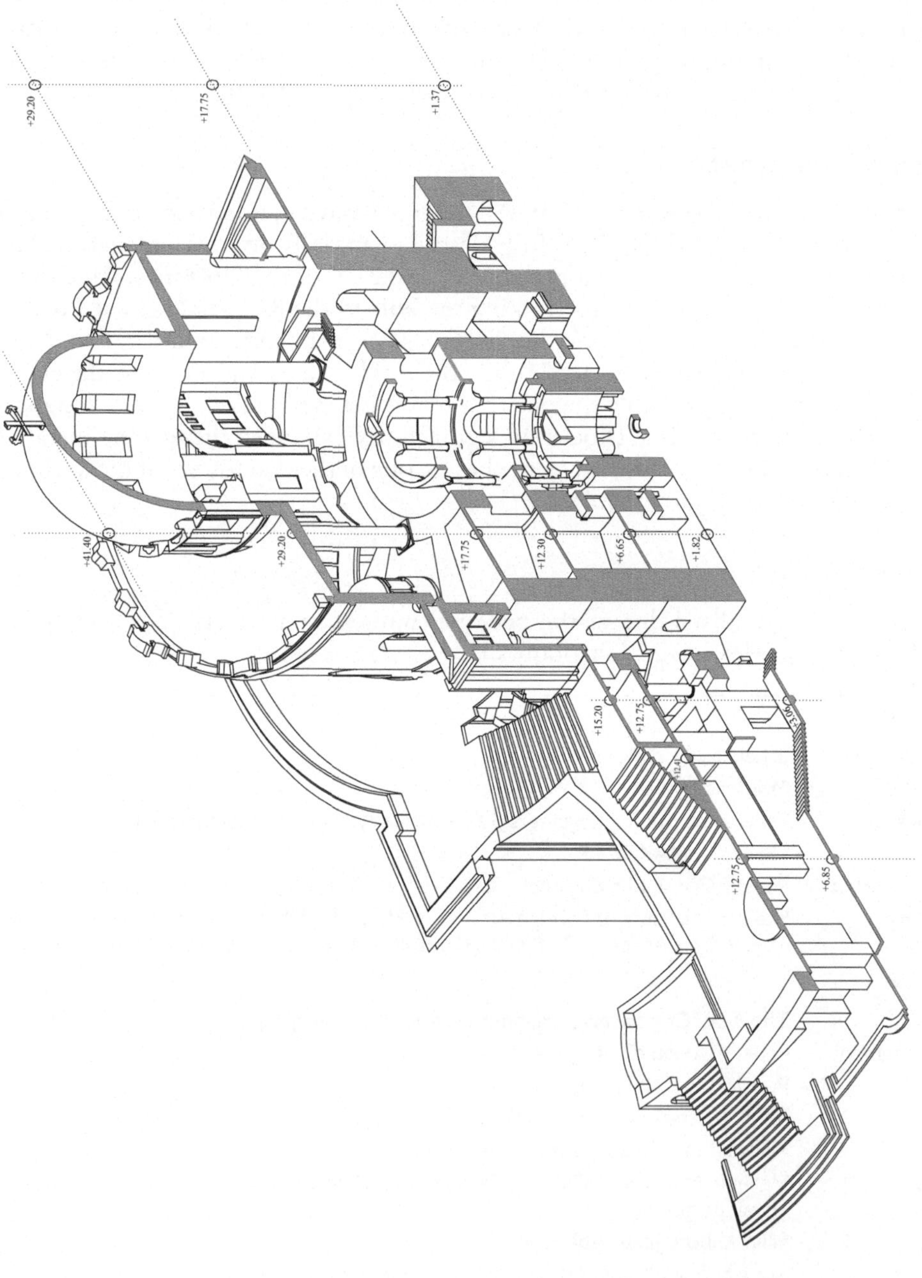

Figure 9.6.2 Axisymmetric representation of section of the church of St. George and the Roman tower. (Adapted from Lekkas, E., Chiotinis, N., and Awad, M. (2008) Protection and revelation of the church of saint George- Cairo. (Archives of the preliminary design); Penelis, S.A. (2016) Archives of final design and construction of the restoration of the Greek Orthodox church of Saint-George, old Cairo.)

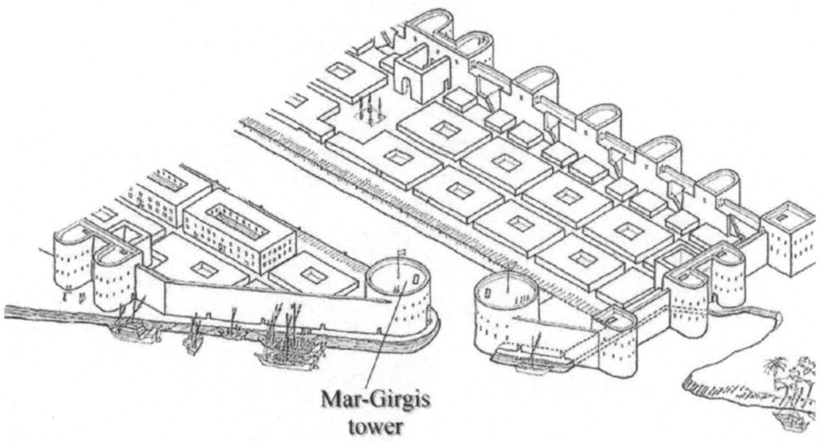

Figure 9.6.3 Axonometric view of the archaeological findings in old Cairo. (Adapted from Sheehan, P. (2010). Babylon of Egypt, Oxford University Press, Oxford, UK. American Research Center in Egypt.)

9.6.5 Pathology

9.6.5.1 Roman tower

The Roman tower had been flooded in its lower level up to 2002, when a massive program of underground tunnels and well points for dewatering parts of old Cairo was completed. The rise of the water table may be attributed to the construction of the metro line (Mar Girgis Station) passing the old Nile bed and hence creating a diaphragm between old Cairo and Nile.

The Roman tower suffered serious damage in the surface of the stone walls of lower level due to the long-term flooding, fully penetrating cracks at the walls, resulting mainly from old settlements and only partially due to the dewatering, while corner stones had been removed at specific locations. The cracks at the tower were stabilised some years after the dewatering as was evident from the glass markers installed (fissurometers). One very significant source of surface damage resulted from the fesses of bats that inhabited the tower in thousands (Figure 9.6.4).

9.6.5.2 St. George's church

The Rotunda presented serious structural issues, as the weight of the Dome created outward thrust forces on the external perimeter wall via the groin vaults. This action had not been taken properly into consideration in the original design, a fact that led to evolving (glass markers) vertical tensile cracks on the perimeter URM wall as well as tensile cracks at the bottom of the groin vaults as the perimeter wall was moving outwards. Additionally, a part of the outer floor of the church which was formed as a terrace was founded beyond the Roman tower, on soft river deposits, which lead to differential settlements (Figure 9.6.5).

Finally, the eight internal circular brick masonry columns suffered extensive cracking at their bottom (spalling of plaster).

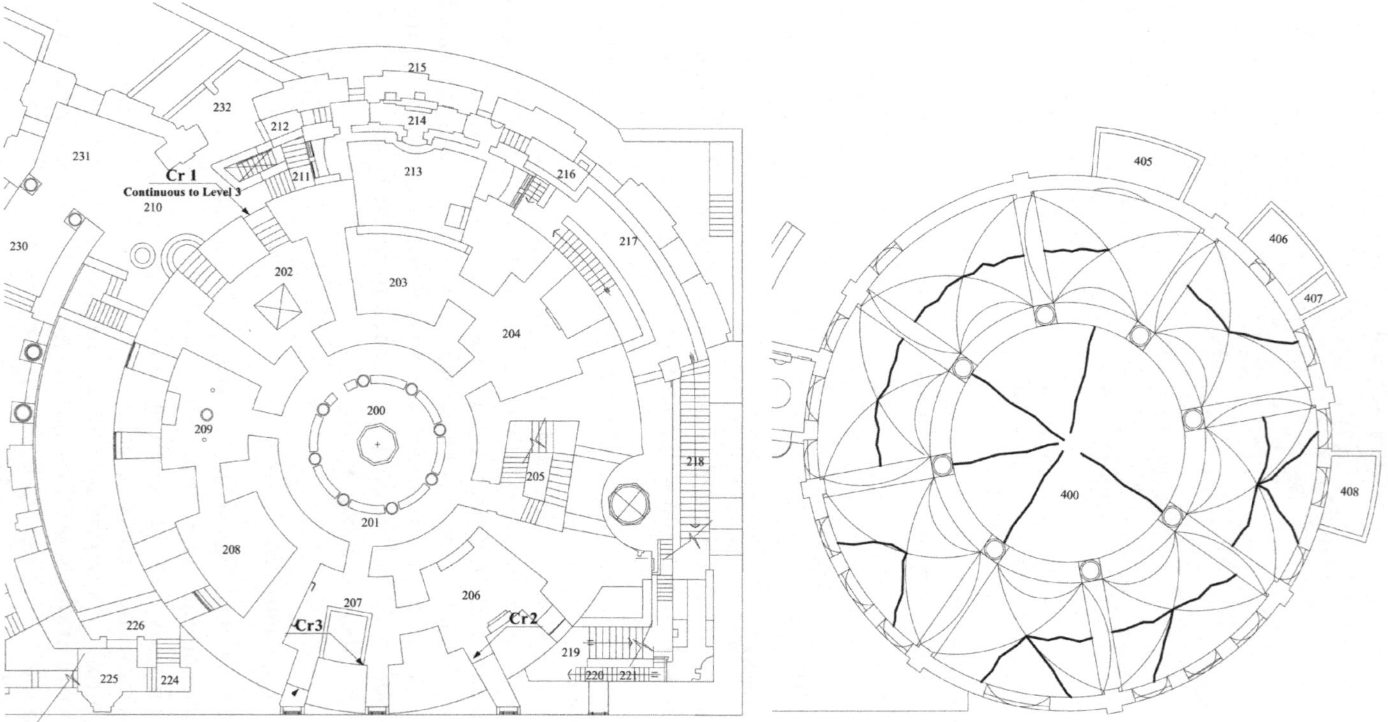

Figure 9.6.4 Plan of the Roman tower and external wall and groin vaults (cracks). (Adapted from Penelis, S.A. (2016) Archives of final design and construction of the restoration of the Greek Orthodox church of Saint-George, old Cairo.)

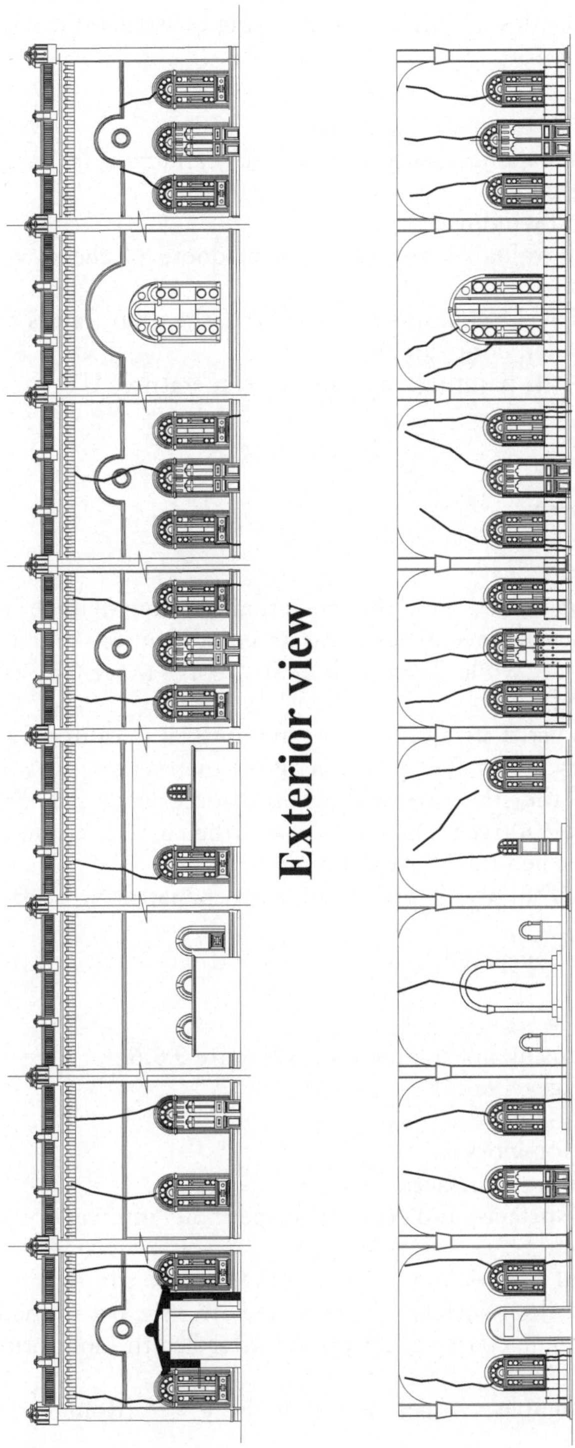

Figure 9.6.5 External and internal wall cracks (ongoing glass markers). (Adapted from Penelis, S.A. (2016) Archives of final design and construction of the restoration of the Greek Orthodox church of Saint-George, old Cairo.)

9.6.6 *In-situ* and laboratory investigations

During the first stages of the design (mainly) and during construction, the following *in-situ* investigations and laboratory tests were conducted:

- Complete architectural survey of the monument
- Chemical and mechanical analyses of stones, bricks and mortars
- Recording of all cracks
- Survey of all doors and windows
- Documentation of all valuable historic wooden doors of the tower (Roman or Islamic style)
- Investigation sections at the columns of the church, the groin vaults and the dome
- Carbon dating of wooden doors and wooden ties
- Investigation of the wall paintings (moisture, temperature, UV cameras, thermic cameras).

9.6.7 Restoration works

9.6.7.1 *Roman tower*

The intervention concept was based upon the thesis that the Roman tower is a significant historic monument including the several additions and modifications and it has sustained during the 2000-year lifetime, while the church of St. George (Mar Girgis) is considered as a recent monument (listed).

The tower is a robust and stiff structure consisting of thick masonry walls with intermediate solid brick masonry layers. The principle of the intervention was the restoration of the bearing capacity of the structure without any addition or strengthening, given that its existence for almost 2000 years is proof of its sufficiency. The static and dynamic analyses of the structure verified the above concept.

As already mentioned, the tower presented fully penetrating cracks that were restored.

The following works were executed:

i. Structural restoration works
   - Crack restoration using pozzolanic grouts (Figure 9.6.6)
   - Crack stitching (Figure 9.6.7)
   - Stone stitching

ii. Architectural restoration works
   - Sandblasting of all wall surfaces
   - Repointing of all surfaces and removal of pervious interventions with cement paste
   - Antidust coating of the brick masonry layers for emphasis
   - Coating of all R/C floor surfaces (constructed after the fire in the restoration of 1909) with handmade coarse grain plaster after health monitoring of the steel beams and rebars
   - Restoration of valuable wooden doors of the tower (Roman, Byzantine and Islamic types)

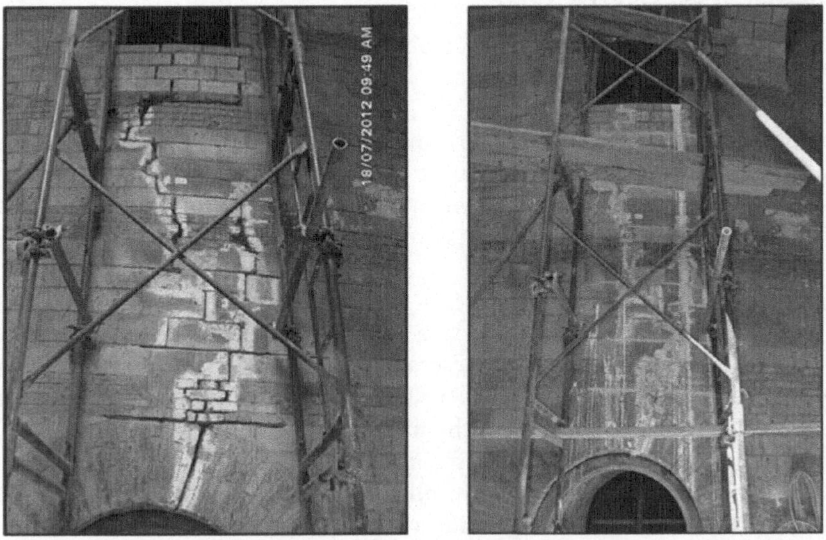

Figure 9.6.6 Crack grouting in the tower before and after the intervention. (From Penelis, S.A. (2016) Archives of final design and construction of the restoration of the Greek Orthodox church of Saint-George, old Cairo.)

Figure 9.6.7 Crack stitching. (From Penelis, S.A. (2016) Archives of final design and construction of the restoration of the Greek Orthodox church of Saint-George, old Cairo.)

- Restoration (supplements) of stone floors
- New architectural lighting

iii. MEP works
- Fire alarm system
- High- and low-voltage cabling.

9.6.7.2 Church of St. George

The Rotunda is a newer monument that has significant structural issues that resulted from its structural form which exerts serious thrusting horizontal forces caused by the weight of the dome and transferred via the groin vaults to the external perimeter wall, pushing it outwards. It has come out from the analysis of the existing situation that the strength of masonry wall is inadequate to carry these thrusts. The principle of the intervention is the structural restoration and strengthening of the Rotunda so that it can adequately receive all static loads as well as have acceptable performance under a significant earthquake.

To determine the intervention scheme, a preliminary elastic dynamic analysis of the Rotunda was performed so that the critical areas could be determined (Figure 9.6.8). This analysis, completely verified the observed crack pattern. Based on the above analysis, the following intervention scheme was chosen: introduction of external posttensioned steel rods at four levels externally and strengthening of the eight internal columns using FRPs. Additionally, the groin vaults had to be strengthened.

The required posttensioning force and all the other interventions at the internal columns, the groin vaults and the dome windows were designed using elastic dynamic analysis and verified using a static nonlinear analysis (Figure 9.6.9) (Penelis, G., et al., 2010; Penelis, Gr., 2006a).

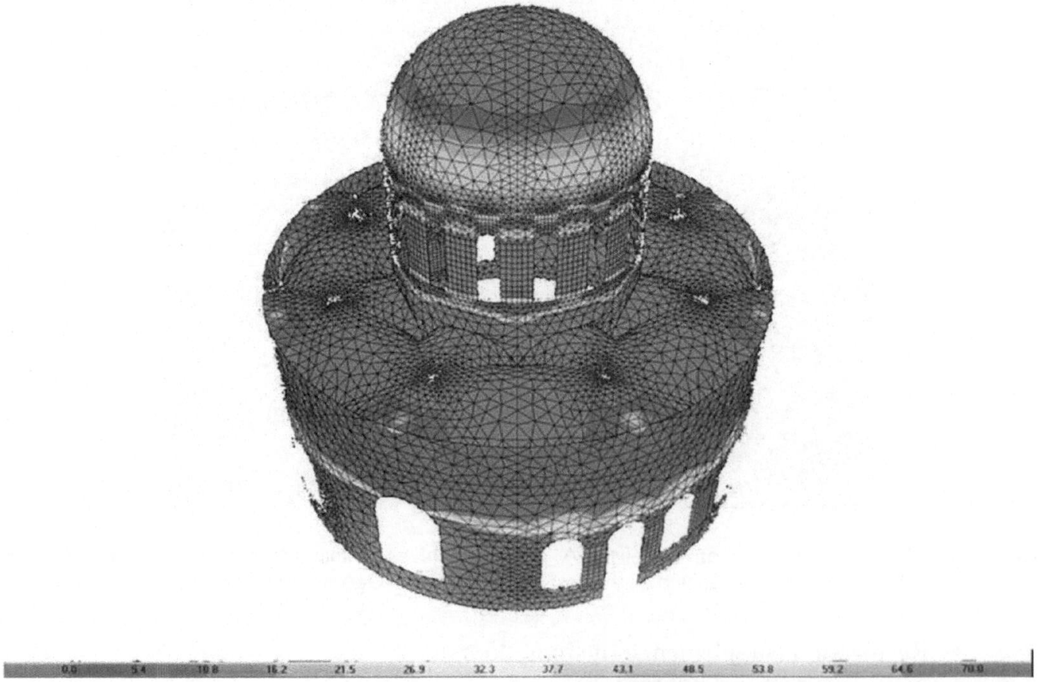

Figure 9.6.8 Elastic analysis of the Rotunda. σ_x>70 [kN/m^2] denotes tensile cracking. (Adapted from Penelis, S.A. (2016) Archives of final design and construction of the restoration of the Greek Orthodox church of Saint-George, old Cairo.)

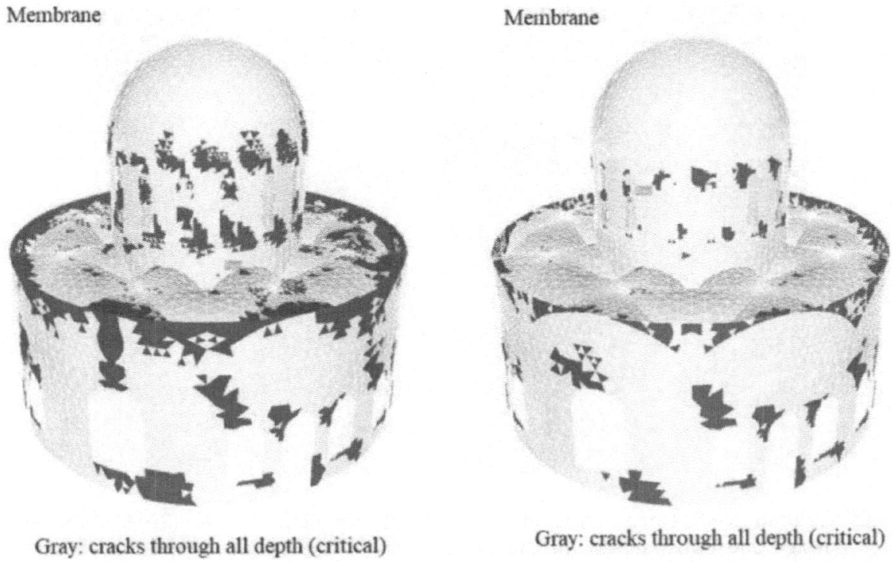

Figure 9.6.9 Static inelastic analysis for dead load plus seismic action 0.30g. Left original structure. Right structure with posttensioned rings. (Adapted from Penelis, S.A. (2016) Archives of final design and construction of the restoration of the Greek Orthodox church of Saint-George, old Cairo.)

The following works were executed in the Rotunda:

i. *Structural intervention works*
- Crack restoration using pozzolanic grouts
- Crack stitching
- Column confinement (Figure 9.6.11)
- External posttensioning at four levels (Figure 9.6.10)
- Steel casing of all the dome window openings
- Groin vaults strengthening

Figure 9.6.10 Posttensioning works. (From Penelis, S.A. (2016) Archives of final design and construction of the restoration of the Greek Orthodox church of Saint-George, old Cairo.)

Figure 9.6.11 Column confinement using FRPs (right). (From Penelis, S.A. (2016) Archives of final design and construction of the restoration of the Greek Orthodox church of Saint-George, old Cairo.)

ii. *Architectural restoration works*
- Parthenis wall paintings restoration
- Icon restoration
- Marble floor removal and conservation
- Wooden windows and doors conservation
- Wooden furniture restoration
- New interior and exterior lighting
- Restoration and repointing of the decorative paintings of the church
- Restoration and retouching of the decorative paintings of the groin vaults
- Restoration of external stone cornices

iii. *MEP works*
- Installation of air conditioning system
- Installation of sound system
- Fire alarm and fire detection system
- Firefighting system
- High- and low-voltage cabling
- Installation of a power pack
- Electrical substation installation
- Fire pumps and a water tank for fire protection.

9.7 THEOFILOU MANSION, THESSALONIKI, GREECE

9.7.1 Introduction

The Theofilou mansion (Figure 9.7.1) is a characteristic example of the Macedonian architecture which is very common in Northern Greece, North Macedonia, Bulgaria and Northwest Turkey.

The project of the restoration was completed in two distinct periods in 1997 and 2016 so that the building was returned to its original use as a high-end residence. The lead members of the design teams of 1997 and 2016 are the following:

- Dimitrios Alexadrou, architect, 1997
- A. Stefanidis, structural engineer, 1997
- Panos Panetsos, architect, 2016
- Dr. Penelis, Gr., structural engineer, 2016.

9.7.2 Description

As already mentioned, the typical construction system that can be found in Macedonia follows general principles which take into account the high seismicity of the region.

The building is assumed to be built in 1840–1860 and was located in the Ottoman (Turkish) quarter of Thessaloniki, currently called Ano Poli. So the three-storey structure has in plan dimensions of 15 m × 12 m and a total height at the base of the roof 10 m (12 m at the top of the roof).

Figure 9.7.1 Front elevation after restoration of 2016.

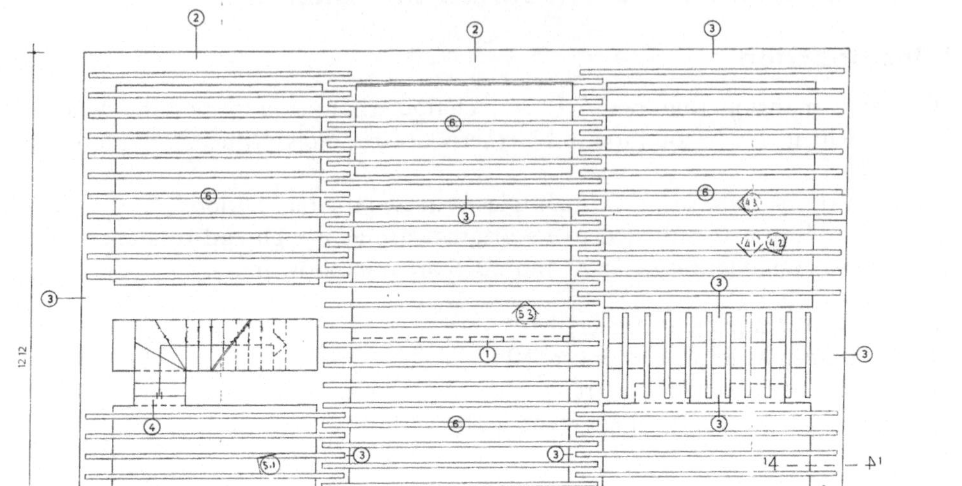

Figure 9.7.2 Lower floor roof plan.

The lower floor (Figure 9.7.2) is very stiff, built from thick, load-bearing stone masonry with a thickness of 70 cm. The first floor (Figure 9.7.3) is of brick masonry (usually solid bricks) and embedded within it are horizontal and diagonal timber beams in order to ensure the proper function of the walls as a unified building element (called 'tsatma'). In the second floor (Figure 9.7.4), timber frames are used with wooden connectors and plaster (called 'bagdati'). All floors are with timber pine beams (8 × 10 or 10 × 10).

The roof is with wooden trusses; while part of it has Byzantine type tiles, the other part has French type tiles.

9.7.3 Structural condition

The pathology recorded prior to the 1997 intervention is to be considered the summation of the building lifespan to date.

In the foundation and lower floor level, no serious damages were located (no cracks or settlements), which is to be expected considering the bedrock foundation of all buildings in the region. Several plaster detachments in the south external wall as well as disorganization of the mortar were recorded there. There was also a cracking mapped at the west façade, while the effect of uprising moisture was present in all lower story walls.

In the first floor, there were significant cracks and deformation in the west wall that had to be temporarily shored. In the south façade wall, the corbels' support caused the wall to deviate from vertical significantly. All roof elements were severely deformed and shored to prevent collapse.

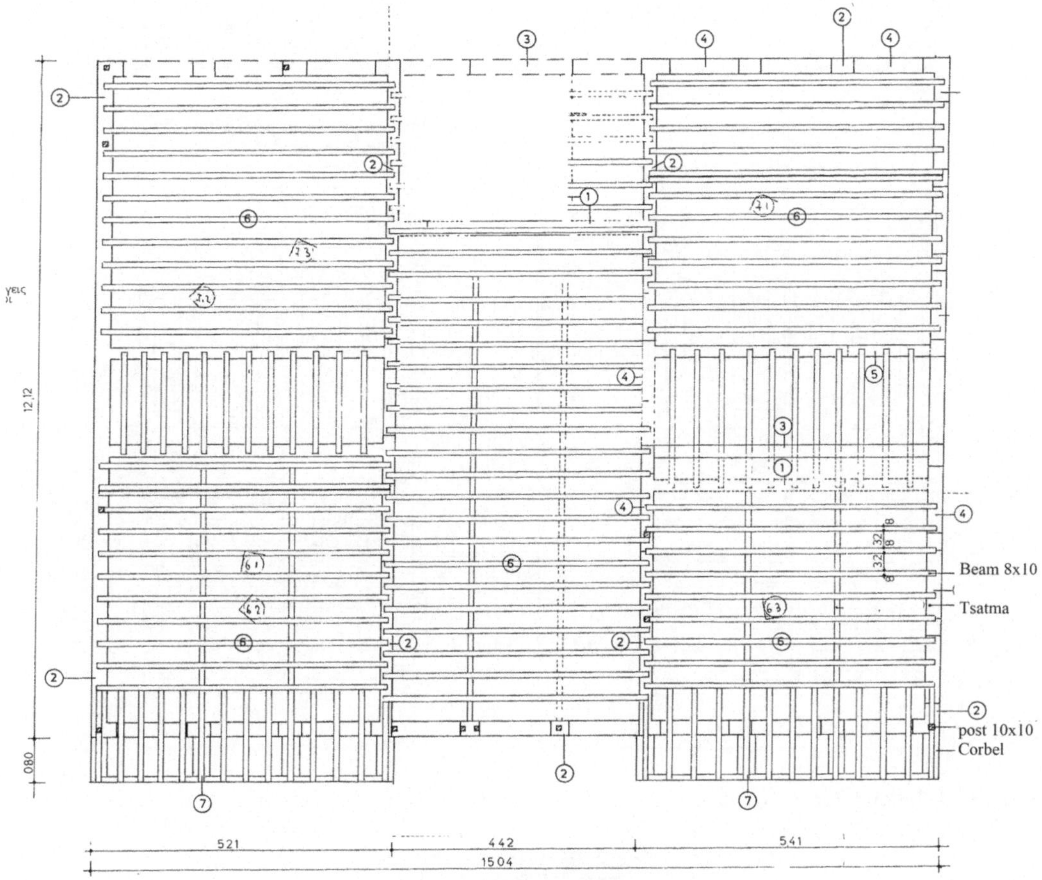

Figure 9.7.3 First floor roof plan.

The second floor (Figure 9.7.5) was the most severely damaged and presented serious cracking and deformations (to complete disarray) of the external west façade wall, as in the floor below, that required immediate shoring. Also the internal walls in the main hall were also damaged probably due to the deformations of the supporting floor.

The roof in the west side presented serious deformations; the roof in the central part also presented deformations, while part of the roof in the east wing had partially collapsed.

9.7.4 Investigations – materials

An architectural survey was conducted using standard survey techniques. To determine foundation conditions, several sections were excavated alongside the external walls. These showed that the stone masonry is resting on the bedrock without the use of large megaliths that would be found in more significant stone masonry monuments, such as churches (Figure 9.7.6).

Regarding building materials, the stones are limestone and dolomites with a minimum compressive strength of 50 MPa. The mortar is a mix of clay and thin sand. It can be removed by hand, does not contain fly ash or trimmed bricks and is of very low

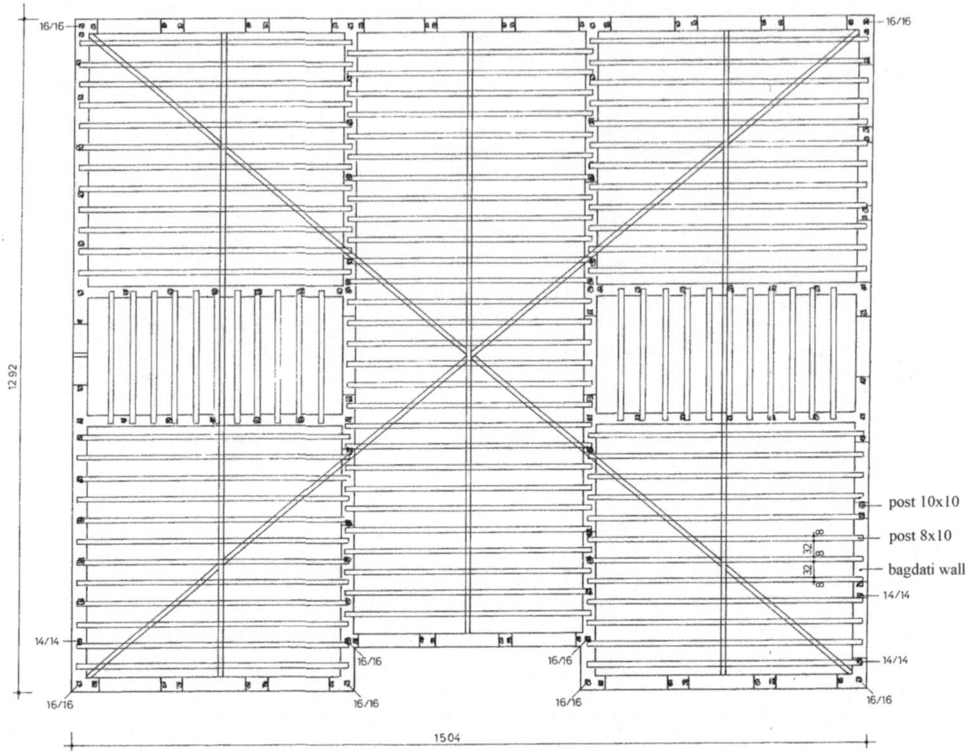

Figure 9.7.4 Second floor roof plan.

Figure 9.7.5 Second floor south façade deterioration.

Figure 9.7.6 Base of stone masonry.

bearing capacity. The considered design compressive strength is 1.20 MPa, and the considered cohesion is 0.12 MPa.

The wood elements both included as ties in the walls or in the roof truss are pine, probably originating from Romania as was custom at the time of construction with a design tensile strength of 8.50 MPa.

9.7.5 Restoration

As already mentioned, the building was fully restored during the 1997 intervention.

During the restoration, the foundation was strengthened by encasing the lower part of the masonry walls with reinforced concrete jackets; the stone masonry walls of the lower floor were grouted, repointed and strengthened with an exterior cement-based plaster and welded wire mesh reinforcement as shown in Figure 9.7.7.

The first floor was repaired and strengthened by replacing all the damaged timber frame elements, strengthening the connections with galvanised steel connectors and

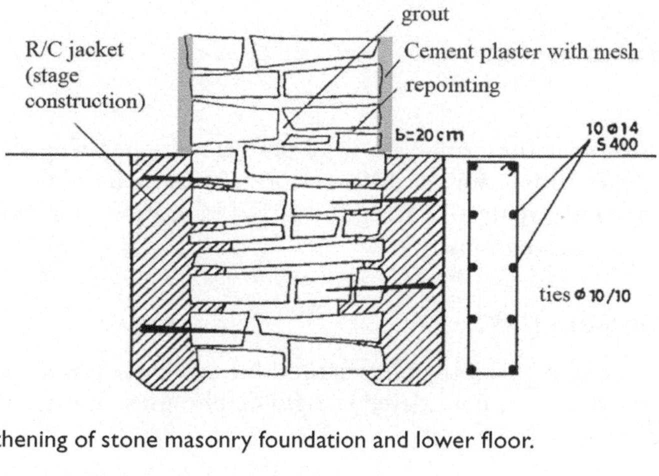

Figure 9.7.7 Strengthening of stone masonry foundation and lower floor.

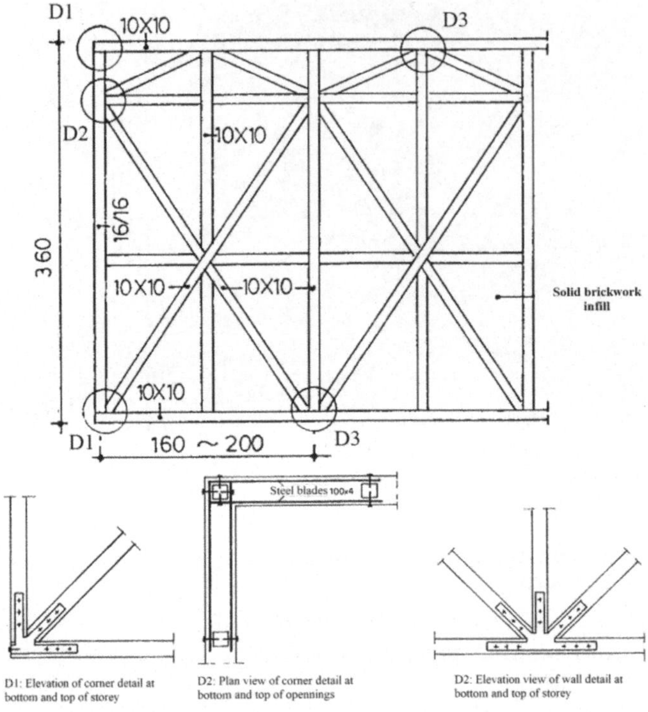

Figure 9.7.8 First floor walls framing (tsatma).

rebuilding the infill solid bricks as shown in Figure 9.7.8. The surface was also covered with a 3–4 cm cement-based plaster and welded wire mesh reinforcement.

The second floor was treated in a manner similar to that of the first floor with the modification that instead of the solid bricks infills, horizontal wooden cladding was applied (bagdati) which was then also covered with a 3–4 cm cement-based plaster with welded wire mesh reinforcement.

The roof was partially rebuilt and repaired using the original structural system with trusses in both directions.

9.7.6 Structural analyses

9.7.6.1 General

Two different sets of structural analyses were performed, one using static approximate approaches which was utilised for the 1997 restoration and a more precise finite element assessment with dynamic analyses during the 2016 restoration that did not require any additional structural interventions.

9.7.6.2 Static approach (1997)

The static analysis of 1997 used the pier method for evaluation of the elastic stresses of the piers; that is, the floors are considered as rigid diaphragms, and each pier participates in the storey shear proportional to its stiffness.

The lower-level piers were considered according to the masonry geometry; however, for the first floor (tsatmas) with brick-filled timber frame and the second floor (bagdati) timber frame, only the timber elements with the diagonals were considered ignoring the brick infills, in the case of the first floor, and the horizontal wooden cladding, in the case of the second floor. This approach is obviously conservative.

The seismic force was applied as a weight coefficient equal to 0.12.

The results from static analysis considering gravity loads and a lateral static load for earthquake were presented in a tabular form, which shows that all piers are adequate regarding elastic compressive and shear stresses. The roof was analysed using a series of 2D trusses and elastic analysis.

9.7.6.3 Dynamic approach (2016)

i. *Modelling*

The building was modelled in its current layout, as it was reformed during the major restoration of 1997. A rigid diaphragm was considered only at the level of the roof base, as the roof was reconstructed during the renovation, resulting in rigid connections and joints with adequate stiffness.

At the other storeys, a rigid diaphragm constraint was not considered, due to the flexibility of the timber floor and the different diaphragm constraints that the timber beams create per direction.

The stone masonry load-bearing walls were modelled as shell elements. The walls of the first and second floors were modelled considering the timber frame structure as well as the brick infill of the walls of the first floor and the cement plaster overlay of the walls of the second floor, using shell elements (Figure 9.7.9).

The timber frame structures of the walls of the first and second floors were verified for a permissible stress $\sigma_{Rd} = 8.5\,\text{MPa}$. The masonry walls of the first and second floors were verified against shear, considering that the timber frame structure resists tension.

The timber members of the roof were verified for a permissible stress $\sigma_{Rd} = 2/3(8.5) = 5.7\,\text{MPa}$, as in the 1997 design (the permissible stress according to DIN1052 was reduced by 33% due to the exposure of the roof members to adverse weather phenomena).

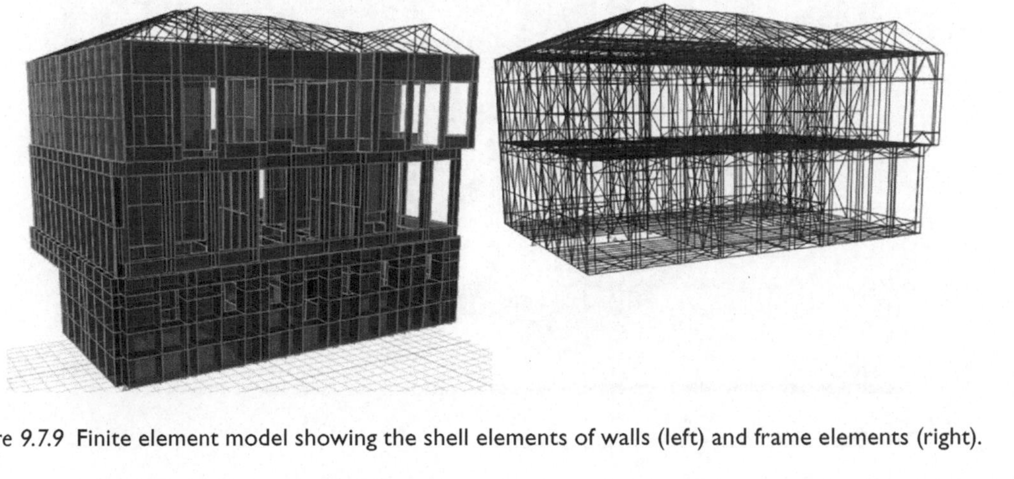

Figure 9.7.9 Finite element model showing the shell elements of walls (left) and frame elements (right).

As a seismic input, the design acceleration spectrum of EN1998-1 code was used with PGA = 0.16 g and a behaviour factor $q = 2.0$.

ii. *Results*

Table 9.7.1 shows the first 10 eigenmodes, while due to the complexity of the modelling, which included the roof, the total modes considered were 100.

The piers were checked for flexure and shear, as per EN1996. For flexure, section analysis considering both axial force and bending was performed, while for shear, the uncracked area of each pier was considered. These checks were performed using the ECtools masonry add-on to Etabs (ECtools, 2013). The results clearly demonstrated that the structure, as strengthened in 1997, is adequate to withstand the design earthquake.

Indicatively the elastic shear stresses of the masonry walls of the first floor are shown in Figure 9.7.10, and the developed stresses are well below the 0.12 MPa cohesion limit.

Table 9.7.1 Modal results

Case	*Mode*	*Period (s)*	*UX (%)*	*UY (%)*	*RZ (%)*	*Sum UX (%)*	*Sum UY (%)*	*Sum RZ (%)*
Modal	1	0.181	2.07	**8.33**	**10.42**	2.07	8.33	10.42
Modal	2	0.151	**11.60**	2.62	0.27	13.67	10.95	10.69
Modal	3	0.128	3.58	0.13	3.15	17.25	11.08	13.85
Modal	4	0.104	0.00	1.83	0.67	17.25	12.91	14.51
Modal	5	0.102	0.03	**4.65**	3.91	17.28	17.56	18.42
Modal	6	0.094	1.98	0.21	0.44	19.26	17.78	18.85
Modal	7	0.093	1.09	0.10	0.00	20.35	17.87	18.86
Modal	8	0.09	0.01	0.65	0.02	20.35	18.52	18.88
Modal	9	0.083	0.55	0.00	0.84	20.90	18.53	19.72
Modal	10	0.078	0.01	0.03	0.33	20.92	18.56	20.05

Note: The values in bold refer to the 3 significant modes of vibration.

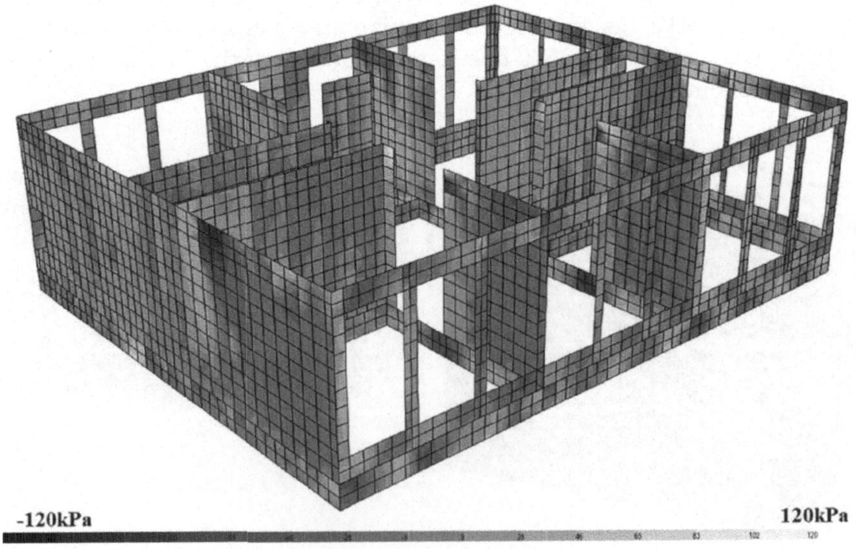

Figure 9.7.10 Elastic shear stresses for first floor.

9.8 RESTORATION OF ST. IRINE SQUARE TOWNHOUSE, ATHENS

9.8.1 Introduction

The building is a characteristic example of the early 20th century Athenian late neoclassical buildings with some neoteric and innovative approaches in its structure. These buildings are usually comprised of unreinforced masonry (URM) bearing walls combined with steel-vaulted floors, steel girders with stone plates and concrete fill or timber floors. They are built in a continuous form; that is, they are surrounded by buildings of the same typology (urban nuclei) as in our case.

The project of the restoration of this building and its conversion to a hotel (Monastery Estate Hotels) started in 2019 and was overseen by architect George Kondilis and structural engineer Christos Drakopoulos, while the authors (Penelis Consulting) acted as the owner's consultants during the design and permitting phase.

9.8.2 Description

As already mentioned, the typical construction system that can be found in early 20th century Athens has been followed. The building is assumed to be built between 1910 and 1930 in its current form and is located in St. Irine Square which is in the central part of Athens. It is a five-storey structure with a basement, ground floor with a mezzanine and first and second floors and is covered by a timber roof with an attic (Figure 9.8.1). It has in plan dimensions of 10 m × 19 m and a total height at the base of the roof 16.50 m (19 m at the top of the roof).

The basement is a rigid box structure with uncut rubble stone masonry perimeter walls with a thickness of 85 cm. The basement roof (ground floor plan) is comprised of a series of main steel beams in the *x–x* direction with double IPN300 steel profiles encased in unreinforced low-quality concrete to increase the stiffness and control deflections (Figure 9.8.2). The secondary beams in the *y–y* direction are IPN80, and the floor is formed by placing stone plates at the bottom of the I-beams and filling with unreinforced concrete (Figure 9.8.5). The final floor is concrete. The novelty of the structure is the use of a composite solution that produces a clear span of 700 m, a solution that has been also used in the ground floor roof structure to create a store without columns.

In the ground floor roof (first floor plan) (Figure 9.8.3), the structural system is built with four large composite steel beams in the *x–x* direction which have planted on them the façade and back load bearing piers of the upper levels as well as the interior load bearing walls. These main beams have a composite concrete steel section with 3xIPN400 for the façade, which has the loads of the piers above and the greater span of 1,000 m and 2xIPN400 for the other beams that have a span of 900 m. The transverse secondary beams are of timber with section of 80 × 150 for the rooms and steel IPN80 for the common areas. Also a steel beam is placed under each interior load-bearing wall of the upper structure. The perimeter walls are of masonry with 50 cm thickness and continue to the roof level.

The structure of the upper levels, first floor, second floor and roof attic are actually comprised of smaller spans by introducing load-bearing brick masonry walls which are planted on the ground floor roof beams (Figure 9.8.4). So the main beams of these levels in *x–x* direction are single IPN360, while the secondary beams in the *y–y* direction are timber beams 75 × 190 with IPN120 in the common areas and back balconies and IPN180

Figure 9.8.1 Front elevation after restoration of 2016.

under the load-bearing brick masonry walls. It is obviously difficult to identify which walls are load bearing, but one should consider the *x–x* interior walls as part of the structure and the *y–y* walls as infill walls that, however, probably withstand some of the weight of the upper-level wall, due to deflections. The duo pitch roof is supported by timber truss.

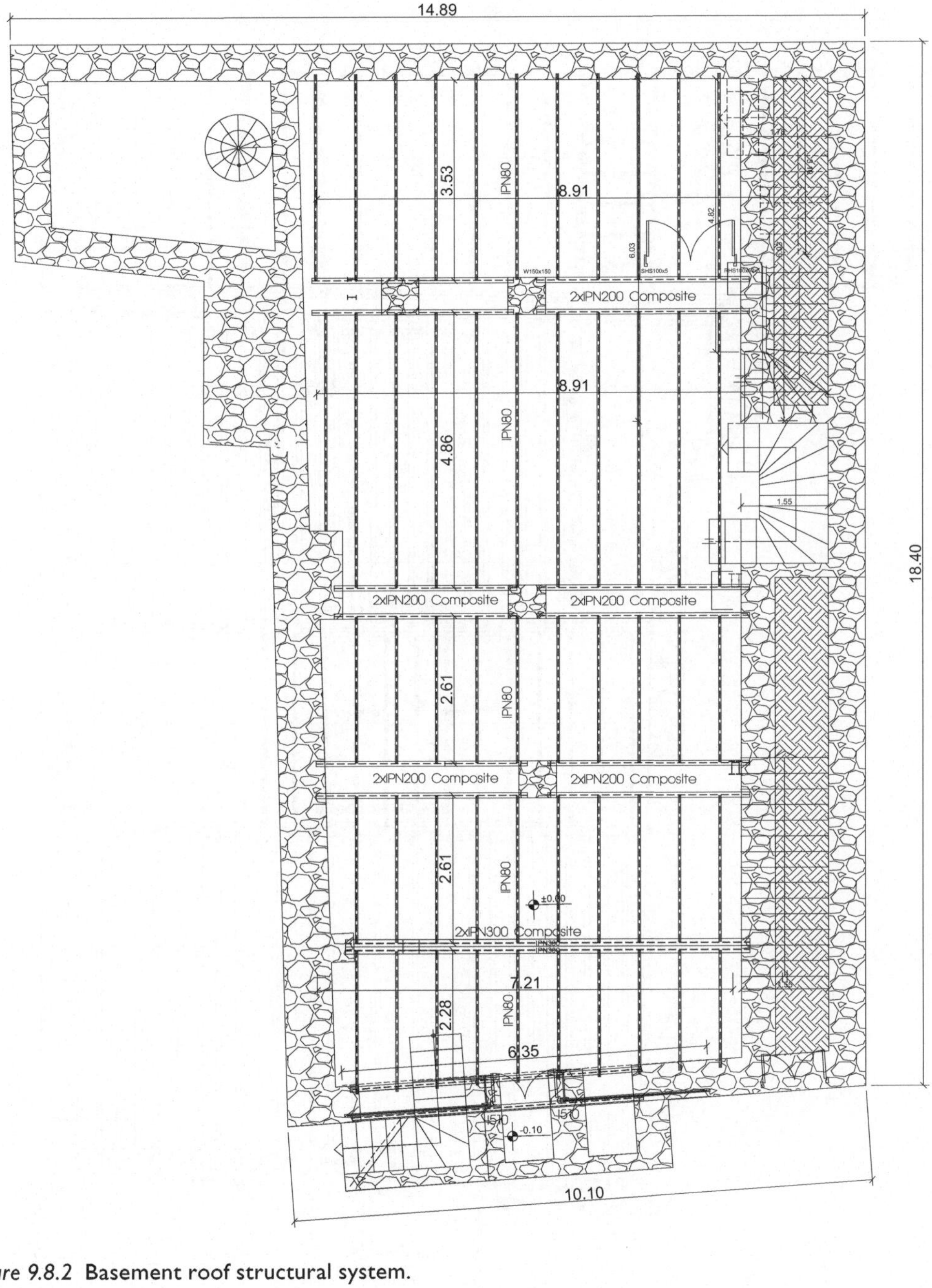

Figure 9.8.2 Basement roof structural system.

9.8.3 Structural condition

During the design period of 2018, the building did not have any pathology from previous earthquakes or ground settlements. Its continuous use for almost a century as a residential building with a ground floor store actually contributed to its preservation. The main

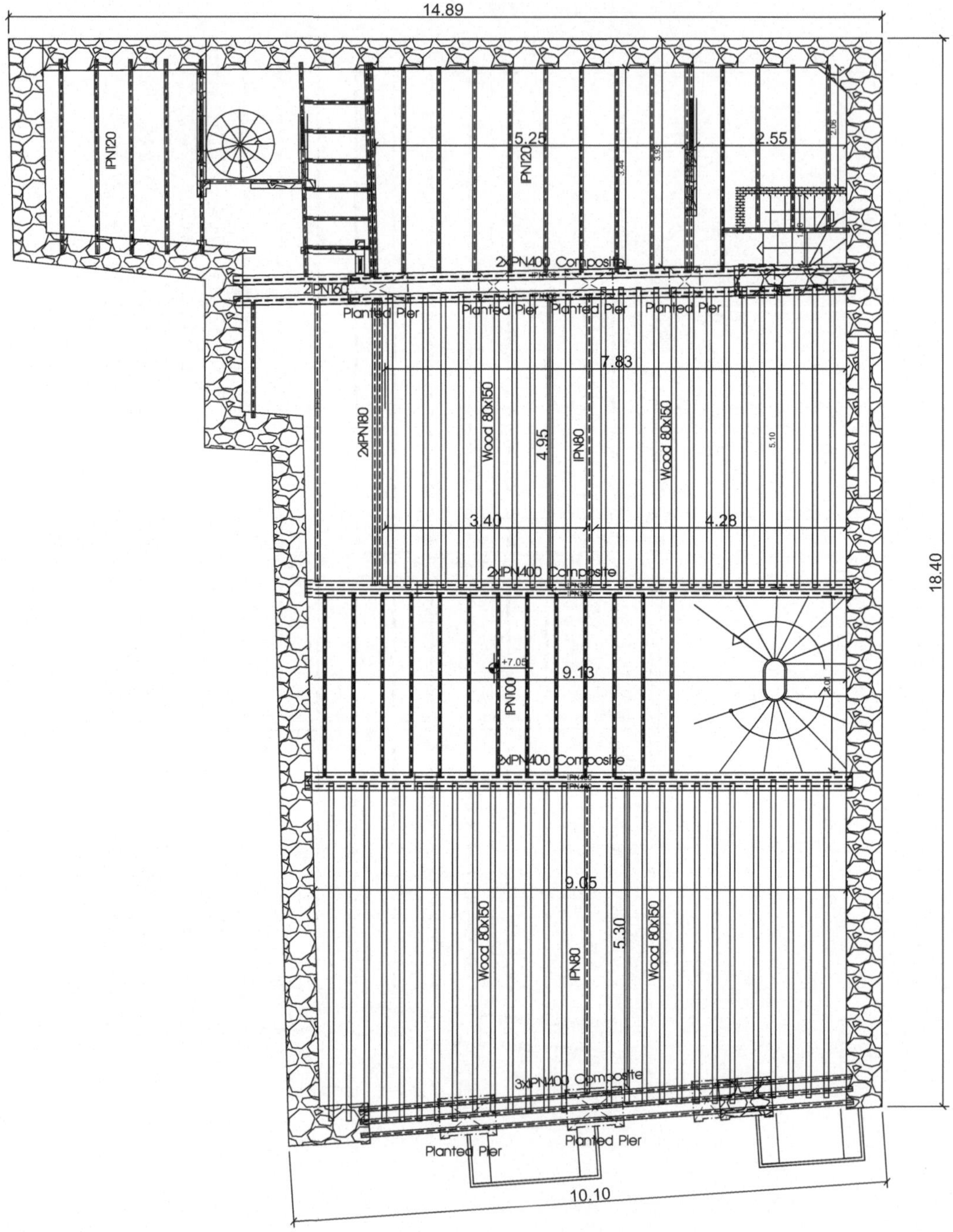

Figure 9.8.3 Ground floor roof structural system.

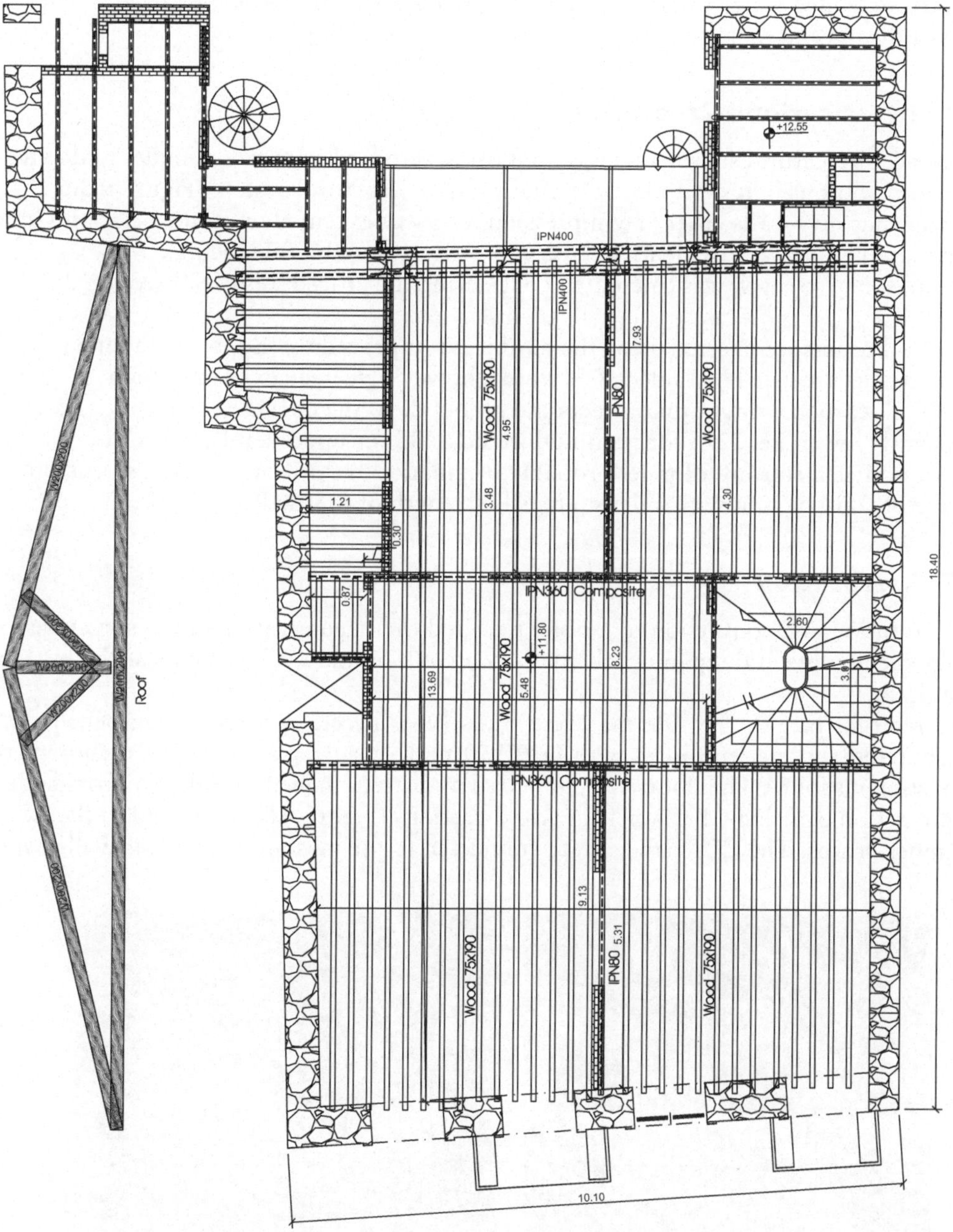

Figure 9.8.4 Typical floor and roof truss sketch.

steel beams of the structure as well as the perimetric rubble stone masonry walls were extremely well preserved and undamaged. The roof was also in good condition.

However, the building presented severe deterioration in the back façade areas which form the structural system, and the plan views were probably a later addition. These areas presented extended water infiltrations, moisture, oxidisation of steel beams and deterioration of the masonry.

9.8.4 Investigations – materials

An extended structural survey was required using standard survey techniques and extensive *in-situ* sections in order to identify the complex structural system (Figure 9.8.5).

For the masonry walls, a set of compression tests was executed on stone, brick and lime mortar samples in order to identify the compressive strength (Table 9.8.1).

The masonry strength was calculated considering the procedures of Subsection 3.2.2.3 (Table 9.8.2)

The steel beams were tested using the hardness test procedure (Equotip rebound tester), and the results are listed in Table 9.8.3, which, however, seemed extremely high, probably due to high carbon content that affects the rebound, so the considered steel grades were significantly lower, i.e. S275 for ground floor ad S235 for upper levels.

The wood elements are of pine, probably originating from Romania as was custom at the time of construction with a design tensile strength of 8.50 MPa.

9.8.5 Restoration

The restoration and conversion to a hotel required the strengthening of the structure for earthquake loading and repairs in the deteriorated floor and wall elements at the back of the building.

The internal face of the side masonry walls (*y–y* direction) were strengthened with 7 cm shotcrete jackets reinforced with D10/150 mesh, ensuring also proper anchorage to the existing masonry. The floors were repaired and strengthened in order to provide diaphragm action by using CLT panels in two directions (Figure 9.8.6) screwed to the existing timber beams. The CLT panels were connected to the masonry perimeter walls using

Figure 9.8.5 Main composite steel beam (left) and secondary steel beams with stone plates and concrete fill (right).

Table 9.8.1 Material tests

Type of sample	*Compressive strength (MPa)*
Stone	
ST01	74
ST02	51.9
ST03	6.4 (outlier)
ST04	67.5
ST05	62.0
ST06	54.5
Brick	
BR1: solid (118 mm × 250 mm × 60 mm)	10.7
BR2: solid (118 mm × 250 mm × 60 mm)	9.9
BR3: solid (115 mm × 245 mm × 40 mm)	21.3
BR4: solid (115 mm × 245 mm × 40 mm)	15.4
BR5: 2× Hollow voids (215 mm × 97 mm × 60 mm)	4.8
BR6: 2× Hollow voids (115 mm × 245 mm × 40 mm)	8.1
BR7: 3× Hollow voids (200 mm × 95 mm × 55 mm)	13.8
BR8: 3× Hollow voids (200 mm × 95 mm × 55 mm)	12.9
BR9: 3× Hollow voids (200 mm × 95 mm × 55 mm)	12.8
Mortar	
M1 (brickwork)	1.25
M2 (brickwork)	1.01
M3 (brickwork)	1.22
M4 (brickwork)	1.56
M5 (brickwork)	2.11
M6 (stonework)	0.83

Table 9.8.2 Masonry properties

Masonry type	*Compressive strength (MPa)*	*Modulus* **E** *(MPa)*	*Poisson's ratio* (ν)
Stone masonry	2.37	2,844	0.376
Solid brickwork	2.94	2,667	0.378
Voided brickwork (two voids)	1.78	2,136	0.384
Voided brickwork (three voids)	2.51	3,012	0.376

Table 9.8.3 Steel properties

Location	*Hardness value*	*Vickers (HV)*	*Tensile strength (MPa)*
GF	561	292	938
GF	565	297	954
First floor	338	99	316
First floor	343	102	325
First floor	380	126	399
First floor	406	143	458

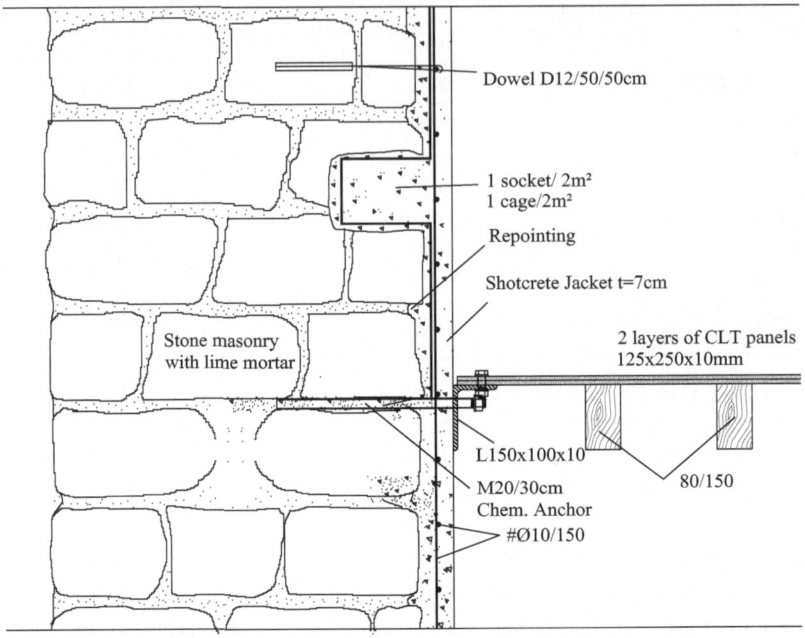

Figure 9.8.6 Wall strengthening details.

steel profiles (L150 × 100 × 10) and chemical anchors. The corners of masonry walls were anchored using anchors from the internal side (Figure 9.8.7). Finally all internal masonry load-bearing and infill walls were strengthened by cement plaster reinforced with welded wire mesh reinforcement.

The existing roof timber rafters were strengthened sideways with new double timber beams, while the attic area was reconstructed using steel girders.

9.8.6 Structural analysis

i. *Modelling*

The building was modelled both in the existing condition as well as in the strengthened condition. The main differences between the two approaches was the consideration of the diaphragm action, which in the existing condition was not present and, in the strengthened condition, was considered according to the in-plane stiffness, and the stiffness and capacity of the masonry walls in their existing and strengthened states (Figure 9.8.8).

As a seismic input, the design acceleration spectrum of EN1998-1 code was used with PGA = 0.16g and a behaviour factor $q = 2.0$.

ii. *Results*

Table 9.8.4 shows the first 10 eigenmodes, while due to the complexity of the modelling, which included the roof, the total modes considered were 50. The modal behaviour is clearly uncoupled and translational.

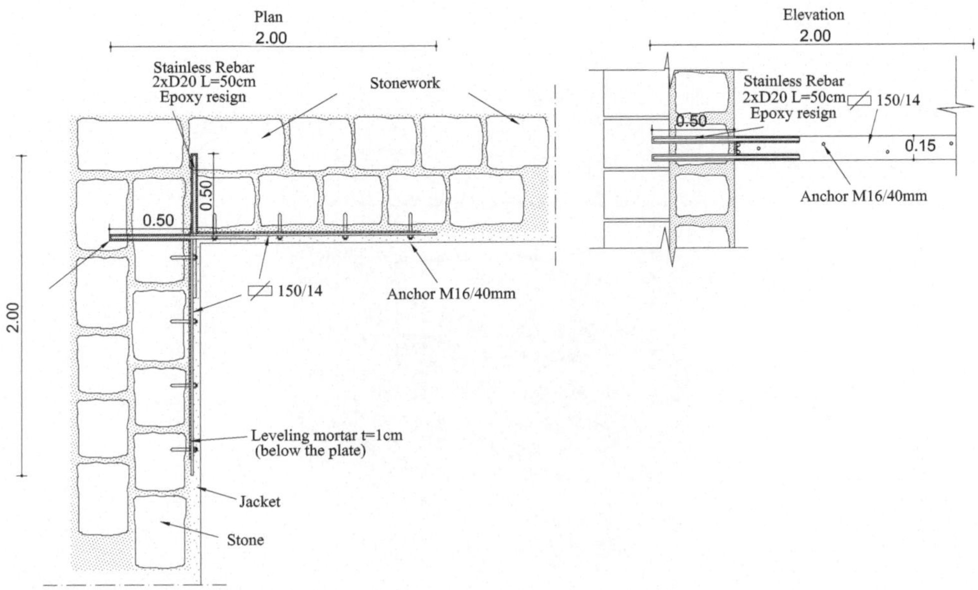

Figure 9.8.7 Corner tie detail.

Table 9.8.4 Modal results

Mode	*Period (s)*	*UX (%)*	*UY (%)*	*Sum UX (%)*	*Sum UY (%)*
1	0.331	**53**	0	53	0
2	0.182	0	**34**	53	34
3	0.164	0	**28**	54	62
4	0.135	0	0	54	62
5	0.106	19	0	73	63
6	0.103	0	2	73	65
7	0.084	1	1	74	66
8	0.079	0	0	74	66
9	0.078	0	0	74	66
10	0.076	0	0	74	66

Note: The values in bold refer to the 3 significant modes of vibration.

The piers were checked for flexure and shear as per EN1996. For flexure, section analysis considering both axial force and bending was performed, while for shear, the uncracked area of each pier was considered. These checks were performed using the ECtools (2013) masonry add-on for Etabs. The results clearly demonstrated that the structure, as strengthened, is adequate to withstand the design earthquake. Figure 9.8.9 shows the section analysis and safety factor of a masonry L-shaped pier strengthened with an unsymmetrical R/C jacket.

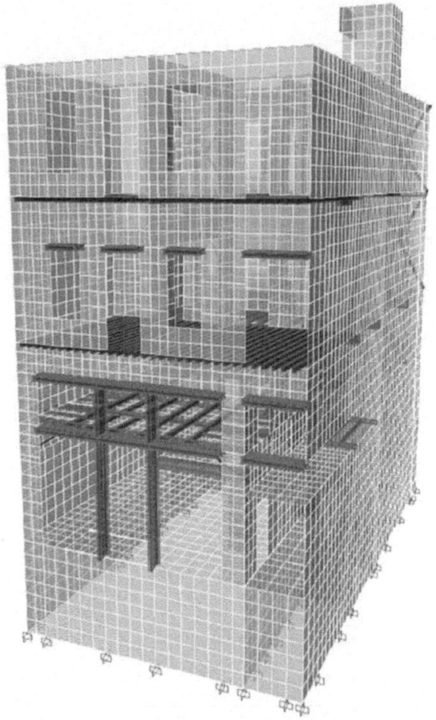

Figure 9.8.8 Finite element model showing the meshing (strengthened structure).

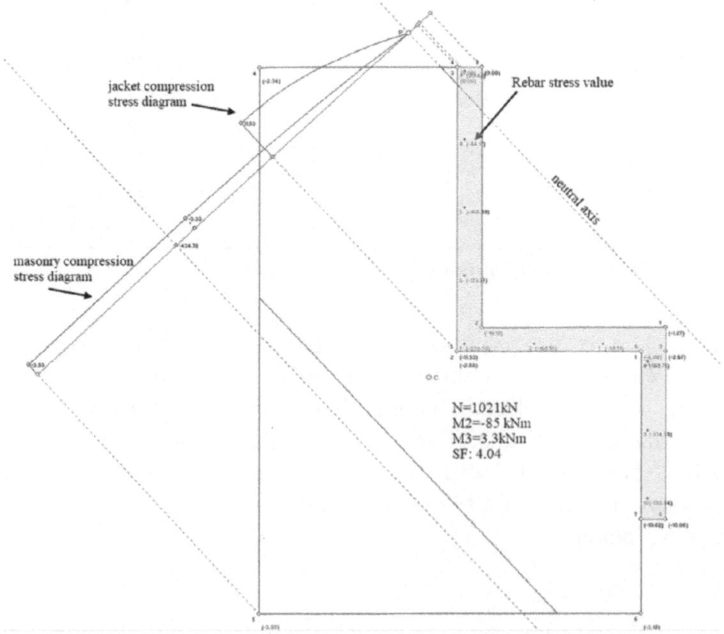

Figure 9.8.9 Section analysis of masonry pier with internal R/C jacket. (ECtools, 2013.) (Penelis S.A. 2019f)

References

Addessi, D. and Sacco, E. (2018). Homogenisation of heterogeneous masonry, *New trends in mechanics of masonry. Meccanica*, Vol. 53, Springer Science & Business. pp. 1699–1717.

Ali, S. and Page, W. (1988). Finite element model for masonry subjected to concentrated loads. *Journal of Structural Engineering of the ASCE*, Vol. 114, pp. 1761–1784.

Al-Kass, R., Hadi, M., Khalil, N., and Ali, F. (1989). Ancient Babylonian bricks: Their technological characterisation and simulation. *Proceedings of International Conference on Structural Repairs and Maintenance of Historical Buildings*. Brebbia, C., (ed.). Birkhäuser Verlag, Berlin. pp. 71–80.

Andrade, Fr. (1995). La construccion romana (II). Revista de Edificación. RE. N°19 Teoriae historia de la arquitectura. p. 69.

Antonnuci, R., Carboni, M., and Cocchi, G. (1989). Analytical model and numerical application of the structural behaviour of masonry walls under in-plane stress. In structural repair and maintenance of historical buildings. Computational Mechanics Publications.

Aristotle (1955b). Mechanical problems, *in Aristotle Minor works,* Hett, W.S. (ed). William Heinemann, Cambridge, UK.

Aristotle (1955a). *Physica, VIII 5, 249b. Translation into English* by Hardie R.P. and Gaye R.K. The Tech Classics Archive of M.I.T, Boston, USA.

ASCE/SEI 41-17 (2017). *Seismic evaluation and retrofit of existing buildings*. American Society of Civil Engineers / Structural Engineering Institute, Reston, VA, USA.

ATC-40 (1996). *Seismic analysis and retrofit of R.C. buildings*. ATC, California.

Athanassiadou, C., Penelis, G. and Kappos, A. (1994). Seismic response of adjacent buildings with similar or different dynamic characteristics. *Journal of Earthquake Spectra,* Vol. 10, No. 2. pp. 273–317.

Athanassiou, Ph., Malama, V., Misa, M., and Sarandidou, M. (2015). *The restoration of the ruins of the Galerian complex (1994–2014). Documentation and Interventions, Thessaloniki.* Ephorate of Thessaloniki Antiquities. Thessaloniki, Greece.

Baker, J., Horne, M., and Heyman, J. (1956). *The steel skeleton Vol II*, Cambridge University Press, London

Baraldi, D., Reccia, E., and Cecch, A. (2018). In-plane loaded masonry walls: DEM and FEM/DEM models: A critical review. *New trends in mechanics of masonry. Meccanica*, Vol. 53, Springer Science and Business. pp. 1613–1628.

Bei, G. and Papayianni, I. (2003). Compressive strength of compressed earth block masonry. *Proceedings of Conference on Structures Studies Repairs and Maintenance of Heritage Architecture VIII WIT press, Bath, UK. pp. 367–375.*

Bertoresi, E., Milani, G., and Casolo, S. (2018). Homogenisation towards a mechanistic Rigid Body and Spring Model (HRBSM) for non-linear dynamic analysis of 3D masonry structures, *New trends in mechanics of masonry. Meccanica,* Vol. 33, Springer Science & Business. pp. 1819–1855.

Binda, L. (2008). *Learning from failure: Long-term behaviour of heavy masonry structures.* WIT Press, Southampton, UK.

Binda, L., Anzani, J., and Saisi, A. (2003). Failures due to long-term behaviour of heavy structures. Pavia civic tower of the Noto Cathedral. in *Structural studies, repairs and maintenance of heritage architecture VIII*, Brebbia, C., (ed). WIT press, Bath, UK. pp. 99–108.

Binda, L., Anzani, A., and Mirabella Roberti, G. (2000). Tall and massive ancient masonry buildings: Long-term effects on loading. *Proceedings of G. Penelis International symposium on concrete and masonry structures.* AUTH, Thessaloniki. pp. 273–284.

Binda, L., Fontana, A., and Anti, L. (1991). Load transfer in multiple-leaf masonry walls. *Proceedings of the* 9th *International Brick Block Masonry Conference.* pp. 1488–1497.

Binda, L., Fontana, A., and Mirabella Roberti, G. (1994). Mechanical behaviour and stress distribution in multiple-leaf walls. *Proceedings of the 10*th *International Brick Block Masonry Conference.* pp. 51–59.

Binda, L., Pina-Henriques. J., Anzani, A., Fontana, A., and Lourenço, P.B. (2006a). A contribution for the understanding of load-transfer mechanisms in multi-leaf masonry walls: Testing and modelling. *Elsevier Engineering Structures*, Vol. 28. pp. 1132–1147.

Binda, L., Saisi, A., and Tedeschi, C. (2006b). Compatibility of materials used for repair of masonry buildings: research and applications, in *fracture and failure of natural building stones: Applications in the restoration of Ancient monuments*, Kourkoulis, K., (ed). Springer, Dordecht, Netherlands. pp. 167–182.

Bouras, Ch. (1965). Byzantine groin-vaults with ribs. Archaeological Bulletin, Athens, Greece.

Brencich, A., Gambarotta, L., and Lagomarsino, S. (1998). A microelement approach to the three-dimensional seismic analysis of masonry buildings. *Proceedings of the XI European Conference on Earthquake Engineering.* Paris, AA Balkemon (Abstract Vol. 2 CD-ROM). p. 602.

Bröcker, O. (1961). Stein Festigkeit und Wand Festigkeit. *Ziegelindustrie*, Vol. 2. Germany. pp. 19–21.

Capecchi, D. (2012). *History of virtual work laws. Science Network Historical Studies*, Vol. 42, Birkhäuser; Springer-Verlag, Italy.

Chiozzi, A., Milani, G., Grillanda, N., and Tralli, A. (2018). A fast and general upper-bound limit analysis approach for out-of-plane loaded masonry walls *New trends in mechanics of masonry. Mecanica,* Vol. 33. Springer Science & Business. pp. 1875–1898.

Choisy, A. (1873). *L'art de batir chez les Romans.* Ducher etc i.e Paris.

Choisy, A. (1883). *L'art de batir chez les Byzantins*, Librairie de la societe anonyme de publications periodiques. Paris.

Como, M. (2019). Thrust evaluation of masonry domes. An application to the St. Peters dome. *International Journal of Masonry Research and Innovation,* Vol. 4, No. ½, Interscience Enterprises Ltd. pp. 32–49.

Couplet, P. (1730). De la poussée des voûtes. Histoire de l'Aca'demic Royale des Sciences, France (1729, p. 79 and 1730, p. 117).

Croci, G. (2007). *The conservation and structural restoration of architectural heritage.* Advances in architecture series. WIT Press, Southampton.

CSIR India (2017). *Brick masonry in mud mortar.* India.

Cucci, N., Sassu, M., Rossi, P. and Palcini, A. (1995). Comparison of non-destructive in-situ techniques for vertical strength assessment in masonry walls. *Transactions of the Built Environment*, Vol. 15, WIT Press. pp. 349–356.

D'Ayala, D. and Speranza, E. (2003). Definition of collapse mechanisms and seismic vulnerability of historic masonry buildings. Earthquake Spectra 19 (3), pp. 479–509.

Diaferio, M. and Foti, D. (2017). Seismic risk assessment of Trani's Cathedral bell tower in Apulia, Italy. *International Journal of Advanced Structural Engineering,* Vol. 9, Springer. pp. 259–267.

Didaskalou, Gr. (2018). Geoter Didaskalou Sp. (Archives) Thessaloniki.

DIN1052 blatt 1 and 2 (1976). Holzbauwerke – Berechnung and Ausführung Dubelverbindungen besonderer art. *Beton – Kalender, Teil II* Verlag von Wilhelm, Ernst & Sohn, Berlin. pp. 382–433.

Drosou, K. (2014) Archives of the Ephorate of Antiquities of Halkidiki, Greece.

Drougas, A. (2009). Investigation of the use of iron in construction from antiquity to the technical revolution. *Master's thesis Universität Politécnica de Catalunya*, Spain.

E.C.8-1/EN 1998-1 (2004). *Design of structures for earthquake resistance: General rules, seismic actions and rules for buildings*. CEN, Brussels, Belgium.

EAK (2000). *Greek Aseismic Code*, Athens, Greece.

ECtools (2013). 'User manual', 3Pi Software ltd, Thessaloniki, Greece, www.ectools.eu, 2013.

Egermann, R. (1993). Investigation on the load bearing behaviour of multiple leaf masonry. *Proceedings of the IABSE Symposium 'Structural Preservation of the Architectural Heritage'*, Rome. pp. 305–312.

EN1052-1 (1999). *Methods of test for masonry*. Determination of compressive strength B.S.I., UK.

EN 1996-1-1 (2005). *Design of masonry structures: General rules for reinforced and unreinforced masonry walls*. CEN, Brussels.

Ephorate of Antiquites of Thessaloniki (1978) Archives of Acheiropeitos church, Thessaloniki.

Ersoy, U. (1989). Diagnosis, assessment and emergency interventions for historic masonry. Structural conservation of stone masonry. *Proceedings of International Technical Conference ICCROM*, Athens. pp. 13–19.

Fajfar, P. (1999). Capacity spectrum method based on inelastic demand spectra. *Earthquake Engineering and Structural Dynamics*, Vol. 28. pp. 979–993.

Fajfar, P. and Dolsek, M., (2000). A transparent nonlinear method for seismic performance evaluation. *3rd workshop of the UK – Japan Seismic Risk Forum*, Imperial College Press, London.

Feilden, M. B. (2003). *Conservation of historic buildings*, 3rd Edition. Architectural Press (Routlege), London & New York.

Galasco, A., Lagomarsino, S., Penna, A., and Resemini, S., (2004). Non-linear seismic analysis of masonry structures. *13th World Conference in Earthquake Engineering*. Paper No. 843.

Gambarotta, G., Lagomarsino, S., and Marbitucci, R. (1996). Brittle-ductile response of in-plane loaded brick masonry walls. *10ECEE*.

Gambarotta, L. (2007). *Mechanics of masonry structures: arches, shear walls and vaults*. Seminari DICAT University of Genova, Italy.

GEOMETRICS (2018). Crosshole seismic testing. www.geometrics.com/applications/geophysical-methods/cro. pp. 1–5.

Giambanco, G., La Malfa Ribolla, E., and Spanda, A. (2018). Meshless mcso-modelling of masonry in the computational homogenisation framework, New trends in mechanics of masonry. Meccanica, Vol. 33, Springer Science & Business. pp. 1673–1697.

Girkmann, K. (1963). *Flachentragwerke 6. Aufl.* Springer-Verlag, Wien, Austria.

Gregorezyk, P. and Lourenço, P. (2000). A review on flat-jack testing. *Engenharia Civil. UM*. Vol. 9, Portugal. pp. 39–50.

Guerra, G. (1979). Principles of earthquake resistance of masonry structures. *7th UNESCO Seminar*, Istanbul.

Harris, J. (1992). Weathering of rock, corrosion of stone and rusting of iron. Masonry Constructions. Mechanica, Vol. 27, No. 3. Calladine (ed.), Springer-Science & Business Media, B.V.

Hart, F. (1965). *Kunst und Technik der Wolbung*. Verlag George Callwey, München, Germany.

Hatziantoniou, H. (1988). Rotunda minaret. *Preliminary technical report. 9th Ephorate of Byzantine antiquities*, Thessaloniki, Greece. In Greek.

Hemeda, S. (2012). Grouting of multiple leaf-masonry walls: Application on some Islamic historical monuments in Cairo, Egypt. *International Journal of Conservation Science*, Vol. 3. pp. 275–288.

Hendry, A. (1981). *Structural brickwork*. The Macmillan Press LTD, London, UK.

Heyman, J. (1966). The stone skeleton. *International Journal of Solids and Structures*, Vol. 2. pp. 270–279.
Heyman, J. (1977). *Equilibrium of shell structures*. Clarendon Press, Oxford, UK.
Heyman, J. (1986). Statical aspects of masonry vaults and domes. *Proceedings of the international symposium of Thessaloniki on restoration of Byzantine and post-Byzantine monuments*. 9th Ephorate of Byzantine Antiquities Thessaloniki Greece. pp. 229–236.
Heyman, J. (1989). The care of masonry buildings: The engineer's contribution. in *Structural repair and maintenance of historical buildings*, Brebbia, C.A., (ed.). Computational Mechanics Publication, Southampton, UK.
Hilsdorf, H. (1969). An investigation in the failure mechanism of brick masonry in axial compression. *Designing, engineering and constructing masonry product*, Johnson, F.B., (ed.). Gulf, Houston, TX, USA. pp. 34–41.
Holstrom, I. (1977). Suitable materials for use in repair of historic structures. *Proceedings of Conference on Structural Conservation of Historic Buildings ICCROM*, Rome.
ICOMOS. (1994). *The Nara document on authenticity*. ICOMOS, symposia, Nara, Japan.
IESEE (Institute of Engineering Seismology and Earthquake Engineering, Greece). (2003). *Strong Motion Data Base of Greece 1978-2003*, Thessaloniki, Greece.
Ignatakis, C. (2000). Mechanical behaviour of Roman masonry, numerical parametric investigation. *G. Penelis International Symposium on Concrete and Masonry Structures*. pp. 285–298.
Ignatakis, C. and Stylianidis, K. (1989). Criteria for the substitution of wooden tie rods in Byzantine churches; St Pandeleimon of Thessaloniki. *Proceedings of Conference on Structural Repair and Maintenance of Historical Buildings*, Florence. Brebbia, C.A. (ed.). Computational Mechanics Publications. pp. 385–393.
Ignatakis, C., Stavrakakis, E., and Penelis, G. (1989a). Analytical model for masonry using the finite element method. *International Conference on Structural Studies, Repairs and Maintenance of Historical Buildings,* Florence Italy.
Ignatakis, C., Stavrakakis, E., and Stylianidis, K. (1989b). A special F.E. axisymmetrical model for the analyses of interventions in domes, including cracking consideration. *Proceedings of the Internat. Technical Conference on Structural Conservation of Stone Masonry*, Athens.
Ignatakis, Ch., Stavrakakis, E., and Penelis, G. (1993). *Mechanical Characteristics of Ancient Roman Masonry*. IABSE symposium. Rome. pp. 433–440.
Ignatakis, Ch., Stavrakakis, E., and Penelis G. (1990). Analytical model for masonry using the finite element method. *Software for Engineering Workstations*, Vol. 6, No.2. pp. 90–96.
Kappos, A., Penelis, Gr. and Drakopoulos, C. (2002). Evaluation of simplified models for lateral load analysis of Unreinforced Masonry Buildings. *Journal of Structural Engineering*, Vol. 128, No.7, ASCE. pp. 1–8.
Karantoni, F. (2004). *Masonry constructions*. In Greek. *Papasotiriou, Athens*.
Karaveziroglou, M., Papayianni, I., and Penelis, G. (1995). Mortars and grouts in restoration of Roman and Byzantine monuments. *Proceedings of the European Seminar on Ancient and Traditional Mortar Technology and Their Importance in Conservation Projects,* Brussels.
Karaveziroglou, M.., Papayianni, I., and Penelis, G. (1995). Mortars and grouts in restoration of Roman and Byzantine monuments. *European Seminar of Ancient and Traditional Mortar Technology and Their Importance in Conservation Projects*, Rethymno, Greece.
Karaveziroglou, M., Papayianni, I., and Penelis, G. (1989). Criteria for selecting materials for repair brick masonry. *ICCROM, Proceedings of International Technology Conference on Structural Conservation of Stone Masonry*, Athens, Greece. pp. 335–342.
Kazakos, P. (2010). *Μηχανικά Αριστοτέλους*. *Ms Thesis NTUA*, Athens, Greece.
Kenit, T., Forest, S., Galliet, I., Mounoury, V., and Jeulin, D. (2003). Determination of the size of the representative volume element for random composites; statistical and numerical approach, *International Journal of Solids and Structures*, Vol. 40, No. 13. pp. 3647–3679.
Korvest Galvanisers (2018). *Atmospheric corrosion resistance*, pp. 1–9

Krautheimer, R. (1965). *Early Christian and Byzantine architecture.* Pelican History of Art, Penguin, UK.

Kupfer, H. (1973). *Das Verhalten des Betons unter mehrachsiger Kurzzeitbelastung unter besondere Berücksichtigung der zweiachsiger Beanspruchung.* Deutscher Ausshuss für Stahlbeton, Berlin.

Lagomarsino, S., Giovinazzi, S., Potesta, S., and Resemini, S. (2003). RISK-UE-EVK4-CT-2000–00014. *An advanced approach to earthquake risk scenarios with application to different European towns.* WP5: *Vulnerability assessment of historical buildings Brussels.*

Lekkas, E., Chiotinis, N., and Awad, M. (2008) Protection and relevation of the church of saint George- Cairo. (Archives of the preliminary design)

Le Seur, T., Jacquier, F., and Boscovich, R.G. (1742). *Parere di tre matematici sopra I danni che si sono trovati nella cupola di S. Pietro sul fine dell' anno 1742, dato per ordine di N.S Benedetto XIV*, Roma. Reprint of the Faculty of Architecture of University of Rome Supienza, Roma.

Lizzi, F. (1982). *The static restoration of monuments.* Sagep Editrice, Genova, Italy.

Lourenço, P. (2018). Technologies for seismic retrofitting and strengthening of earthen and masonry structures: Assessment and application. *Recent advances in earthquake engineering in Europe*, Pitilakis, K., (ed.). Springer, London; New York. pp. 501–518.

Magenes, G. and Calvi, M. (1997). In-plane seismic response of brick masonry walls. *Journal of Earthquake Engineering and Structural Dynamics*, Vol. 26. pp. 1011–1112.

Magenes, G. and Della Fontana, A. (1998). Simplified non-linear seismic analysis of masonry buildings. *5th International Masonry Conference Proceedings of the British Masonry Society.*

Mainstone, R. (1975). *Developments in structural form.* Penguin Books Ltd, England.

Mainstone, R. (1989). Characteristics and assessment of old masonry structures. Structural conservation of stone masonry. *Proceedings of International Technical Conference ICCROM*, Athens. pp. 63–76.

Malinowski, R. (1981). Ancient mortar and concrete durability aspects. *Proceedings of Conference on Mortars, Cements and Grouts used in the Conservation of Historic Buildings*, Rome ICCROM. pp. 341–350.

Mango, K. (1975). *Byzantinische Architecture.* Beiser/Electa Verlag Stuttgart, Mainland.

Mann, W. (1989). Diagnosis and assessment of existing structures. Structural conservation of stone masonry. *Proceedings of International Technical Conference ICCROM*, Athens. pp. 1–12.

Marotta, A., Liberatore, D., and Sorrentino, L. (2016). Estimation of unreinforced tuff masonry compressive strength based on mortar and unit mechanical parameters. in *Brick and block masonry – Trends innovations and challenges.* Modena, C., da Porto, F., and Valluze, M.R. (eds.). Taylor and Francis Group, London. pp. 1715–1722.

Martinez, G., Roca, P., Caselles, O., and Clapés, J. (2006). Characterisation of the dynamic response for the structure of Mallorca Cathedral, in *Structural analysis of historical constructions*, Lourenço, P.b., Roca, P., Modena, C., and Agrawal, S., (eds.). Springer, New Delhi, India.

Mavropoulou, Ch. (1986). The earthquake of 1978 and the monuments of Thessaloniki. The positive aspect. *Proceedings of the International Symposium of Thessaloniki.* 9th Ephorate of Byzantine antiquities. Thessaloniki. pp. 9–24.

Miltiadou-Fezans, A., Papakonstantinou, E., Zambas, K., Panou, A., and Frantzikinaki, K. (2005). Design and application of hydraulic grouts of high injectability for structural restoration of the column drums of the Parthenon Opisthodomos. *Proceedings of STREMAH IX*, WIT Press. pp. 461–471.

Miltiadou, A. (1990). Contribution a l'etude des coulis hydrauliques pour la reparation et le renforcement des structures et des monuments historiques en maconnerie. These de doctorat ENPC, LCPC, Rapport de laboratoires, Series ouvrages d'art A.A.8

Moropoulou, A., Biscontin, G., Bakolas, A., Michailidis, P., and Basiotis, J. (1997). Historic mortars in Mediterranean monuments. *Proceedings of the 4th International Symposium on the Conservation of Monuments in the Mediterranean,* Vol. 3. Technical Chamber of Greece, Greece. pp. 213–231.

Moutsopoulos, N. (2013). *The rotunda of St George in Thessaloniki.* In Greek. Kyriakidis editions, Thessaloniki, Greece.

Mühler, K. (1976). *Holzbau, Beton Kalender 1976 Teil II,* Verlag von Wilhelm Ernst & Sohn. Berlin. pp. 573–626.

Nguyen, V.P., Stroeven, M., and Sluys, L.J. (2011). Multiscale continuous and discontinuous modelling of heterogeneous materials a review on recent developments. *Journal Multiscale Model,* Vol. 3, No. 04. pp. 229–270.

Oliveira, C. and Costa, A.,(2010). Reflections on the rehabilitation and the retrofit of historical constructions, in *Earthquake engineering in Europe*, Garevski, M., Ansal, A., (ed.). Springer, London, New York. pp. 199–221.

Oliveira, D., Silva, R., Lourenco, P., and Garbin, E. (2012). Strengthening of three-leaf stone masonry walls. An experimental research. *Materials and Structures,* Vol. 45, No. 8. pp. 1259–1276.

Omer, U. (2019). Ulker Consulting Engineers. Archives of Galata Greek School renovation and reuse in Istanbul.

Ordinanza OPCM 3274. (2003). *Modified according to OPM 3431, 2005. Allegato 2, Norme tecniche per il progetto, la valutazione e l'adeguamento sismico degli edifici.* Consiglio dei Minister Rome.

Orlandos, A.K. (1995) *The structural materials of the ancient Greeks Part I, Athens 1955 and Part II 1958.* In Greek. Athens, Archaeological foundation.

Page, W. (1981). The biaxial compressive strength of brick masonry. *ICE, Part 2,* Vol. 7. pp. 893–906.

Page, W. (1982). An experimental investigation of the biaxial strength of brick masonry. *Proceedings of the 6th International Masonry Conference Roma.* pp. 3–15.

Page, W. (1983). The strength of brick masonry under biaxial compression. *International Journal of Masonry Construction,* Vol. 3, No.1. pp. 26–31.

Page, W., Kleeman, W., and Dhanasekar, M. (1985). An in-plane finite element model for brick masonry. *New Analysis Techniques for structural masonry ASCE.* New York. pp. 1–18.

Page, W., Samorasinghe, W., and Hendry, W. (1982). The in-plane failure of masonry – a review. *Proceedings of British Ceramic Society. Load Bearing Brickwork,* Vol. 7, No. 30, pp. 90–100.

Palladio, A. (1965). *The four books of architecture.* Dover Publications, New York.

Papasouliotis, G. (2018). Restoration activities at Hagios Dimitrios church of Thessaloniki, 1930. Thessaloniki.

Papayianni, I. (2014). Retrofitting and strengthening masonries of heritage structures: materials used. in *Encyclopaedia of Earthquake Engineering,* Beer, M., Kougioumtzoglou, I.A., Patelli, E., and Au, S.K. (eds.). Springer-Verlag, Berlin. pp. 1–18.

Papayianni, I. and Pachta, V. (2012). High performance lime-based grouts for repair of historic masonries. *SAHC 8th International Conference on Structural Analysis of Historical Constructions.* Warsaw, Poland.

Papayianni, I. (1997). A holistic way of studying mortars and bricks of ancient masonries for manufacturing compatible repair materials. *Proceedings of the 4th International Symposium on the Conservation of Monuments in the Mediterranean,* Vol. 3. Technical Chamber of Greece, Greece. pp. 265–274.

Papayianni, J. and Karaveziroglou, M. (1993a) Aggregate gradation of ancient mortars. Relationship of strength and porosity. *Proceedings of the International RILEM/ UNESCO Congress. Conservation of Stone and Other Materials.* Published by E&FN SPON, Paris. pp. 493–496.

Papayianni, J. and Karaveziroglou, M. (1993b). Compressive strength of masonry with thick mortar joints. *Proceedings of the International RILEM/UNESCO Congress. Conservation of Stone and Other Materials.* Published by E&FN SPON, Paris. pp. 212–219.

Papayianni, J. and Batsouka, C. (2017). *Unfired mud brick masonries.* MSc Thesis of Batsouka C. and supervised by Prof. Papayianni. J.

Parkin, G. and Phinn, A. (1977). Rock support in stainless steel, Stainless steel mining July/August. pp. 26–34.

Pasvantis, St. (2019) Photo files of Thessaloniki.

Pela, L., Aprile, A., and Benedetti, A. (2009). Seismic assessment of masonry arch bridges *Engineering Structural*, Vol. 31. pp. 1777–1788.

Penelis, G. (1965). *Membrane state of stress of shells with undulating radial symmetry. Phd Thesis*, Aristotle University of Thessaloniki, Greece.

Penelis, G. (1976). Load bearing capacity of R/C 1-D structures. *Seminar on modern trends in analysis and design of R/C structures.* In Greek. Penelis, G., (ed.). Thessaloniki, Greece. pp. 1–41.

Penelis, G. (1982). *The Masonry Roman and Byzantine Monuments in Greece. UNDP/UNIDO, PROJECT (RER/79/015) National Report of Greece.*

Penelis, G. (1992). The Rotunda of Thessaloniki: Seismic behaviour of Roman and Byzantine structures. in *Hagia Sophia from the age of Justinian to the present*, Mark, R. and Çakmak, A. (ed.). Cambridge University Press, Cambridge. pp. 132–157.

Penelis, G. (1995). Mortars and grouts in restoration of roman and byzantine monuments. *European seminar on Ancient and traditional mortar technology and their importance in conservantion projects.* Rethymnon, Creta.

Penelis, G. (1998). Assessment of residual bearing capacity of masonry structures. *1st International Course on 'the conservation of ancient masonry in Byzantine Architecture'* (CAMBA-98) ICCROM-AUTH Thessaloniki, Greece.

Penelis, G. (2002). Structural restoration of historical buildings in seismic regions. *Progress in Structural Engineering Materials*, Vol. 4, No.1, John Wiley & Sons Ltd. pp. 64–73.

Penelis, G. (2014). Structural restoration of monumental buildings in seismic areas, *MARE.14 materials and techniques for the restoration of monuments lifelong learning unit. AUTH*, Thessaloniki.

Penelis, G. and Penelis, Gr. (2001). The structural restoration of the National Library of Greece in Athens. *Proceedings of the 3rd International Congress of Studies in Ancient Structures.* Istanbul. pp. 857–865.

Penelis, G. and Penelis, Gr. (2015). *Restoration of the historic church of Saint George in Cairo and the underlying Roman Tower. Special publication for the church of Saint George in old Cairo.* Patriarchate of Alexandria. pp. 257–263.

Penelis, G. and Penelis, Gr. (2018). Restoration of the MarGirgis Church and Roman tower in Cairo. *Proceedings of the 4th Architectural and Urban Restorations BRAU4,* CI COP Italia. pp. 163–174.

Penelis, G. and Penelis, Gr. (2019). *Concrete buildings in seismic regions.* CRC Press (A SPON Press Book), London.

Penelis, G. and Stylianidis, K. (1994). *Strengthening of the minaret of Rotunda in Thessaloniki. Final report of a research programme.* Aristotle University of Thessaloniki, Greece. In Greek.

Penelis, G. and Stylianidis, K. (2001). Structural restoration of the Acheiropoietos basilica in Thessaloniki – Greece. *Proceedings of the 2nd International Congress on Studies in Ancient Structures.* (ed. Arun, G. and Seçkin, N), Istanbul. pp. 847–856.

Penelis, G., Karaveziroglou, M., and Papayianni, I. (1989). Materials used in the restoration of historical buildings and monuments. *Seminario per Conservao Recuperaçäo e consolidacäo de monumentos e edificios histöricos*, Conselho Directivo National (ed) Lisbon.

Penelis, G., Karaveziroglou, M., Stylianidis, K., and Leohdaridis, D. (1980c). *Restoration and Strengthening of Vlatades Monastery.* Archives of the Ephorate of Antiquities of Thessaloniki.

Penelis, G. Karaveziroglou, M., Stylianidis, K., and Leohdaridis, D. (1980a). *Restoration and strengthening of Rotunda of Thessaloniki.* Archives of Ephorate of antiquities of Thessaloniki.

Penelis, G. karaveziroglou, M., Stylianidis, K., and Leohdaridis, D. (1980b). *Restoration and strengthening of Hagios Andreas, Peristera.* Archives of the Ephorate of Antiquities of Thessaloniki.

Penelis, G., Karaveziroglou, M., Stylianides, K., and Leondaridis, D. (1984). The church of Hagios Andreas, Peristera case study in the *UNDP/UNIDO manual for repair and strengthening of historical monuments and buildings in urban nuclei.* UNDP/UNIDO, project RER 79/015 Vol. 6, Vienna.

Penelis, G., Karaveziroglou, M., Stylianidis, K., and Leondaridis, D. (1985). Repair and Strengthening of Rotunda, Thessaloniki. *Proceedings of the International Symposium of Thessaloniki on Restoration of Byzantine and Post-Byzantine Monuments.* 9th Ephorate of Byzantine Antiquities. pp. 25–48.

Penelis, G., Karaveziroglou, M., Sytlianides, K., and Leontarridis, D. (1983). Strengthening of the Rotunda Monument in Salonica. *Proceedings of the IABSE Symposium Strengthening of Building Structures – Diagnosis and Therapy.* Venezia, Italy.

Penelis, G., Papayianni, I., Stylianides, K., Ignatakis, C., and Athanassiadis, L. (1993). Strengthening of a 400-year-old Ottoman Minaret. *IABSE Proceedings of the IABSE Symposium 'Structural prevention of the architectural heritage'*, Rome, Italy. pp. 629–630.

Penelis, G. and Stylianidis, K. (1998). Structural restoration of Acheiropoietos church in Thessaloniki. In Greek. *Journal KTIRIO: Scientific Review.* A/1998 Thessaloniki. Vol. 1, pp. 33–40.

Penelis, G. and Stylianidis, K. (2001). Structural restoration of the Acheiropoietos basilica in Thessaloniki. *Proceedings of the 2nd International Congress on Studies in Ancient Structures (ed. Arun, G. and Seçkin, N) Istanbul,* pp. 847–856.

Penelis, G., Stylianidis, K., and Ignatakis, C. (1997). Strengthening of the Ottoman minaret of Rotunda in Thessaloniki using external metal strips. *Proceedings of the International Conference on studies in ancient structures.* Istanbul, Turkey. pp. 595–604.

Penelis, Gr. (2006a). An efficient approach for pushover analysis of unreinforced masonry (URM) structures. *Journal of Earthquake Engineering,* Vol. 10, No. 1. Imperial College Press, UK. pp. 359–379.

Penelis, Gr. (2006b). An efficient approach for pushover analysis of URM structures *International Journal of Masonry Construction*, Vol. 3, No. 1. pp. 26–31.

Penelis, Gr. and Penelis, G. (2001). The structural restoration of the National Library of Greece in Athens. *Proceedings of the 2nd International Congress on Studies in Ancient Structures (ed. Arun, G. and Seçkin, N)* Istanbul, pp. 857–863.

Penelis, G.Gr., Penelis, Gr.G., and Papanikolaou, V. (2010). Quasi-inelastic static analysis of URM monuments. *Proc. of 14th European Conference in Earthquake Engineering*, pp. 1133–1140

Penelis, Gr., Kappos, A., and Stylianidis, K. (2003). Assessment of the seismic vulnerability of unreinforced masonry buildings. in *Structural studies, repairs and maintenance of heritage architecture VIII*, Brebbia, C., (ed.). WIT Press, Bath, UK. pp. 575–584.

Penelis, G., Karaveziroglou, M., Stylianides, K., and Leontarides, D. (1984). The Rotunda of Thessaloniki. Case study in the *UNDP/UNIDO Manual for Repair and strengthening of historical monuments and buildings in urban Nucler*: UNDP/UNIDO, RER 79/015, Vol. 6, Vienna Austria. pp. 165–187.

Penelis, S.A. (2001a). Archives of Restoration of the National Library of Greece, Athens. In Greek.

Penelis, S.A. (2001b). Archives of Restoration of Hagia Anastasia monastery, Vasilika Thessaloniki.

Penelis, S.A. (2009). Restoration of bell tower of Zografos monastery, Mount Athos Penelis S.A., Archives. In Greek.

Penelis, S.A. (2014a). Archives of restoration project of Margirgis Tower, Old Cairo.

Penelis, S.A. (2014b). Archives of restoration project of Zardinides, mansion Thessaloniki, Greece.

Penelis, S.A. (2014c). Archives of Restoration of Mangos mansion in Athens.

Penelis, S.A. (2016) Archives of final design and construction of the restoration of the Greek Orthodox church of Saint-George, old Cairo.

Penelis, S.A. (2017). Archives of Hagia Irene, Instabul.

Penelis, S.A. (2019a) Archives of Acheiropoietos of Thessaloniki.

Penelis, S.A. (2019b) Archives of Atreus Tomb, Mycine, Greece.

Penelis, S.A. (2019c). Archives of Hagios Pandeleimon church, Thessaloniki Greece.

Penelis, S.A. (2019d). Archives of photographic material of Rotunda of Thessaloniki.

Penelis, S.A. (2019e). Archives of photographic material of Red House of Thessaloniki.

Penelis S.A. (2019f). Restoration of St. Irine square townhouse, Athens.

Peroni, S., Tersigni, C., Torraca, G., Cereu, S., and Forti, M. (1981). Lime-based mortars for the repair of ancient masonry and possible substitutes. *Proceedings of Conference on Mortars, Cements and Grouts used in the Conservation of Historical Buildings*, ICCROM, Rome. pp. 63–100.

Poleni, G. (1748). *Memorie istoriche della gran cupola del tempio vaticano e de'donni di essa, e de' ristoramenti loro. Padova. Included: Breve Pavere di Vanvitelli L.* Roma: reprint of the Faculty of Architecture of University of Rome Sapienza, Roma.

Ramalho, M., Taliercio, A., Anzani, A., Binda, L., and Papa, En. (2008). A numerical model for the description of the nonlinear behaviour of multi-leaf masonry walls. *Journal of Advances in Engineering Software*, Vol. 39. No. 4, UK. pp. 249–257.

Raptis, K. and Zombou, A. (2010). The consolidation and restoration project of Acheiropoietos basilica in Thessaloniki. *Proceedings of the 8th International Symposium on the Conservation of Monuments in the Mediterranean Basin Monubasin 8*, Patras, Greece, pp. 411–428.

Rehm, G., Lämke, A., Nürnberger, U., Rieche, G., Martin, H., and Rauen, A. (1974). *Verhalten von Verzinkten Spannstählen und Bewehrung Stahlen D.A.f.* St. B. Heft 242, Wilhelm Ernst and Sohn.

Robertson, E. (1982). *Physical properties of building stone.* The National Academies Press, Washington, D.C.

Roca, P., Cervera, M., Gariup, G., and Pela, L. (2010). Structural analysis of masonry constructions. Classical and advanced approaches. *Archives of computational methods in engineering, 2010/17*: pp. 299–325.

Roca, P., Massanas, M., Cervera, M., and Arun, G. (2004). Structural analysis of Küçük Ayasofya Mosque in Istanbul, in *structural analysis of historical constructions IV* Balkema, Amsterdam. pp. 679–686.

Roca, P., Molins, C., and Mari, A.R. (2005). Strength capacity of masonry wall structures by the equivalent frame method. *Journal of Structural Engineering, ASCE*, Vol. 131, No. 10. pp. 1601–1610.

Ruth, W. and Schuller, M. (1996) *Masonry repair by grout injection.* The Aberdeen Group, USA.

Sacco, E., Addessi, D., Sab, K., (ed.) (2018). *New trends in mechanics of masonry. Meccanica*, Vol. 53. Springer Science & Business, Berlin. pp. 1565–1569.

Salonikios, T., Theodoulidis, N., Morfidis, K., Zacharopoulou, G., and Raptis, K. (2017). Efficiency investigation of structural interventions based on in-situ ambient vibration measurements on Acheiropoietos early byzantine basilica, Thessaloniki, Greece. *Journal of Civil Structural Health Monitoring.* Springer-Verlag GmbH. Germany.

Samara Singhe, W. and Hendry, W. (1982). The strength of Brickwork under biaxial tensile and compressive stress. *Proceedings of British Ceramic Society, No 30,* UK. pp. 120–139.

Saxiabani, C. (1986). Injections at Rotunda application. *Proceedings of International Symposium of Thessaloniki on Restoration of Byzantine and Post-Byzantine Monuments.* 9th Ephorate of Byzantine Antiquities. pp. 49–58.

Schnabel, B., Lysmer, J., and Seed, B. (1972). SHAKE - *A computer program for earthquake analysis of horizontally layered sites, Rep. No EERC 72-12, University of California,* Berkeley.

Selby, Ar. and Wilson, Jm. (1997). The structural safety and acceptability of bell towers. *Transactions of the Built Environment*, Vol. 26, 1997, WIT Press, pp. 321–330.

Sextos, A., Pitilakis, K., and Kappos, A. (2003). Inelastic dynamic analysis of R/C bridges accounting for spatial variability of ground motion, site effects and soil-structure interaction phenomena Part 1: Methodology and Analytical tools. *Earthquake Engineering and Structural Dynamics*, Vol. 32, No. 4. pp. 607–627.

Shan, J. and Toth, C. (2018). *Topographic Laser Ranging and Scanning: Principles and Processing*, 2nd Ed. Taylor and Francis, UK.

Sheehan, P. (2010). *Babylon of Egypt*, Oxford University Press, Oxford, UK.

Shing, B., Lofti, R., Barzegarmehrabi, A., and Brunner, J. (1992). Finite element analysis of shear resistance of masonry wall panels with and without confining frames. *10 WCEE, Rotterdam.*

Shuller, M., Atkinson, R., and Borgsmiller, J. (1994). Injection grouting for repair and retrofit of unreinforced masonry. *10th 132 MaC, Galgary, Canada.* pp. 549–559.

Shultz & Barnsley (1917). Photos of Hagios Demetrios of Thessaloniki, British School of Athens.

Skoulikidis, T. (1977). Report on physio-chemical problems; Solutions proposed. *Proceedings of International Meeting on the Restoration of Erechtheion.* Athens.

Stavrakakis, E., Ignatakis, C., and Penelis, G. (1995). Behaviour of Roman masonry under biaxial loading analytical assessment of complete failure criterion. In Brebbia, C.A. (ed.). *Proceedings of the 4th International Conference on Structural Studies of Historical Buildings,* Southampton, Boston; computational mechanics. pp. 339–347.

Stevens, G.P. (1927). *The Erechteum, measured, drawn and restored.* Harvard University Press, Cambridge, MA, 1927

Stylianidis, K. (2000). Strengthening of the minaret of Rotunda in Thessaloniki. G. Penelis. *International Symposium of Concrete and Masonry Structures*, Stylianidis, K. and Kappos, A. (eds.). Thessaloniki, Greece pp. 335–348.

Stylianidis, K. and Ignatakis, C. (1995). Experimental investigations of the mechanical properties of original and strengthened masonry of the Russian Hospital of Thessaloniki. In Greek. Journal 'Monument and Environment' Special publication. pp. 119–128, Greece (special publication).

Stylianidis, K., and Ignatakis, Ch. (1997). *Masonry structures.* In Greek. Xirographia, Thessaloniki, Greece.

Swallow and Carrigton (1995). Limes and lime mortars-part I. *Journal of Architectural Conservation*, Vol. 1, Issue 3 C.R.C (Taylor and Francis) UK. pp. 7–25.

Swallow and Carrington (1996). Limes and Lime mortars-part II. *Journal of architectural conservation*, Vol. 2, Issue 1 CRC (Taylor and Francis) UK. pp. 7–22.

Tassios, T. (1986). On selection of 'modern' techniques and materials in structural restoration on monuments. *Proceedings of International Symposium of Thessaloniki. Restoration of Byzantine and Post-Byzantine Monuments.* 9th Ephorate of Byzantine Antiquities. pp. 357–363.

Tassios, T. (2001). Interdisciplinarity in studying ancient structures, *Proceedings of the 3rd International Congress on Studies in Ancient Structures*, Istanbul, pp. 3–11.

Tassios, Th. (1992). *Masonry mechanics.* In Greek. Athanassopoulos – Papadonis Ltd, Athens, Greece.

Tassios, Th. (2004). Rehabilitation of three-leaf masonry. In *evoluzione nella sperimentazione per le construzioni, seminario internationale, 26th Sept -30th Oct Centro Internationale di Aggiornamento Sperimentale - Scientifico* (CIAS), Italy.

Theocharidou, K. (1988). The architecture of Hagia Sophia, Thessaloniki. *B.A.R International Series 399*, Great Britain.

Theocharidou, K. (1992). The structure of Hagia Sophia in Thessaloniki from its construction to the present, in *Hagia Sophia from the age of justinianian to the present*, Mark, R. and Çacmak, A. (eds.). Cambridge University Press, Cambridge, USA. pp. 83–99.

Theocharidou, K. (1985). Acheiropoietos: Problems and criteria of restorative interventions. *Proc. of the International Symposium on Restoration of Byzantine and PostByzantine Monuments.* In Greek. 9th Ephorate of Byzantine antiquities. Thessaloniki. pp. 61–70.

Tomazevic, M. (1997). Seismic resistance verification of buildings, following the new trends. *in Seismic methodologies for the next generations of codes.* Fajar, P. and Krawinkler, H. (eds.). Balkema, Rotterdam.

Tomazevic, M. (1999). *Earthquake – resistant design of masonry buildings.* Imperial College Press, London, UK.

Tomazevic, M. and Apih, V. (1993). The strengthening of stone-masonry walls by injecting the masonry friendly grouts. European Earthquake Engineering, Vol. 7, No. 2. Patron, Bologna. pp. 10–20.

Toumbakari, E. (2002*). Lime-pozzolan-cement grouts and their structural effects on composite masonry walls.* Doctor thesis, Catholic University Leuven.

Tsintziras, G. (2018). *Ten Brinke Group (Archives) Unknown*, Thessaloniki.

Ulrich, R. (2006). *Roman woodworking.* Yale University Press, New Haven, USA.

UNDP/UNIDO (1984). *Repair and Strengthening of Historical Monuments and Buildings in Urban Nuclei*, 6, Vienna.

Valluzi, R., da Porto, F., and Modena, C. (2001). Behaviour of multiple leaf stone masonry wall strengthened by different intervention techniques. *Proceedings of the 3rd International Seminar on Structural Analysis of Historical Constructions* (EA. P.B. Lourenco, P. Roca). pp. 1023–1032.

Valluzzi, M., da Porto, F., and Modena, C. (2004). Behaviour and modelling of strengthened three-leaf stone masonry walls. *Materials and Structures,* Vol. 37. pp. 184–192.

Van Gemert, D. and Toumbakari, E. (2003). *Structural consolidation and strengthening of masonry.* Historical overview and evolution. KU Leuven, Belgium.

Varoufakis, G. (2005). *The history of iron from Homer to Xenophon.* In Greek. Hellinika Grammata, Athens.

Vintzileou, E. (2006). Grouting of three-leaf stone masonry: Types of grouts, mechanical properties of masonry before and after grouting, in *Structural analysis of historical constructions*, Lourenço, P.B., Roca, P., Modena, C., Agrawal, S. (eds.). New Delhi. pp. 41–58.

Vintzileou, E. (2011). Three-leaf masonry before and after grouting: A review of literature. *International Journal of Architectural Heritage*, Vol. 5, Taylor and Francis Group, LLC. pp. 513–538.

Vintzileou, E. (2018). Unreinforced masonry walls subjected to in-plane shear: From tests to codes and vice versa. *Recent advances in earthquake engineering in Europe*, Pitilakis, K., (ed.). Springer, London, New York. pp. 439–458.

Vintzileou, E. and Tassios, Th. (1995). Three-leaf stone masonry strengthened by injecting cement grouts. *Journal of Structural Engineering*, Vol. 121, No.5. pp. 848–856.

Vintzileou, E., Miltiadou-Fezans, A., Palieraki, V., and Delinicolas, N. (2005). The use of radar techniques and endoscopy in investigating old masonry. in *Structural analysis of historical constructions.* Modena, C., Lourenço, P.B., and Roca, P. (eds.). Taylor & Francis Group, London, UK, pp. 351–359.

Vitruvius (1998). *On architecture. Books 1–10 LOEB Classical Library,* Harvard University Press, London, England.

Vrouwenvelder, A. (2003). *The plastic behaviour and the calculations of beams and frames subjected to bending.* Technical University Delft, faculty of civil engineering and geosciences Delft, Holland.

Wenzel, F. and Maus, H. (1992). Repair of masonry structures. in *Masonry construction; structural mechanics and other aspects*, Calladine, C., (ed.). Meccanica, Vol. 27, No.3, Springer Science & Business, Berlin. pp. 223–232.

Wood, C. and Burns, C. (2013). Grouting solid masonry walls. *English Heritage website, Conference of Exeter* UK. pp. 1–6.

Zacharopoulou, G. (1998). *The renascence of lime-based mortar technology: an appraisal of a bibliographic study. S.P. publication on compatible materials for protection of European Cultural Heritage*, PACT 55/1998, INCOMA-RECH-RAPHAEL 97/E/412 Technical Chamber of Greece. pp. 89–114.

Zambas, K. (1982) *Articulated ancient Greek and Roman buildings. UNDP/UNIDO, PROJECT (RER/79/015) National Report of Greece.*

Index

H

I

L

P

Q

R

S

T

U

V

W

X